Exploring Biology Today

Exploring Biology Today

ANN O. WILKE
Adjunct Senior Lecturer
Formerly, Director of Undergraduate Laboratories
University of Missouri, St. Louis

with 422 illustrations and 30 in color plates

Mosby
St. Louis Baltimore Boston Chicago London Philadelphia Sydney Toronto

Mosby
Dedicated to Publishing Excellence

Editor-in-Chief: James M. Smith
Editor: Robert J. Callanan
Developmental Editor: Jean Babrick
Project Manager: Carol Sullivan Wiseman
Production Editor: Pat Joiner
Designer: Susan Lane

Printed in the United States of America

Mosby–Year Book, Inc.
11830 Westline Industrial Drive
St. Louis, Missouri 63146

ISBN 0-8016-6855-7

93 94 95 96 97 CL/VH/CD 9 8 7 6 5 4 3 2 1

To the student

Have you ever wondered how your body works or what different organisms are like and how they affect each other? Are you interested in how our living world changes? Welcome to the study of life! During this term, you will have opportunities to explore these and many other questions about life. As you start to understand how biologists learn about the living world, you can apply their methods to your life to help you understand your own body and the world around you.

During this term, you will be asked to think about your thinking. How do you think through problems and make sense out of a jumble of information? We can learn a lot about thinking and learning if we watch small children interacting with their environment. They investigate new objects and invent ways of doing things that may turn their surroundings into chaos. They handle, manipulate, and frequently explore objects in ways quite different from the ways that adults intended the objects to be used. From these investigations, children learn to understand their world. Exploring and manipulating new materials and examining processes when trying to understand them can be as useful to you now as it was when you were a child.

The laboratories associated with science classes give you the opportunity to explore materials, to get involved with them, and from these explorations to develop an understanding of materials and their processes—just as you might explore a puzzle such as Rubik's Cube.

You may find the laboratories that you work with in this book are considerably different from others you have encountered previously. One of the most striking differences may be that you will be working with materials in the laboratory *before* you attend a lecture or read about them in your textbook. There is an important reason for this arrangement: actual experience (handling, manipulating, and exploring materials) is one of the easiest ways to learn. Working with materials in a laboratory *after* you hear or read about them probably means that you will find what you were told you would find and observe processes that you were told would happen. This is not the way that scientists work. They become curious, get excited, and are challenged by the unknown as they seek answers to their questions about the natural world. If you only look for something that you have already accepted because you were told it is there, you lose the opportunity to challenge the unknown and to become fascinated and involved as you make new discoveries.

Sudents differ in the ways they handle information. Some students incorporate bits of information and draw relationships among a collection of facts and processes more quickly than others. Some students have been encouraged by their former teachers to wrestle with understanding material rather than just memorizing it. Those who have considerable experience building a framework of concepts from isolated experiences and information may find it easier to make sense of many pieces of information and use them as a foundation on which to incorporate new information. If you have not had such experiences in the past, you will learn these skills quickly in the investigative laboratory setting.

The laboratories in this manual use a teaching strategy called the *learning cycle*. There is nothing mystical or magical about the learning cycle. Its format is related to the way we learn. It helps students who previously just memorized facts to move to a more effective level of thinking—to an understanding of the facts and how they are related. It provides opportunities for students with some understanding to expand their knowledge. When you understand, you can use the information you synthesize from your classes and apply it to your daily life. As you know, facts that you memorize for a test are soon forgotten. When you put facts together in meaningful relationships, however, you remember and understand them and can use them in new situations.

The learning cycle strategy has three components or phases, exploration, concepts/terms development, and application. Here is a brief description of the learning cycle components as used in this manual:

1. *Exploration* You begin to learn about new topics by exploring materials and processes. You will have the opportunity to manipulate, examine, investigate, and do many different things with the materials. Through questions in the manual, I will nudge you into observing and thinking about the materials and your activities in ways that biologists might think about them. Your teacher will invite you to "dig deeper" by asking additional questions. Sometimes, these explorations and your consideration of them may fit neatly with your previous experiences. On other occasions, you may find that they are different and confusing—they do not fit with previous information and relationships that you have constructed in your mind. For these new and confusing experiences, you have to build a new set of relationships.
2. *Concepts/terms development* After exploring materials and while you are thinking about the results, I provide some of the relationships and descriptions of terms that biologists use for these materials and processes in the concepts/terms sections. You can compare these to your interpretations of your exploration experiences.
3. *Application* After you have a relationship or concept tentatively built in your mind, you can solidify it as you apply it to other situations. In the application phase you will have one or more opportunities to apply the concepts you have developed.

In this manual, the parts of the learning cycle do not rigidly follow the sequence of one exploration section, one concepts/terms section, and one application phase. Instead, you may have three suggested exploration activities before you reach the concepts/terms and application sections for those explorations; or may have two sets of exploration and concepts/terms sequences before you apply the concepts from both of these sets in an application section. Look for all three learning cycle phases that invite you to explore materials, form concepts, and apply new knowledge.

An integral topic throughout the manual is science. You will have many opportunities to develop a deeper understanding of science as a way of learning about our world and the way that scientists use information obtained through their work to explain our natural world. Frequently, you will investigate questions by designing and conducting experiments in the way that scientists do. Sometimes, you will use these experiments to test newly formed concepts; at other times, you may conduct experiments in an exploration section.

In these laboratories the role of your teacher is to help you understand the material you are exploring. The teacher is an enabler—a facilitator—not a source of knowledge with a multitude of facts to pour into your mind. Because learning is a construction process in and by each individual, your teacher cannot do it for you. Only you can make sense out of your experiences. Your teacher will help you develop understanding by *answering* your questions about your experiences *with* questions. This may by very frustrating if all you want is the correct answer to memorize. Because you have to build your own understanding, your teacher is giving you maximum assistance by encouraging you to develop your own answer step by step. You will quickly become more comfortable with this type of help, and you will make progress in developing an understanding of biology and biology as a science.

To you—one of my current students—I say, "Welcome to these explorations." I hope that you will enjoy your experiences and, above all, that you will accept the challenge of exploring our fascinating living world.

ANN O. WILKE

To the teacher

Ann Wilke has created an innovative laboratory manual based on current learning theory, exploratory in nature, and structured in a way that is easy to use. The instructional approach of *Exploring Biology Today* is based on the learning cycle, which was developed by Robert Karplus at the University of California at Berkely during the late 1950s and the early 1960s. (Professor Wilke has included a bibliography of readings in the instructor's manual for those interested in this theory.)

The learning cycle has three phases: exploration, concept/term development, and application. During the exploration phase, students explore and manipulate materials in a guided manner that leads to their understanding of one or more concepts. In the concept/term introduction phase, the concepts that the students have explored and have come to understand are articulated and named. In the application phase, students apply this knowledge in a new situation.

In *A Theory of Instruction: Using the Learning Cycle to Teach Science Concepts and Thinking Skills* (1989), Lawson, Abraham, and Renner list and describe three types of learning cycles: descriptive, empirical-abductive, and hypothetical-deductive. The primary difference among the three is the degree to which students test alternative hypotheses. Professor Wilke makes greatest use of descriptive learning cycles, in which students discover patterns in the natural world, name these patterns, and then look for similar patterns elsewhere. As students become more comfortable with the learning-cycle approach and progress in their reasoning abilities, she incorporates facets of the empirical-abductive approach, asking students to generate ideas regarding possible causes of a pattern and to test these ideas. The semester project and other questions bring students to the hypothetical-deductive learning cycle, in which students pose a question, generate alternative explanations, and design and conduct an experiment to test these hypotheses. During the course of these laboratory explorations, Professor Wilke also asks students to think about their thinking. She provides clear descriptions of the various reasoning patterns that students use during their laboratory activities and asks them to reflect on and identify the reasoning patterns that they use in specific situations.

Although learning cycles are often open-ended in their approaches, Professor Wilke has constructed learning cycles that are highly structured. She has done this so that these laboratories can be used by a variety of persons teaching the introductory biology laboratory who may be unfamiliar with the learning-cycle approach. She has created laboratories in which *she* is there, guiding students during their explorations and applications. She also presents the concept/term introductions or provides activities for this phase of the learning cycle. The job of the teacher in the laboratory then becomes that of guide and questioner, helping students learn by challenging their ideas and guiding their thought processes through questions.

We have chosen the learning cycle as the foundation of this manual because it is an instructional strategy that closely follows the way in which scientists learn about the natural world. Exploring questions and uncovering patterns as scientists do is inherently more interesting and exciting to students than simply going through a series of steps to verify what someone has already told them. The learning-cycle approach also provides a framework in which students are expected to ask questions and to figure out the answers themselves. To become educated, thinking citizens and consumers, students must practice thinking and problem-solving skills so that they can use these skills and apply new knowledge in their daily lives.

Another reason for choosing the learning-cycle approach is that many students in introductory classes learn best by first manipulating concrete objects and puzzling out questions rather than having abstract knowledge "delivered" to them. As students construct their new knowledge in an introductory biology course, they build on the concrete experiences in which they engaged in the laboratory. Therefore to maximize learning with this approach, Professor Wilke and I suggest that laboratory exercises precede lecture, not vice-versa.

The teacher's guide that accompanies *Exploring Biology Today* should be an extraordinary help in implementing this laboratory course. Professor Wilke not only provides lists of all the materials you will need, but also suggests from her 20 years of experience in designing, conducting, and implementing introductory laboratories how to tailor activities to either a 2- or 3-hour laboratory session. With this information, you can plan the laboratory and sequence it effectively with lecture within your own scheduling framework. In addition, Professor Wilke describes each laboratory and provides the teacher with valuable hints and background information. I'm sure you will find the teacher's guide a valuable adjunct to the laboratory manual for teachers, laboratory prep personnel, and teaching assistants.

All of us involved in the development of *Biology Today* are interested in your comments. We will strive to continue to produce pedagogically sound, error-free materials that best meet the needs of today's students and faculty. We hope you enjoy using *Exploring Biology Today*.

SANDRA S. GOTTFRIED

ACKNOWLEDGMENTS

No one is in the world alone, and certainly the profession of teaching is fully rooted in interactions among people. Teachers learn, are stimulated, and glean information and experiences from many sources. So it is with me and the writing of this manual. A part of every one of my students is represented within these pages in the things I have learned from them and the ways they have taught me to help them. I treasure these interactions.

Also, I value the years of interactions with my faculty colleagues and teaching assistants at the University of Missouri-St. Louis. From these people, I have received encouragement, stimulation, information, new perspectives, and the challenge of expanding the boundaries of understanding in biology. Among these colleagues is Dr. Sandra Gottfried, who convinced Mosby that learning cycle–based laboratories would be an important auxiliary to her textbook, *Biology Today*. She has graciously and generously given her time and shared her expertise in providing many valuable suggestions and continuing encouragement for this project.

Another group of colleagues who are deeply interested in quality teaching in biological laboratories, the members of the Association for Biology Laboratory Education (ABLE), have readily shared their solutions to laboratory problems and ideas for interesting laboratory activities. Among the ABLE members to whom I am particularly indebted are two creative sisters, Drs. Marsha Fanning of Lenoir-Rhyne College and Sheryl Shanholtzer of DeKalb College. They introduced me to deshelled eggs as osmometers (Laboratory 3) and termites as organisms for exploring aspects of the process of science (Laboratory 1). Many suggestions for improving other laboratories have come from teaching assistants. I am indebted to one, Rich Nolte, for his imaginative use of dominoes to simulate nerve cell activity (Laboratory 18).

Credit for the readability of this manual and dispatching of the many details of its production goes to the staff at Mosby. Their guidance was important in making this text easier to read and more visually interesting. Particularly, people such as Robert Callanan (Acquisition Editor), Jean Babrick (Senior Developmental Editor), Jennifer Collins (Editorial Assistant), Linda McKinley (copy editor), and Pat Joiner (Senior Production Editor) were important to the production of this manual. Eileen Draper created the needed illustrations to clarify topics and techniques. To all of these people—students, the publishing staff, and my colleagues—I give my heartfelt thanks. Without their input, this manual would not have been written.

ANN O. WILKE

Contents

APPENDIXES

LABORATORY 1

Aspects of the scientific process

Is science a common or exotic activity?

Learning about things that make up the world is part of your daily life. At an early age, you learned that if you pushed a glass of milk off the highchair tray, the glass would fall and the milk would make an interesting pattern on the floor. Later, you learned that if you turned the room thermostat up to 90° F (32° C), you would become too hot. Determining cause-effect relationships such as these is a major part of your everyday life. In fact, you do it so often that you are seldom aware of the process that determines these relationships. They become part of a body of knowledge, an understanding of your world that is frequently called **common sense.** All of us develop common sense knowledge by drawing relationships among objects we encounter, activities in which we participate, and phenomena (happenings) we observe.

Is the way we acquire scientific knowledge an exotic activity? Or is it related to the way we acquire common sense knowledge? Most people would agree that scientists are human and the scientific process, the way scientists inquire about our world, is a human activity. Therefore the process of science cannot be too far removed from what people do in their normal lives. In fact, some scientists and scholars of science say that developing scientific knowledge is just a more formal version of obtaining common sense knowledge. For them, **science is a refinement of common sense.**

What do these scholars mean by the qualifying term *refinement?* If science and its methods are generally similar to common sense and its methods, what aspects of science are summarized by the term *refinement?* Developing an understanding of refinement is an important part of what you will be doing in the activities planned for this laboratory class.

Let's get on with it! Life and science are fascinating, and experience is a much more effective teacher than the "couch potato" process of reading about biology and science.

EXPLORING

The scientific process—termites

With your partner(s), get the following materials from the source indicated by your teacher:

1 hand lens
2 sheets of white paper
1 set of different colored ink pens
1 paintbrush
1 container with two termites
1 index card

As you begin exploring with the termites, be aware that worker termites cause millions of dollars in damage to homes each year. Using the hand lens for magnification, visually observe the structures of these organisms. Use your other senses as well. Can you detect a termite odor? Are their bodies soft or hard? As a **naturalist** (an observer of nature) you may also wonder about other questions such as the following:

How many legs does it have?
Are the legs better for walking or digging?
How many sections are in the body? Are they specialized or all alike?
Does it have structures that detect sounds, odors, and light?
What kind of structures are present for ingesting (taking in) food?
What other questions about termites might you explore?

1 ____

____

On the lines below, describe the form, or structure, of the termites.

2 ____

____

____

Now expand your exploration. Instead of just observing the termites' form, *quietly* watch their behavior as they are placed on the paper. You can move them around with the paintbrush. Observe their responses to ink lines drawn on the paper.

After a short period of casual checking, seek information a bit more scientifically—a bit more formally. Based on your observations and any other available information, propose a hypothesis about one type of termite behavior. In science, a **hypothesis** is a general statement—a tentative explanation—that links a possible cause to an effect. (For example, you might form a hypothesis that dogs scratch because they have biting fleas.) In your exploration, the effect is the termites' behavior. Because time is limited, pick an easy-to-change environmental factor that may be the cause of this behavior. Each of you should state your hypothesis about this proposed cause-effect relationship. What are these hypotheses?

3 ____

____

SCIENCE NOTE

Using all the information they can gather, scientists propose explanations for questions they have about events and things in our natural world. These tentative explanations are called hypotheses.

Scientists do more than simply propose hypotheses, they also set up **experiments** to test them. They test a hypothesis by testing predictions that apply the hypothesis to specific situations. (For example, you might predict that if you remove the fleas from your dog, Waggles, he will stop scratching.) A **prediction** states an anticipated change in a specific response (stop scratching) as the proposed cause factor is changed (remove fleas). In an experiment, scientists manipulate the cause factor and measure the response. If the response corresponds to the prediction, the hypothesis is supported. If the measured response fails to correspond, the hypothesis is rejected.

With your partner(s), set up a test of one of the proposed hypotheses. State your prediction (the expected change in the termite's response as the possible cause factor is manipulated) below. Then describe the experiment you design.

4 Prediction

Experiment

CONCEPTS/TERMS

Useful terminology

Two components of your hypothesis and experiment are known by conventional scientific terms. One is the factor that you manipulate in your experiment to *cause* a change in the termites' behavior. This cause factor is the **independent variable.** What is the independent variable in your experiment?

5

The second component is the *proposed effect* on the termite that may change because you manipulate the independent variable. The effect is the **dependent variable.** What is the specific dependent variable in your experiment?

6

The relationship between the independent variable and the dependent variable is a **cause-effect relationship.**

EXPLORING

An experiment with termites

Carry out your experiment and record the results on the following lines or in an appropriate table.

7

Once the results are obtained, scientists examine the data (the results) and compare them to the prediction based on the hypothesis.

Examine your data and compare them to your prediction. What change did you predict in the termites' behavior as you changed the independent variable?

8

Describe how the data support or reject your prediction and its hypothesis.

9

SCIENCE NOTE *Scientists design experiments to test specific predictions that are based on hypotheses. These tests provide data that are analyzed to yield information about each prediction and its hypothesis. As information is gathered, our knowledge about the phenomena of the natural world increases.*

Now that you and your collaborating investigator(s) have tested your prediction and drawn conclusions about your hypothesis from your experiment, share the information with other members of the class. On the chalkboard, put your information in Table 1-1. If directed by your teacher, record all class information in Table 1-1 of the manual.

Carefully look at the information from all the experiments and group together those with the same independent variable.

Table 1-1 Factors affecting termite behavior

Investigator initials	Independent variable	Dependent variable	Conclusion
1.			
2.			
3.			
4.			
5.			
6.			
7.			
8.			
9.			
10.			
11.			
12.			
13.			
14.			
15.			

Examine the group that has the largest numbers of experiments and answer the following questions: What is the independent variable for this group?

10 ______________________________

Within this group, did all investigators test for the same dependent variable?

11 ______________________________

If not, what are the dependent variables within this group?

12 ______________________________

____

Compare the experiments that have the same independent *and* dependent variables. Do all investigators of these experiments come to the same conclusion?

13 ______________________________

If all investigators working on the same question did not produce the same results and conclusion, there is a problem. How do you know if or how an independent variable is affecting the termites' behavior when investigators have different results and conclusions?

Try to determine the factors that might be contributing to these different conclusions. In a class discussion, suggest reasonable explanations for these different conclusions. What are these explanations?

14 ______________________________

____

____

____

____

Does one explanation refer to changing factors *other than* the planned independent and dependent variables?

15 ______________________________

If experimental conditions were not always constant, how can you determine a cause-effect relationship between the independent variable and the behavior?

16 ______________________________

CONCEPTS/TERMS

Increasing understanding of the scientific process

The factors other than independent and dependent variables must remain the same throughout the experiment. These are the **controlled** or **standard conditions.** They are controlled conditions because they *remain the same* throughout the experiment. They are standard conditions because they are used as a base for comparison. If any of these other factors changed in addition to the independent variable, you do not know which of the changing factors caused the change in the responding variable (termites' behavior). Only when all other factors are unchanged, or controlled, is the experiment a fair test of the effect of the independent variable on the dependent variable. Only then can comparing the measured value of the dependent variable before and after a change in the independent variable support the cause-effect relationship.

SCIENCE NOTE *Scientists design experiments to examine cause-effect relationships between the manipulated independent variable and the responding dependent variable. To develop a cause-effect relationship between the independent and dependent variables, all other factors that may affect the dependent variable must remain unchanged throughout the experiment. All unchanging factors are termed* standard *or* controlled conditions *of the experiment.*

Keeping in mind your experiment with termites, consider all scientific experiments in general. Even if the investigator makes a great effort to plan and conduct a well-controlled experiment, is it possible to be absolutely sure that *only* the intended independent variable is changing? Scientists have to face a problem when answering this question because the answer is always "no." No one can be absolutely sure that at least one more changing factor is not affecting the data. How then can scientific information be determined?

In general, scientists handle this problem in the following ways:

1. Scientists frequently try to reject a null hypothesis. A **null hypothesis** proposes *no* change in the response as the independent variable is manipulated. If a null hypothesis is rejected, the scientists conclude that the independent variable does affect the dependent variable. Then other hypotheses (alternatives) are proposed.
2. Scientists make a great effort to uncover all factors that may affect the results of the experiment and establish controlled conditions for them.
3. Investigators are never satisfied with running the experiment only one time. The experiment is repeated until they are convinced that the independent variable is the cause of the dependent-variable response.

The process of science requires two levels of agreement or verification. The first level consists of the original experimenters agreeing that the data are accurate and support the conclusions. For science, this is not sufficient. It also requires that other scientists interested in the question agree. Information produced by investigators must always be examined and verified, or confirmed, by others. Interested scientists run experiments to satisfy themselves that the cause-effect relationship proposed and tested is reliable. Only after this **verification** process is the information accepted into the body of scientific knowledge as an explanation of one small phenomenon in the natural world. This requirement makes science *a community affair.*

How do the processes of acquiring common sense knowledge and scientific knowledge compare in the requirement for controlled experimental conditions and repetitions?

17 ______________________________

____

How do common sense and science compare in the requirement for verification of information by a community of people?

18 ______________________________

____

If science is defined as a refinement of common sense, to what does *refinement* refer?

19 ______________________________

____

SCIENCE NOTE

Because scientists can never be absolutely sure that only the intended independent variable is the cause for the observed change in the dependent variable, they frequently do the following: (1) Scientists work to reject null hypotheses, (2) they repeat the experiments under controlled conditions, and (3) other scientists verify the hypotheses before these cause-effect explanations are accepted as scientific information.

APPLYING

The scientific process

Effects of heat on river fish (a hypothetical experiment)

Now put the scientific process into practice. Design an experiment to test a hypothesis. First, you need a problem situation about which there are some questions.

Suppose a northeastern resort town's new power plant is just beginning operations. The plant is designed to convert a source of energy into electricity that can be used by people in the town. This energy conversion involves a process that releases some of the energy as heat. River water is used to cool the equipment. The plant pumps 14° C (about 59° F) water from the town's river into the plant for cooling and returns 22° C (72° F) water to the river. The volume of water in the river remains the same as before the construction of the plant.

The resort owners and the fishermen of the area claim that the extra heat in the river water will harm the river's organisms and result in smaller native fish. They also claim the heat will even change the types of fish that live in the river. They believe that changes in the fish populations will damage the resort businesses and the fishermen's favorite pastime.

A class at the local college decides to run a fish growth experiment in their laboratory tanks to test the claim of the resort owners and fishermen. The class obtains native fish (brook trout) and two kinds of nonnative fish (rainbow and brown trout) that have been stocked in the river (Figure 1-1 and Plate 1). The class decides to test the hypothesis, *The warmer water decreases the growth of the*

A

C

B

Figure 1-1
A, Brook trout.
B, Rainbow trout.
C, Brown trout.

native fish. What is the null hypothesis for the question about the effect of temperature on the fish?

20 ____

Based on the null hypothesis, what is a specific prediction that can be experimentally tested?

21 ____

Help the class determine the following:

Dependent variable

Independent variable

Standard conditions

Procedure for testing

The class obtained the data in Table 1-2 from their experiment. Examine the data. From the table, determine how the class designed their experiment. How is their design different from your suggested experiment?

22 ____

What should the class conclude from their experiment? In your answer, include the particular observations that you think are most important and how you linked them together.

23 ____

Table 1-2 Increase in length of fish grown at 14° and 22° C

Fish type/identification number	Initial length (cm)*	Length after 8 months (cm)*	Percent change	Average percent change ($\bar{X}$)
GROWTH IN WATER AT 14° C				
Brook trout 1	11	19	72	
2	10	16	60	
3	12	20	67	62
4	11	16	45	
5	9	15	67	
Rainbow trout 1	12	19	42	
2	13	20	53	
3	10	14	40	46
4	12	18	50	
5	11	16	45	
Brown trout 1	13	17	31	
2	14	19	36	
3	11	16	45	45
4	9	15	67	
5	13	19	46	
GROWTH IN WATER AT 22° C				
Brook trout 1	13	17	31	
2	11	16	45	
3	12	16	33	41
4	13	19	46	
5	10	15	50	
Rainbow trout 1	14	22	57	
2	12	18	50	
3	10	17	70	56
4	12	19	58	
5	14	20	43	
Brown trout 1	10	17	70	
2	12	21	75	
3	12	19	58	71
4	11	19	73	
5	10	18	80	

*Standard length measured from anterior tip to base of tail fin.

How convinced are you that the conclusion from the class experiment is correct?

24 ______________________________

Explain your answer.

25 ______________________________

____

____

In science, the results of experiments are statistically analyzed to determine if changes in the response are big enough to conclude that there is a difference.

Now think for a few minutes about the next step that should be taken by the townspeople. What do you recommend?

26

In what ways do you agree *and* disagree with the following statements:

1. Life in the river will not be harmed by the small increase in heat at the power plant outlet because the heat soon dissipates in the flowing water of the river.
2. There should be no restrictions on the temperature of the outflow water from the plant.

27

Effects of an environmental factor on growth of organisms (a semester-long project)

On the demonstration table is a display of organisms that have been grown under a variety of environmental conditions. Members of the same kind of organisms, such as begonias, were all of similar size and health at the beginning of the special treatments.

Some of the environmental conditions under which these organisms were grown are given on a card beside them. For plants, this information may include temperature, amount of light (measured in foot candles), type of soil (½ clay and ½ sand or ⅓ clay, ⅓ peat moss, and ⅓ sand), amount of water (number of milliliters [ml] given at specific times each week), and type of water (city, distilled, pond water, acid water, 10% alcohol). In some cases only part of this information is available.

For fungi, environmental information may include temperature, amount of water, type of water, amount of light, and growth substrate (bread, orange, newspaper, or leaves).

For animals, the environmental condition information may include temperature, type of food, type of soil, amount of water, and type of water.

Observe the organisms in the display. Prepare a list of characteristics you want to compare for each group of similar organisms. For example, a few characteristics you may observe for plants are size of leaves, color of leaves, number of flowers, and length of branches. List these and other characteristics on the appropriate line.

Plants

Fungi ______________________________

____

Animals ______________________________

____

In Table 1-3, indicate the environmental conditions given for each plant, fungal mass, or animal group. Describe their growth and apparent health by noting the state of observed characteristics.

Now that you have collected some information and a lot of unanswered questions from these display organisms, plan a long-term experiment. For this experimental project, work in groups designated by your teacher. Together, plan an experiment using one of the display organisms as the experimental organism. The following steps will assist your group in planning the experiment:

Step 1. Using observations of the display organisms and any other information your group members may have, choose an experimental organism and an

Table 1-3 Growth and health of organisms under varying environmental conditions

Type of organism	Light (ft cd)	Temperature (° C)	Water (ml)	Water Type	Soil Type	Substrate Type	Growth and health
1.							
2.							
3.							
4.							
5.							
6.							
7.							
8.							
9.							
10.							

independent variable from one of the environmental conditions used for the display. (Hint: Did you notice that in comparing two organisms or groups of the same kind that more than one environmental factor differed between them? Perhaps this knowledge will help you ask a question about one of those environmental factors.) If your teacher gives permission, other environmental factors not included in the display may be chosen as your independent variable.

Step 2. Propose a relationship between the independent variable and the growth of your organism. State this cause-effect relationship in the form of a hypothesis. Predict the specific effects that changes in the independent variable will have on the growth of the organism.

Step 3. Determine how and when you will measure and record the dependent variable. For example, if your organism is a plant, you may want to measure the number and length of the leaves every third day.

Step 4. Determine the levels of the independent variable your group will test. For example, if the independent variable is the amount of water for begonia plants, one plant might be watered with 20 ml of water, a second plant with 40 ml, and a third plant with 60 ml. (For help with the metric system, see Appendix D.)

Step 5. Determine the standard conditions for the experiment.

Step 6. Design the experiment so that each group member will raise one complete set of experimental organisms. For example, if your group is testing the effects of water volume on begonias, each member of your group is responsible for raising three begonia plants, and each one will be watered with a different one of the specific water volumes.

After your group has completed the experimental design, discuss it with your teacher. Be prepared to provide the following:

1. Name of the experimental organism
2. Independent variable
3. Dependent variable
4. Hypothesis
5. Prediction
6. How the dependent variable will be measured and recorded
7. A list of standard conditions
8. A list of all required equipment and materials with the required amounts

Following the directions of your teacher, obtain your organisms and supplies. Pot the plants or set up the fungus and animal containers. Each group member will take responsibility for the care, measuring, and recording for one set of organisms.

Near the end of the term, a group report of the experiment will be required. Include the following in the report:

1. An introduction that includes the problem investigated, the hypothesis, the prediction, and information about the organism and the independent and dependent variables
2. Description of the experimental procedure
3. Presentation of the data
4. Discussion of several topics, including the following:
   a. The relationship of averaged (mean) group data to the prediction and hypothesis
   b. The relationship of the data obtained by each individual and how it is similar or different from total group data
   c. Application of the information you have gathered to current problems such as waste disposal, air quality, crop production yields, and animal growth
   d. The conclusion

LABORATORY

2

Properties of the molecules of life

Hydrophilic and hydrophobic interactions, pH, and buffers

PART ONE

Hydrophilic and hydrophobic interactions

EXPLORING

The interaction of water and oil

Two important materials in living things are water and oils. Water is the primary liquid of life. Lipids, such as oils, are major components in the structure of the basic building blocks of living things. See what you can discover about the interactions of these two types of materials through in vitro explorations. (*In vitro* means "in glass" and refers to being outside of living things.) From in vitro explorations you begin to understand how these substances might interact within a very complex living system such as your body.

Let's begin by exploring how these materials behave when they are placed together.

From the location identified by your teacher, pick up the following items:

1 bottle of water
1 bottle of vegetable oil
A few *small* crystals of rock salt
2 Pasteur pipets (long-pointed droppers)
1 piece of wax paper (about 20 by 30 cm, or 8 by 12 inches)
2 magnets

Place a few drops of water in a small puddle on the wax paper. Make another water puddle next to the first so that the two puddles touch. What happens as the water molecules in these two puddles touch?

1 ______________________________

Make two puddles as above but use oil instead of water. What happens as the oil in these two puddles touch?

2 ______________________________

Now that you have discovered how water interacts with water and oil with oil, test how these two materials interact with each other. Again, make two adjacent puddles, but make one puddle of water and the other of oil. What happens in the interaction between water and oil puddles that is different from the interactions between puddles of the same material?

3 ______________________________

How do you think that water and oil might interact within your body?

4 ______________________________

EXPLORING

The interaction of salt and water or oil

Now proceed one step further. Make two new separate puddles: one of water and the other of oil. At the edge of each puddle, place a crystal of salt so that it touches the surface of its puddle. Describe the interaction between the water and salt.

5 ______________________________

Using this short-term observation and your common knowledge, predict what the interaction will be between the salt and the water over time.

6 ______________________________

Describe the interaction between the oil and salt.

7 ______________________________

Do you think that oil will interact with salt in the same way as water does over time?

8 ______________________________

Discuss your answer with your neighbor; describe your reasoning process that linked your observations to your predictions. If your neighbor disagrees with your answer, ask your teacher for guidance.

EXPLORING

Magnet interactions

If you have not previously explored interactions of magnets, manipulate the two magnets until you can describe their interactions. What are the interactions?

9 ______________________________

____

Do the interactions depend on the orientation (position) of the magnets relative to each other?

10 ______________________________

Does a magnet have any attraction for your skin or for the metal rings of your notebook?

11 ______________________________

Describe how a magnet interacts with three other objects in the laboratory.

12 ______________________________

Some substances such as water and salt behave as two magnets when their areas of unlike charges attract each other. Using this information, give a possible explanation for the merging of water puddles and the interaction between water and salt crystals.

13 ______________________________

____

CONCEPTS/TERMS

Some useful terminology from chemistry

All physical things in the universe are made of matter. Chemistry is one science that studies matter. When biologists study living things, they study the matter or materials of life and use some of the terminology of chemistry. A few terms from chemistry will be useful to you. The smallest chemical unit of matter is an **atom.** There are many different kinds of atoms such as oxygen, hydrogen, gold, and calcium. Most atoms are attracted to other atoms, forming combinations that are **molecules.** Some molecules have only one kind of atom. Other molecules such as those of water have more than one kind of atom. Each molecule of water has two hydrogen atoms and one oxygen atom.

Thinking about your observations

The properties of molecules arise from their structure. A water molecule (chemically symbolized as H_2O) has weak positive charges *(+)* on its structure at the hydrogen (chemical symbol, H) locations, and a weak negative charge *(−)* at the oxygen (symbol, O) area (Figure 2-1, *A,* and Plate 2). Because water has charged areas like the poles of a magnet, water is a **polar molecule.** Thus like the model of the magnets, water molecules attract one another and hold together (Figure 2-1, *B,* and Plate 2).

This "hanging together" gives water its special properties that are so important to life. Because water molecules are attracted to one another, they form a surface where water and air meet or interface. This surface is strong enough to support some water insects and allows you to float a sewing needle. (If you have not observed this before, try it with a needle from the supply table.) The making and breaking of these attractions between water molecules result in a gain or a loss of heat energy; therefore water absorbs or releases a relatively

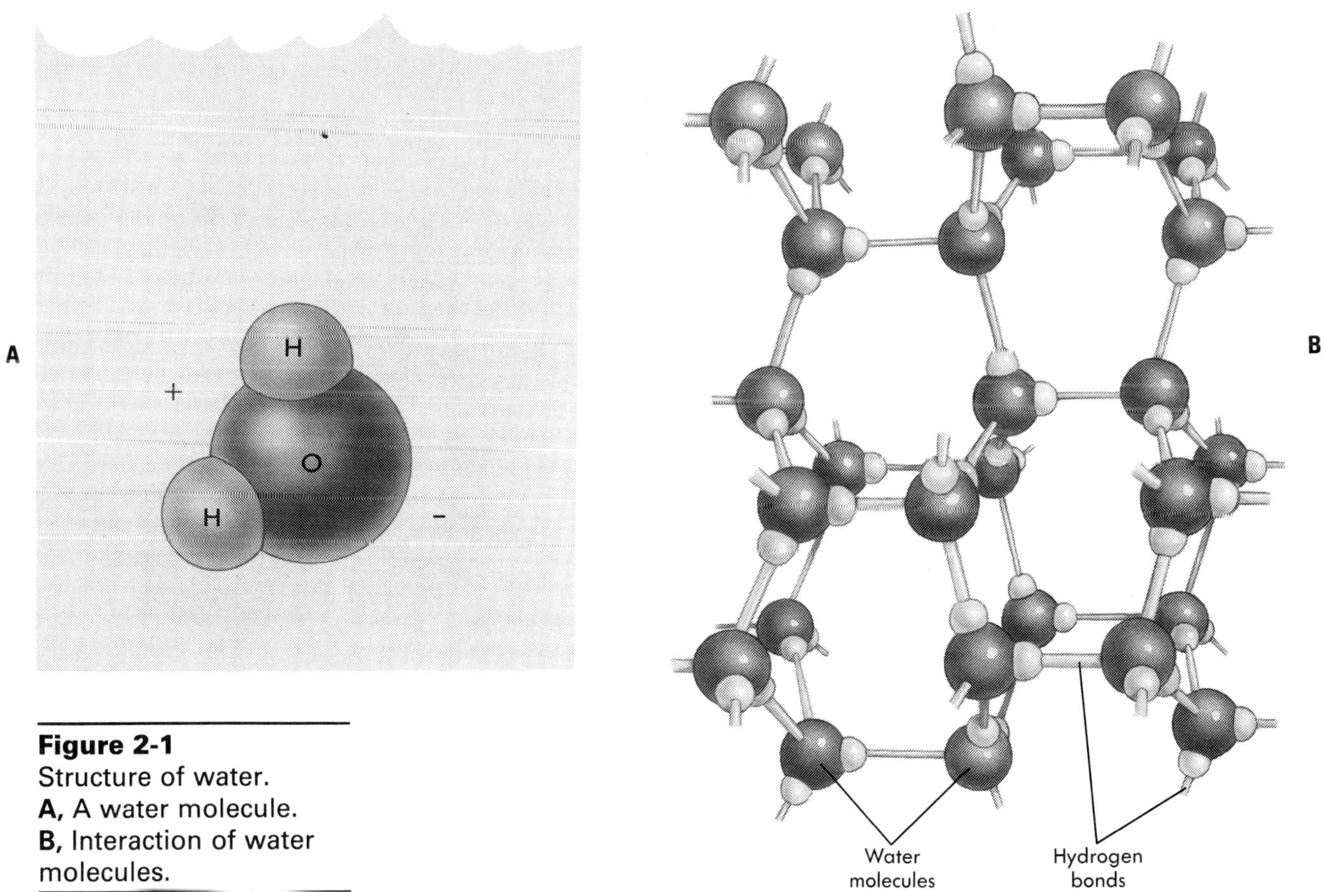

Figure 2-1
Structure of water.
A, A water molecule.
B, Interaction of water molecules.

large amount of heat while changing its temperature only a few degrees. This property is the reason a body of water moderates large swings in air temperature and provides a more constant environment for organisms.

In water, the atoms of some kinds of molecules separate, producing single atoms or groups of atoms with positive or negative charges. These charged atoms are **ions.** Table salt, sodium chloride, has two ions: sodium (symbol, Na^+) and chloride (symbol, Cl^-). In water, the positively charged sodium and the negatively charged chloride ions of salt each attract weakly charged water molecules. As water molecules surround Na^+ and Cl^-, the sodium and chloride ions are separated from each other.

The surrounding of individual particles by many molecules of a liquid such as water is the process of **dissolving.** The liquid is the **solvent** and the surrounded particles are the **solute.** Together the solvent and the solute form a **solution.** Salt water and syrup are examples of solutions. The salt and the sugar are the solutes in the respective solutions. Salt and other polar materials that attract water because they have charges are **hydrophilic** (water-loving). Hydrophilic substances dissolve in water.

As you observed, the oil behaved differently than the water. Oil was excluded from the water and failed to dissolve the salt. Oil molecules have no charged portions; therefore water molecules are not attracted to the oil and the two types of molecules fail to mix. Because substances such as oil are without poles, these substances are **nonpolar.** Because nonpolar substances are rejected by the bonding water molecules, they are **hydrophobic** (water-fearing) substances.

Considering the explanations for water-salt and water-oil interactions, give an explanation for the salt and oil interaction.

14 ______________________________

SCIENCE NOTE

Are magnets like molecules? The magnets you handled do not have the same size, shape, feel, or most of the other properties that are generally associated with a molecule such as water. However, charged portions of magnets do interact with charged areas of other magnets—much like the water molecules attracting one another. From your experience with two magnets, you can imagine how two water molecules position themselves with areas of opposite charges closely associated and with areas of like charges away from each other. In this way the magnets serve as models, *or hypothetical representatives, for the interactions among unseen water molecules.*

A model is not an exact replica of a phenomenon and therefore has limitations. For a model to be used by scientists, it must have a structure, property, or process that can be measured or observed that helps explain some of the characteristics of a phenomenon. A model is discarded if a better model is developed. Scientists also frequently use more than one model at a time because one is insufficient to explain all properties and activities of the phenomenon.

APPLYING

Interactions of stains and solvents

Now that you have a basic understanding of hydrophilic and hydrophobic interactions, try applying it in other situations. Obtain the following materials:

7 test tubes
1 test tube rack
1 bottle of oil
1 bottle of water
2 10-milliliter (ml) graduated cylinders
1 wax marker

1 dropping bottle of Sudan IV (a stain or biological dye)
1 dropping bottle of methylene blue (a stain)
1 Pasteur pipet
1 bottle of colored dish detergent

Number five tubes from one to five, then add the materials listed in the first column of Table 2-1 to the appropriately marked tube. (If you need help with the metric system of measure, see Appendix D.) Next, briefly shake the tubes and return them to the rack. In the second column, record your observations of the interactions among the materials in each tube. In the third column, give an explanation for the interactions of each stain and the solvents—water, oil, or both—in each tube.

Think back to your puddles of water and oil on the wax paper. Does the wax paper have a hydrophilic or a hydrophobic surface?

15 ______________________________

Which observations in the original experiment were important for your decision about the wax paper surface?

16 ______________________________

Table 2-1 Interaction of stains and solvents

Tube number and materials	Description of interactions	Explanation for interactions
1. 3 ml water and 1 drop Sudan IV		
2. 3 ml water and 1 drop methylene blue		
3. 3 ml oil and 1 drop Sudan IV		
4. 3 ml oil and 1 drop methylene blue		
5. 3 ml water 3 ml oil 1 drop Sudan IV and 1 drop methylene blue		

Name two common substances that are hydrophilic.

17 ______

APPLYING

A more complex interaction

Now apply hydrophilic and hydrophobic concepts to a slightly more complex interaction. To the two remaining test tubes add 3 ml of water and 3 ml of oil as before. With the Pasteur pipet, add 1 drop of colored dish detergent to the first tube where the water and oil meet or interface. Describe the interaction of the detergent with the water and oil.

18 ______

Briefly shake both tubes equally. Observe the tubes immediately after shaking and at 1, 2, and 5 minutes after shaking. Record your observations in Table 2-2.

In the third column, explain the differences in the interactions you observed over the 5-minute interval. Base these explanations on your understanding of hydrophilic and hydrophobic interactions.

Using these observations, explain how detergents remove greasy spots from your clothes.

19 ______

How do you think the concept involved in detergent action might be applied to oil spills in water and on land?

Table 2-2 Interactions of detergent with water and oil

Tube contents	Observations	Explanation
Water, oil, and detergent	0:	
	1 min:	
	2 min:	
	5 min:	
Water and oil	0:	
	1 min:	
	2 min:	
	5 min:	

20 ____

____

Describe any effect this environmental application might have on organisms living in the area.

21 ____

____

PART TWO Acids and bases

EXPLORING

pH and hydrogen ion concentrations

From the supply area, obtain the following materials:

1 wedge of red cabbage or a handful of solid-red coleus leaves
1 knife or single-edge razor blade
1 funnel
1 double-layered square of cheesecloth (about 10 × 10 cm)
2 250-ml beakers
1 bottle of distilled water
1 hot plate
1 pair of eye protective goggles for each person
1 oven mitt
1 10-ml or 25-ml graduated cylinder
1 dropping bottle of 1 N hydrochloric acid (HCl)
1 dropping bottle of 1 N sodium hydroxide (NaOH)
2 50-ml beakers
Access to a pH meter

CAUTION

The 1 N HCl is a very strong acid; 1 N NaOH is a very strong base. Both of these chemicals can burn your skin, damage your eyes, and make holes in your clothes. Handle with care! Wear goggles to protect your eyes. Follow your teacher's advice for cleaning up any spills.

Wearing protective goggles, chop enough coleus or cabbage leaves into thin shreds to fill one of the beakers to the 100-ml mark. Add enough distilled water to barely cover the leaves and place the beaker on the hot plate. Bring the cabbage and water to a boil and boil the leaves for 5 minutes at a moderate rate to extract the leaf pigments (colored substances). Using the oven mitt, move the beaker from the hot plate and let the beaker cool. Then strain the liquid into a clean beaker through two layers of cheesecloth laid in the funnel. Discard the boiled leaves. Add additional distilled water to bring the extract's total volume back to 100 milliliters. This is your "stock" extract.

The pH values of a solution can be measured by using an instrument called a pH meter. Read Appendix F, "Using a pH Meter."

What will happen when you add acid or base to red leaf extracts? Accurately measure 10 ml of your extract solution into a small beaker. Wearing goggles and following instructions in Appendix F, carefully measure the pH of your 10-ml extract using the pH meter.

CAUTION

If you have any questions about the operation of the pH meter, ask for assistance from your teacher; the electrode is easily damaged and is expensive to replace.

In Table 2-3, record the pH of the extract and its color. Add one drop of 1 N HCl to your extract. Gently swirl the beaker to mix the extract and acid, measure and record the pH of the extract, and note its color. Calculate the change in the pH units on adding one drop of acid by subtracting the new pH reading from the previous one. Keeping track of the number of drops, continue to add HCl, one drop at a time, until you have added a total of ten drops. After each addition, mix, measure, record, and calculate as before. After you have added the first 10 drops individually, add 10 at one time (for a total of 20 drops). Mix, measure, and record as before. Because you added 10 drops this time, when you calculate the change in pH divide the pH change by 10 to get an average pH change per drop of acid added.

Table 2-3 Effects of adding HCl and NaOH to red leaf plant extract

Drops HCl	pH	Change in pH/drop	Color	Drops NaOH	pH	Change in pH/drop	Color
0		X		0		X	
1				1			
2				2			
3				3			
4				4			
5				5			
6				6			
7				7			
8				8			
9				9			
10				10			
20				20			

Compare the change in pH units on adding each of the individual drops of HCl. Are these changes in pH the same for each drop of acid or do they differ?

22 ______________________________

If these changes differ, describe how they are different.

23 ______________________________

Set this beaker aside.

Measure 10 ml from your stock extract into another beaker. Using the pH meter, measure the pH of this portion of extract. If it is not the same as the initial pH of the first 10-ml portion, notify your teacher. Add drops as before to this new portion of extract, but this time use the 1N NaOH in place of HCl. As you add each NaOH drop, record pH and color changes of the extract in Table 2-3. Also calculate the change in the pH units for each drop of NaOH added to the extract.

CONCEPTS/TERMS

Relating observations to hydrogen ion concentrations

Materials that ionize, or break apart, producing positively charged hydrogen ions (H^+) are **acids.** All the molecules of some acids such as HCl ionize completely, producing a lot of H^+. These are strong acids. In other acids, only some of the molecules ionize, producing fewer H^+. These are weak acids.

Materials that can incorporate hydrogen ions, removing them from the solution, are **bases.** There are also strong and weak bases. Strong bases ionize readily and remove a lot of H^+ from the solution; weak bases do not. NaOH (a strong base) ionizes to Na^+ and OH^-; the OH^- combines with H^+ in the solution forming water and thereby removes the H^+ from the solution.

The pH units represent the concentration of hydrogen ions in solution. However, the pH scale does not have a direct linear relationship to the hydrogen ion concentration. Examine your data. In general, what happened to the pH as you increased the hydrogen ions of the extract by adding HCl?

24 ______________________________

____

In fact, the hydrogen ion concentration has an inverse relationship to pH. How would you define an inverse relationship?

25 ______________________________

Since there is an inverse relationship between pH and H^+, what is the relationship between pH and the strength of acids?

26 ______________________________

This inverse relationship is not the only relationship between pH and hydrogen ion concentration. Review the changes in the pH units as you added each drop of HCl to the extract. Would you agree or disagree that there is a greater difference in the amount of hydrogen ion represented by a pH change at low pH units than at higher pH units?

27 ______________________________

Check your answer with your neighbors and describe how you related these different pH changes to form your answer. If there is any disagreement, talk with your teacher.

This unequal relationship between the change in pH and the change in hydrogen ion is a logarithmic relationship. It is the same relationship that is used in the Richter Scale to measure the strength of earthquakes. In the logarithmic relationship (base 10) each unit represents a tenfold difference. Therefore a pH of 2 has 10 times the H^+ as does a pH of 3 (remember also the inverse relationship). The hydrogen ion concentration is expressed mathematically in the following equation:

$$\text{pH} = \log \frac{1}{H^+ \text{ concentration}}$$

The pH units range from 1 to 14. (For information on the ionization of water and more about the specific relationships among the hydrogen ion concentration, pH, acidity, and alkalinity [which refers to a base], see your textbook.)

Figure 2-2
The pH of some common solutions.

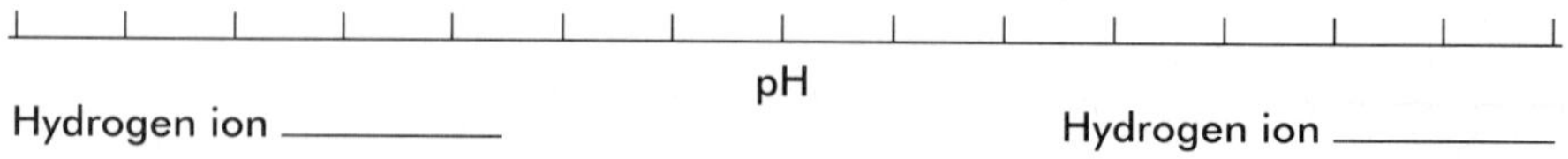

CONCEPTS/TERMS

Beneficial use of the relationship between pH and color changes

There are a number of environments besides well-equipped laboratories in which biologists may want to determine the pH of a solution. Because most pH meters are expensive and delicate, it is not practical to carry pH meters into a woods, field, or other rugged locations. At these times, biologists may determine the pH of a solution by adding a substance that changes to specific colors as the hydrogen ion concentration changes. By knowing the relationship between the color of that substance and its pH, the investigator can roughly determine the pH of materials without a pH meter. These compounds are called **pH indicators.** The red substance in red cabbage leaves is an indicator. Solutions of indicators are sometimes impregnated into slips of paper. One example is litmus paper.

APPLYING

Concepts of pH

The fluids we use every day have widely varying pHs. Your teacher has asked you to bring several liquids for pH testing to class. Other fluids will be available to you in the laboratory. Obtain the following:

3 small beakers, each with 10 ml of a test-fluid
Containers of pH papers
Access to a pH meter

Test the pH of these fluids using the pH meter and the pH indicator papers available in the laboratory. After determining each fluid's pH, pool your information with the pH information provided by other members of your class. Rank these fluids in an order based on their pH, from low to high. On the line in Figure 2-2, add numbers representing the pHs and indicate the relative hydrogen ion concentrations at opposite ends of this pH scale. At the appropriate pH, place the names of fluids tested by class members.

Compare the pH meter readings to the pH determined by the pH paper. Which method of determining pH is the most accurate?

28 ______________________________

Which observations were the bases for your answer to this previous question?

29 ______________________________

____

Table 2-4 Effects of adding HCl and NaOH to red leaf plant extracts with $NaHCO_3$

Drops HCl	pH	Change in pH/drop	Color	Drops NaOH	pH	Change in pH/drop	Color
0		X		0		X	
1				1			
2				2			
3				3			
4				4			
5				5			
6				6			
7				7			
8				8			
9				9			
10				10			
20				20			

PART THREE Buffers

EXPLORING

Modifying changes in hydrogen ion concentrations

Return to the effect of acids and bases on red leaf extract but with a slight modification. Repeat the one drop at a time addition of HCl and NaOH to 10 ml of stock extract as before, but this time add 10 drops of sodium bicarbonate solution ($NaHCO_3$) to the extracts *before* you start measuring and recording. Sodium bicarbonate is also known as baking soda. Record these data in Table 2-4.

Table 2-5 Effects of adding HCl and NaOH to egg white solution

Drops HCl	pH	Change in pH/drop	Drops NaOH	pH	Change in pH/drop
0			0		
1			1		
2			2		
3			3		
4			4		
5			5		
6			6		
7			7		
8			8		
9			9		
10			10		
20			20		

Compare the pH changes as acid or base was added to the extract with $NaHCO_3$ to the pH changes in the extract without $NaHCO_3$ (see Table 2-3). What differences occurred, and what do you think is happening?

30 **Differences** ______

Explanations ______

______

CONCEPTS/TERMS

Thinking about buffers

$NaHCO_3$ is one of a group of compounds called *buffers*. You are familiar with the action of buffers in stomach antacids; they modify or counteract acidity of the stomach. Buffers minimize the changes in the H^+ and OH^- concentration of solutions by combining with hydrogen ions when they are in excess and by donating hydrogen ions when they are low.

To understand the way buffers work, examine the activity of $NaHCO_3$. In solution, the dissolving $NaHCO_3$ ionizes, forming a bicarbonate ion (HCO_3^-):

$$NaHCO_3 \rightarrow Na^+ + HCO_3^-$$

When H^+ is available, as when adding HCl, the bicarbonate ion combines with it, forming a weak acid, carbonic acid (H_2CO_3). (A weak acid fails to ionize as much as a strong acid; therefore it does not contribute as many hydrogen ions to the solution.)

$$HCO_3^- + H^+ \rightarrow H_2CO_3$$

This reaction removes the excess hydrogen ions from the solution and decreases the change in hydrogen ion concentration.

When hydrogen ions are unavailable, instead of forming carbonic acid, the reaction reverses; hydrogen and bicarbonate ions are produced from the carbonic acid:

$$H_2CO_3 \rightarrow HCO_3^- + H^+$$

This reaction produces hydrogen ions, increasing their concentration in the solution.

Attaching to hydrogen ions when they are in excess and releasing hydrogen ions when they are in low supply is the mechanism by which buffers modify changes in pH.

APPLYING

Do egg whites have buffers?

See if you can determine whether or not egg white contains buffers. Obtain two 10-ml volumes of diluted egg white solution in separate beakers. Measure the solution's pH using the pH meter. Add drops of either HCl or NaOH to the beakers of egg white following the same procedure used to observe pH changes in the red leaf extract, except, there will be no color changes to observe. Record your data in Table 2-5.

Examine your data. Do egg whites have buffers?

31 ______________________________

Describe the observations that served as the bases for your decision.

32 ______________________________

If you think that there are buffers in egg white, answer the following question: At what pH level is buffering activity most prominent?

33 ______________________________

LABORATORY 3

Properties of the molecules of life

Diffusion and osmosis

PART ONE Diffusion

EXPLORING

Diffusion through gases

In one corner of the laboratory, at an identified time zero, your laboratory teacher will pour solution A on a surface at room temperature. As soon as you smell it, record the number of seconds from time zero and identify the solution in Table 3-1. Compare your time with others in the class. Do students sitting closer to the source of solution A smell it more or less quickly than you?

1 ______________________________

When did students farther from the source smell it?

2 ______________________________

At this time, what is the best explanation you have for your nervous system suddenly being able to detect the odor of solution A when you were not able to detect it before it was poured?

3 ______________________________

What explanation do you have for students farther from the source smelling it after you did?

4 ______________________________

Now try it with a different solution; at an identified time zero, the teacher will pour solution B onto a heated surface. Record the time for detection and name of the solution.

Table 3-1 Diffusion rates through air

Solution/name	Temperature (° C)	Time (sec)
A/ ________	Room:	
B/ ________	Hot:	

Inspect your data. Which of the solutions did you smell more rapidly?

5 ______________________________

You could smell solutions A and B shortly after they were poured on a surface; before that time you had no evidence either solution was in the room. Overall, did the molecules of solutions A and B move from an area of their high concentration or an area of a low concentration?

6 ______________________________

How do you think the molecules of solutions A and B moved?

7 ______________________________

Explain the thought processes you used to analyze and make a decision regarding the questions. In other words, what kind of information did you use and how did you put it together to arrive at your decision?

8 ______________________________

____

____

CONCEPTS/TERMS

Diffusion

Molecules move, gradually spreading to less concentrated areas. In fact, molecules move constantly except at the lowest possible temperature of absolute zero (0° K or −273° C). This spreading is similar to the scattering of ping pong balls after you drop them in a small area in the center of a table—more of the balls randomly move away from the drop area than stay in the drop area or move back into the area after colliding with other balls. This random movement is the mechanism, or way, by which molecules spread from areas of higher concentration to areas of lower concentrations. This process is **diffusion.**

Alternative hypotheses

Look back at your data. You detected solution B more quickly than solution A. What is a possible explanation for this difference in detection time?

9 ______________________________

Are you sure that your explanation is the only possible one?

10 ______________________________

Check with your neighbors; do they all have the same explanation?

11 ______________________________

In thinking about the experiment, did you notice that the solutions were not the only factors that differed in the procedures? What other factors could cause the time difference?

12 ______________________________

Perhaps the different detection time was caused by the chemistry of the two different materials. But it could also have been caused by the temperature difference. Do you have enough evidence now to decide which of these possibilities, if either, was the cause of the difference in detection times?

13 ______________________________

State at least two possible hypotheses for the speed with which you detected the two solutions.

14 1. ______________________________

2. ______________________________

(others) ______________________________

In this situation you have two or more **alternative hypotheses,** and you need to find out which hypothesis gives the best explanation for the observation, in this case the detection time difference.

SCIENCE NOTE

Scientists frequently propose more than one hypothesis as they consider all the information that is currently understood about a phenomenon and any new unassimilated information. After forming alternative hypotheses, scientists decide which provide the best explanations for the phenomenon and begin experimentally testing them.

EXPLORING

Types of solutions and temperatures

For this exploration, focus on two possible causal factors, the type of solution and/or the temperature of the solutions. Joining forces with two or three of your neighbors, form a specific prediction from each hypothesis concerning these causal factors. Design one or more experiments to test the predictions. Can you do this with just one experiment or do you need two or more?

15 ______________________________

Explain your answer to your group members. Below, state the prediction(s) and describe the procedures of the experiment(s).

16 ______________________________

____

____

____

____

____

____

____

____

After your group describes the procedures to your teacher, carry out the experiment(s). Record your results in a table and attach it to this manual.

After collecting data, review them. Summarize your group's data below, and describe how you linked them to arrive at a conclusion.

17 ______________________________

____

____

____

CONCEPTS/TERMS

Thinking about cause and effect

The community of scientists have verified the information that increasing the temperature of a substance increases the speed of its molecules. (This information is accepted as a scientific fact.) Think about the definition of diffusion and this statement. What does this statement combined with the definition of diffusion imply about the difference in detection times of solutions?

18 ______________________________

Does the conclusion of your experiment logically support or dispute this implication?

19 ______________________________

Describe how your conclusion relates to this implication.

20 ______________________________

____

If the conclusion of your experiment disputes the statement about temperature and molecular movement, discuss this with your teacher.

APPLYING

Concepts of diffusion to other materials

Working with your group members, obtain the following materials:

1 hard boiled egg
2 squares of unflavored gelatin (about 4 × 4 cm)
a cork borer, 3 mm
1 single-edge razor blade or scalpel
2 petri dishes
1 dropper bottle of methylene blue stain, 1% in water
1 ruler marked in millimeters
1 wax marker

Make wells that will hold fluid in the egg white and in the gelatin. To do this, cut the egg in half lengthwise. Discard the yolk. Using the cork borer, carefully cut one hole *halfway* through the white near each end of the two halves (Figure 3-1). Once the egg white is cut with the borer, carefully flip out the plugs of the wells with the corner of the razor blade or point of the scalpel. Also make two wells about 2 centimeters (cm) apart in each gelatin square.

Place one egg white half and one gelatin square in each of two petri dishes. Put one drop of methylene blue in each well. Cover and mark with your names. Set one of the petri dishes in the refrigerator; place the other in the warm area designated by your teacher.

At the end of the laboratory period, retrieve your plates and cut down through the middle of one well in the egg white and one in the gelatin—

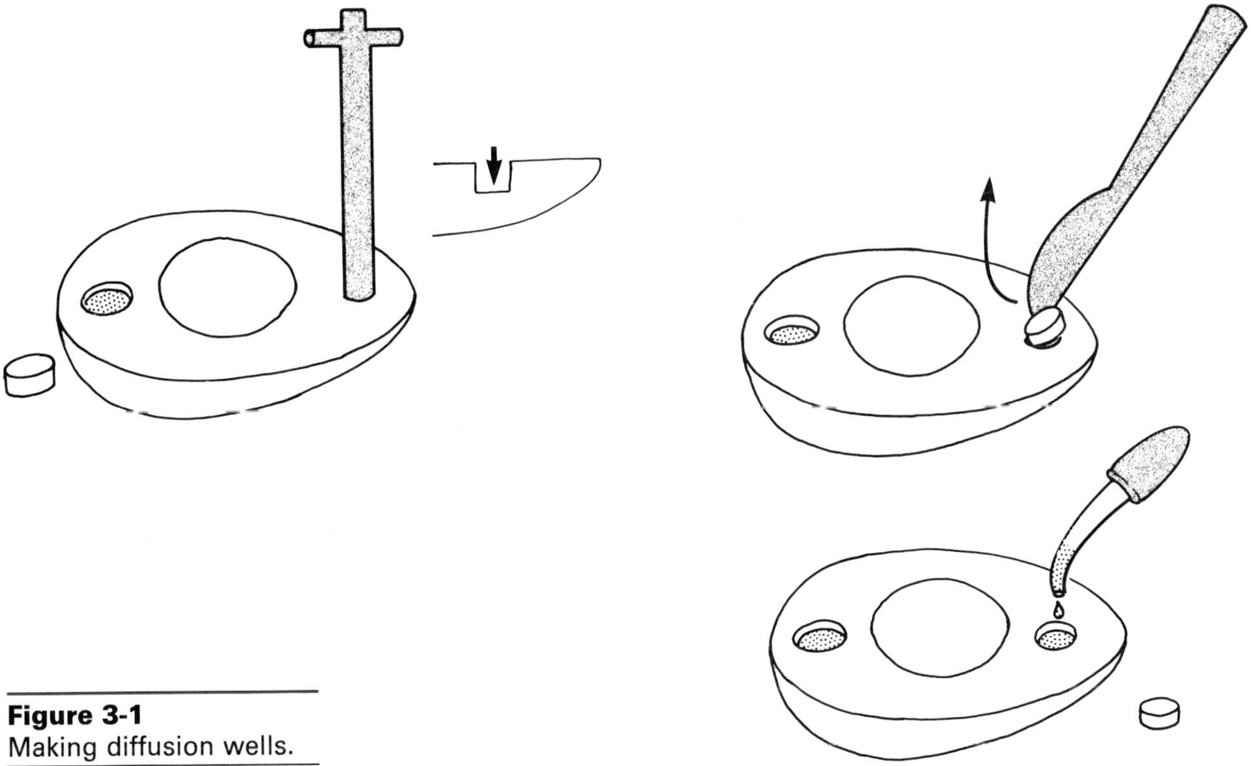

Figure 3-1
Making diffusion wells.

Table 3-2 Diffusion of methylene blue through solids

Material	Temperature (° C)	Distance travelled (mm)	
		Time: ____ hr	Time: ____ hr
Egg white	Cold ________		
	Warm ________		
Gelatin	Cold ________		
	Warm ________		

taking care not to spread the stain with your cutting tool. Measure the distance the dye moved into the egg white and gelatin from the edge of each well in millimeters (mm). Record your measurements in the third column of Table 3-2 and record the time since you put the stain in the wells at the top of the column. Return your plates to the specific temperature areas. Tomorrow or at a time indicated by your teacher, measure the diffusion distance for methylene blue from each of the other wells in your egg and gelatin, and complete Table 3-2.

Examine your data. What are your conclusions about the effects of temperature on diffusion of molecules through solid materials?

21 ____

____

Compare the molecules' speed of diffusion through solids to the diffusion of molecules through gases such as air.

22 ______________________________

____

Compared to speed of diffusion of molecules in gases and solids, what is your prediction of diffusion speed in liquids?

23 ______________________________

PART TWO Osmosis

EXPLORING

A special case of diffusion

Working in a group, obtain the following:

5 fingerbowls—each with 250 ml of one of the following solutions:
- Distilled water (0% sucrose)
- 10% sucrose (table sugar) solutions
- 20% sucrose
- 30% sucrose
- 40% sucrose

5 raw, deshelled eggs that are soaking in a sucrose solution
1 fingerbowl (to transport the eggs)
Paper towels or Kim-wipes
1 balance
1 weighing boat

With a paper towel, gently blot visible fluid from the surface of an egg. Place the egg in the weighing boat. Weigh the egg to the nearest 0.1 gram using the balance. (If you need instructions on use of the balance, ask your teacher.) Record the initial weight* of the egg in the second column of Table 3-3 at one of the five sugar concentrations. As soon as you record the weight of the egg, place the egg into its sugar solution. Weigh the remaining eggs, record their weight, and place them in the appropriate sucrose solution. At 30-minute intervals from the time eggs are placed in the solutions, reweigh each egg and record the weight in the designated column of Table 3-3.

After you complete the weighings for a specific time, calculate the percent change in the weight of the egg by the following equation:

$$\text{\% change in weight} = \frac{\text{Current egg weight} - \text{Initial egg weight}}{\text{Initial egg weight}}$$

You may wonder why it is important to calculate percent change in weight for each egg. Eventually, you want to compare the gain or loss in weight of each egg to each of the other eggs. Because the eggs initially have unequal weights, the change in actual weights presents a comparison problem, since a 1-gram change in a small egg will be a greater change for that egg than a 1-gram change in a big egg. (It's as if a race to determine the fastest student is run with contestants starting at different locations on the raceway.) To negate the effects of initial weight differences, you calculate the gain or loss as a percent-

*The terms *mass* and *weight* are not equivalent. Mass is the quantity of matter in an object as measured by its inertia (its tendency to remain stationary or to not change its movement if it is moving). Weight measures the force of gravity acting on an object. An example of the difference is: the mass of an object is the same on the moon as on the earth, but its weight is much less on the moon than on the earth. In this manual, the mass unit of grams will be used as the measuring unit associated with the process of weighing materials in the laboratory.

Table 3-3 Change in weight of deshelled eggs

Solution	Initial weight	Weight (gms) at specific time intervals: 30 min — Weight	30 min — % change	60 min — Weight	60 min — % change	90 min — Weight	90 min — % change
0% sugar							
10% sugar							
20% sugar							
30% sugar							
40% sugar							

age of the initial weight. If you need help understanding this idea, ask your teacher for an additional explanation. Did any of your eggs lose weight?

24 ______________________________

If an egg lost weight, will its percent change from its initial weight be positive or negative?

25 ______________________________

To present the data, plot the percent change on the graph. In biology it is customary to have the horizontal **X-axis** represent the independent variable (in this case, the length of time that the eggs were in the solution) and the vertical **Y-axis** represent the dependent variable (the percent change in weight). Because eggs could increase or decrease in weight, the "0" point on the Y-axis should be in the middle of that axis, with a decrease from the initial weight plotted below the "0" level and gain in weight plotted above the "0" level. Because five different concentrations of solutions were used, plot the data points for the egg in each specific solution separately and connect only those data points with a line. Use distilled water as meaning no (0%) sugar in the solution. The initial weight is the comparison weight; therefore it has no change and all eggs start at 0% change at time zero. When the graph is completed, it should have five lines, one for each egg. Label each line with the appropriate concentration level for that egg. Use the space of Figure 3-2 for your graph.

Can the data from your set of eggs can be applied to all deshelled eggs?

26 ______________________________

How could you—with information available in the class—increase your confidence about the change in weight of eggs in sucrose solutions?

27 ______________________________

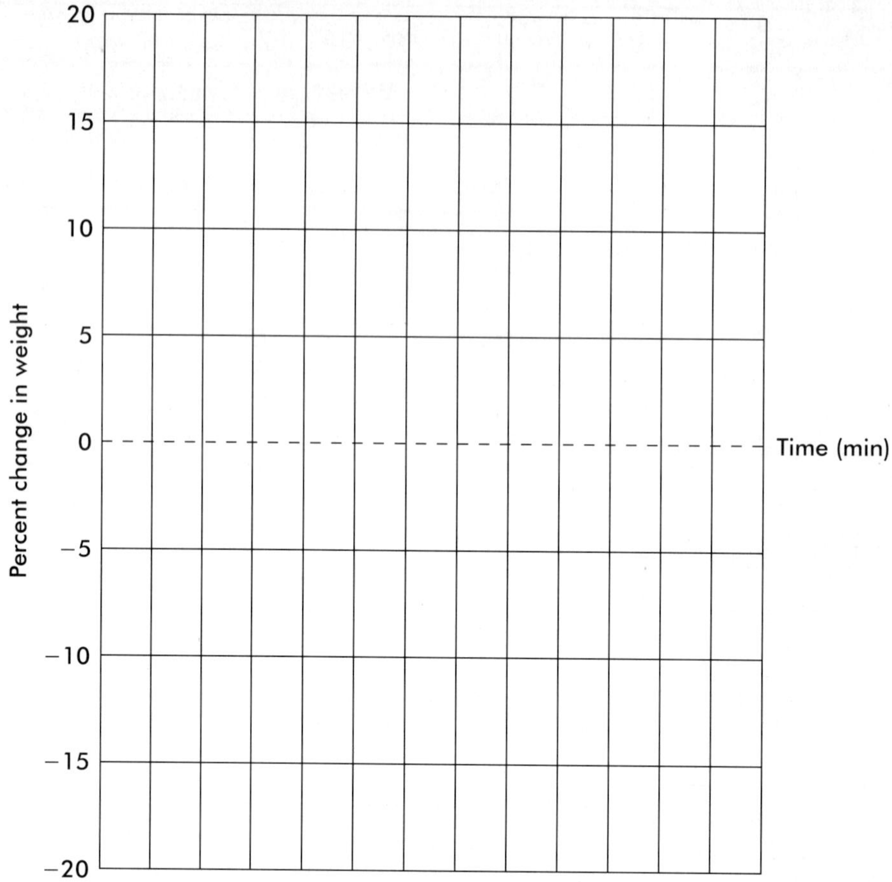

Figure 3-2
Effect of sucrose concentrations on mass of deshelled eggs.

CONCEPTS/TERMS

Diffusion, eggs, and sucrose

Now consider where diffusion fits into the picture of the eggs and sucrose solutions. The egg membrane is a differentiating structure; it permits some substances to pass through it but not others. Biologists call this type of membrane a **differentially permeable,** or **selectively permeable, membrane.** Water passes through this membrane but sucrose molecules cannot pass. In your experiment, which egg had the greatest amount of water diffuse into it?

28 ______________________________

Which observations formed the basis of your answer?

29 ______________________________

Using the concept of diffusion, explain the gain or loss of weight in each of the eggs:

30 0% ______________________________

10% ______________________________

20% ______________________________

30% ______________________________

40% ______________________________

As you might guess, biologists have terms for these conditions. These terms are based not on the differences in water concentrations but on differences in solute concentrations. When the solution around the egg has more water available to diffuse into the egg through the selectively permeable membrane than there is water inside the egg to diffuse out, biologist say that the solution is **hypotonic** to the egg. This term refers to the relative low (*hypo,* meaning "under") solute concentration outside the membrane (where water is in high concentration). When the solution has less water available to diffuse through the selectively permeable membrane than there is water to diffuse out, the solution is **hypertonic** to the egg. This term refers to the relative high (*hyper,* meaning "over") solute concentration outside the membrane (where water is in low concentration). When the diffusable water concentrations and the solute concentrations are equal on each side of the membrane, the solution is **isotonic** (*iso,* meaning "same") to the egg. This condition results in an equal volume of water moving in as out. Return to your explanations for the weight gain or loss and add these terms in the margin for each appropriate solution.

Water moving through a selectively permeable membrane from an area of higher concentration to an area of lower concentration is osmosis.

APPLYING

Osmosis and red blood cells

At the demonstration table, observe the commercially prepared slide of non-human blood with the high power objective of a light microscope. The vast majority of cells in view are red blood cells; in humans there are about five million of these in each cubic millimeter of blood. Diagram the shape of the red blood cell (RBC) on Table 3-4.

CAUTION

You will not be using human blood in this activity. However, it is best to handle all blood using safety procedures designed for human blood. The procedures include these safety techniques. Follow them. If you have any questions, ask your teacher.

Obtain the following materials:

3 clean glass slides
3 glass cover slips
1 wax marker
3 dropper bottles, each with one of the following solutions:
  5.0% salt
  0.9% salt
  Distilled water
3 toothpicks
1 pair of plastic gloves
1 pair goggles/student
1 bottle of laboratory bench disinfectant
1 bottle of hand soap

Table 3-4 Effects of salt solutions on red blood cells

Salt solution	Diagram of RBC	Explanation of RBC shape change
Demonstration slide		This is the standard shape, compare other slides to it.
5.0% salt		
0.9% salt		
0.0% salt		

CAUTION
Dispose of the toothpicks and any other blood-contaminated material, except your slides, in the designated disposal bag.

CAUTION
Place the blood slides in the container with disinfectant that is designated by your teacher. Put the gloves in the disposal bag, scrub your laboratory bench, and wash your hands.

Scrub the laboratory bench with the disinfectant and wash your hands. With a wax marker, identify the slides as follows: (1) 5.0% salt, (2) 0.9% salt, and (3) distilled water. Using the solution corresponding to each of the slides, apply one drop of the appropriate solution to the middle of the slides. Using plastic gloves and goggles, obtain a container of nonhuman blood from its storage area.

Using three different toothpicks, add a small portion of one drop of blood to each drop of fluid on the slides. Gently stir the blood into the fluid with its toothpick. Place a clean cover slip over each drop of fluid. Return the container of blood to its storage area.

Observe each slide with the compound light microscope using high power magnification. Compare the shape of the RBCs in each drop with the RBCs on the prepared slide. In the second column of Table 3-4, diagram a representative RBC from each solution.

Compare the RBCs in the slide on the demonstration table to the RBCs in your slides. Which fluids contain RBCs that appear enlarged or swollen?

31 ______

Which fluids contain cells that have broken and released their contents?

32 ______

Which fluids contain cells that appear shriveled?

33 ______

Which fluids have cells that are the shape and size of the RBCs in the demonstration slide?

34 ______

Do you think that the membrane, or covering, of a RBC is selectively permeable?

35 ______

If so, can water diffuse through its membrane?

36 ______

Can salt diffuse though its membrane?

37 ______

Using your understanding of osmosis, explain your RBC observations for each salt solution in the third column of Table 3-4. In your explanations, include the relative concentration of salt and water on both sides of the RBC's membrane and one of the terms *isotonic*, *hypotonic*, and *hypertonic*.

What are the similarities and/or differences between the phenomena observed in the RBCs and the deshelled eggs?

38 ______

LABORATORY 4

Cells

Building blocks of life

EXPLORING

Parts to understand the whole

What mental strategy do you use to thoroughly understand how a very complex thing is built and how it works? If you want to understand how a car works or why a especially good chocolate cake is so tasty, do you look only at the final product? Or do you start by asking questions about how parts of each of these things are built and work together?

How do scientists try to understand a complex system? It is common practice in science, just as in everyday life, to seek information about parts of the system. As we gain an understanding of these simpler parts, we have a base of information for building an understanding of the complex system. For example, if you understand how pistons, disk brakes, carburetors, cooling systems, and all the other components of a car work, you have a base of knowledge to develop a thorough understanding of how these parts work together to move the car. In investigating why a cake is so delicious, if you understand the function of each ingredient such as the flour, eggs, chocolate, baking powder, salt, and sugar, you have a base of information on which you can develop an understanding of how they work together to make the tasty product.

Life has many layers of complexity. Biologists regularly seek information about the "parts" of life; in fact, they seek information about parts of parts.

Think momentarily about some of life's layers of complexity. If we want to understand how a human population functions, we try to understand how the individual functions. Once we know this, we can study the more complex interactions among people.

If we want to understand how a human functions, we study the components, such as the circulatory or reproductive system. Then we investigate how these human systems interact to give us the emerging characteristics we associate with the complete human.

What is the simplest layer of life that is the basic organizational unit? Does it give information important to all life's complex levels? If so, what are the identifying characteristics of these basic "building blocks" of life? How are they structured and how do they work? Are all of the building units exactly the same? Or are there some differences that help explain differences in living organisms? These are some of the questions we will be exploring in this laboratory period.

CAUTION
Compound light microscopes are expensive and require proper use for achieving maximum resolution. Therefore if you have questions about the proper use of your instrument, seek help from your instructor.

Microscopic study

Is there a building block common to all living material, whether it is an oak tree, a cat, bacteria, or a mushroom, that is visible to the naked eye? Probably not. In fact, there are many organisms that are too small for your unaided eyes to see or resolve. When you explore living material seeking answers to the building block questions, you have to work at the microscopic level. Therefore this exploration requires a working knowledge of the compound light microscope. For information about this valuable tool, see Appendix A.

EXPLORING

For building blocks

Obtain your assigned microscope and materials for constructing a wet mount preparation. At the preparation table, prepare a wet mount of two to three long strands of the water organism *Spirogyra* (pronounced as SPY-row-GI-ra).

Using the lowest magnification, focus on one of the strands. The name *spirogyra* comes from the Latin words *spiro* and *gyr* meaning "spiral" and "round." From your observations, describe how this organism got its name.

1 ____

Carefully using the higher power objectives of your microscope, observe these strands. Adjust your light controls to obtain maximum visibility of the structures that have little contrast in color or density with neighboring structures. Practice fine focusing up and down through the organism.

Is the organism subdivided into parts or units?

2 ____

If so, how are these units spatially arranged to form the long strand or filament of *Spirogyra?*

3 ____

Compare the width of the unit to its length and depth. Using its width as a measuring unit, how deep is this unit?

4 ____

Perhaps some of the following questions can help you expand your explorations:

1. Are all units similar?
2. If there are similar units, what is the general shape of the units?
3. What are the dimensions of the units?
4. Do these units appear to have internal structures?
5. If there are structures within the units, what are their shapes, colors, and sizes?
6. Are boundaries of all units similar?

To estimate dimensions of a single unit, use your knowledge of proportional reasoning and the diameter measurement of the microscope's field of view. You can estimate this proportional relationship by the following steps:

1. Estimate the number of units arranged end to end in a line that could lay across the diameter of the field of view.
2. Set up two proportional relationships: the number of units across that field of view to the diameter of the field of view (you have this measurement from previous work on the microscope) and the number one (for one unit) to the unknown length, x.
3. Set these two proportional relationships equal to each other.

   Once your proportional relationships are set up, solve for x, the length of one unit, by cross-multiplying. Mathematically this can be expressed much more concisely as the following:

$$\frac{\text{Number of units}}{\text{Diameter of the field of view}} = \frac{1}{x}$$

Set up your proportional relationships in the space below and calculate both the length and width of the units. Solve for width in the same way as you solved for length. (See Appendix D for metric system conversions.)

Proportional reasoning

A short-cut method of determining dimension is to divide the diameter of the field of view by the number of units that fit across it. That is actually what you do as you solve for *x* in your equation.

In Table 4-1, draw two units of *Spirogyra* and fill in the second and forth columns from your observations.

Table 4-1 Building blocks of life

Organism	Size (μm)	Diagram of units(s)	Description
Spirogyra without stain			
Spirogyra with stain			
Hematococcus without stain			
Cheek lining without stain			
Cheek lining with stain			
Paramecium without stain			

NOTE ON BIOLOGICAL DRAWINGS

Drawings are a representation of the material you have observed. A drawing should be large enough to present all the details you have observed. Represent boundaries of structures with narrow lines. If there is repetition of structure, it is sufficient to draw the details in only one of the units or sections. Indicate depth only when it is essential to your observations. Shading is used sparingly and is done by stipling, that is, small dots placed close together.

Now stain the *Spirogyra* using the following procedure (Figure 4-1): Remove the slide from the microscope. Using a dropping bottle of malachite green, a biological dye, add a small drop at the edge of the cover slip. Touch a piece of paper towel to the opposite side of the cover slip, drawing the stain under the slip. Thoroughly observe your organism again.

To find building blocks common to all life, you need to explore other organisms. Prepare an unstained wet mount of the alga *Hematococcus* (he-MA-toe-COCK-cus) as you did with *Spirogyra* except add a small drop of a slowing fluid, methyl cellulose, to the slide before you add a drop of fluid containing the organisms. *Hematococcus* is a motile organism that is difficult to observe unless its movement is slowed. Asking the same questions as you did for *Spirogyra,* observe carefully. If this organism is constructed as only one complete unit and not several attached units, apply the questions (except those asking about interrelationships among the units) to the entire organism. Record your findings in Table 4-1. In what ways are *Hematococcus* and *Spirogyra* similar and different from each other?

5 Similarities ______________________________

____

Differences ______________________________

____

Prepare an unstained wet mount of the material that lines the inside of your cheek. Gently rub the broad end of a toothpick against the inside of your cheek and stir the material into a drop of water on a slide. Cover it. With the light intensity of the microscope decreased, observe this material carefully. Again ask the same questions as you did with the *Spirogyra* and *Hematococcus.* Add a drop of malachite green to this material and pull the stain underneath the slip as before. Record all of your findings in Table 4-1.

Explore one more organism before you generalize from your observations. Prepare a wet mount of *Paramecium* (PEAR-a-ME-see-um) with a drop of slowing fluid. Observe and record as you have previously.

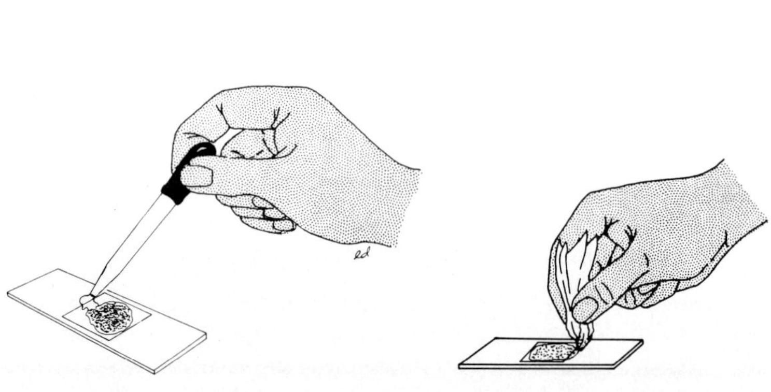

Figure 4-1
Staining a wet mount of cells.

CONCEPTS/TERMS

Generalizing from observations to concepts

List the names of those organisms that have attached units or building blocks.

6 ______________________________

Each individual unit is called a **cell.** List the organisms that lack cells.

7 ______________________________

Compare structures of the organisms that lack cells to the structures of individual cells in the many-celled (multicellular) organisms.

8 ______________________________

____

Do you think that the structure of organisms lacking cellular units is similar to the structure of individual cells in the multicellar organisms?

9 ______________________________

Do you think that all living organisms are composed of one or more cells?

10 ______________________________

On what observations did you base your answer?

11 ______________________________

____

____

Since the 1800s and the efforts of Schleiden and Schwann, biologists have agreed that all living things are composed of a basic, living building block—a cell. This generalization, or concept, coupled with a generalization of the scientist Virchow that living cells can only be produced by other living cells, is the **cell theory**. The structural components of cells that provide specific functions are **organelles.**

EXPLORING

Linking specific organelles to observed structures

The name, structure, function, and a general size range of some cell organelles are given in Table 4-2. These organelles vary greatly in size. Some can be seen with the compound light microscope; others are too small. (If used to its maximum potential at 400 magnification, the light microscope can resolve objects about 1 μm in size.) On Table 4-2, indicate with a check in the margin those structures that can be seen with your microscope. Use this general information about the size of the organelles to determine the identity of the internal structures you observed with your light microscope. Return to your record of observations in Table 4-1 and try to match the structures found in your recent exploration with the name of a particular organelle. Add these names to your diagrams in Table 4-1.

Do all cells have all organelles?

Did you observe all of the organelles listed in Table 4-2 that are large enough to see with the light microscope?

12 ______________________________

Table 4-2 Organelles: components of cells

Organelle	Structure and function	Size (μm)
Nucleus	Large membrane-bound area containing the genetic material; it governs the cell's activities	5-10
Nuclear envelope	Double membrane sac that holds the genetic material	0.02-0.03 (thickness)
Nucleolus	An area of the nucleus that produces ribosome components	1-3
Cytoplasm	All living parts of the cell except the nucleus	
Plasma membrane	A sheet of protein, fat, and other chemicals covering the cell; it is the cell's contact with its external environment	same as cell 0.007-0.011 (thickness)
Chloroplast	Double membrane structure, with stacks of internal membranes that use sunlight to produce sugars	0.8-20 (great variability)
Mitochondrion	Double membrane sac with inner folds that produce usable chemical energy	1-2
Ribosome	A set of two subunits that makes proteins	0.02-0.03
Smooth endoplasmic reticulum	Network of membranes forming channels in the cytoplasm; it modifies proteins and produces lipids	0.02-0.03 (diameter)
Rough endoplasmic reticulum	Network of membranous channels with attached ribosomes that makes proteins for export and lysosomes	0.06-0.15 (diameter)
Golgi apparatus	A series of pancake-shaped membranes that modifies, stores, and packages materials for export	0.3-0.7
Lysosome	Membrane-bound bag of digestive enzyme	0.04-0.4
Microtubules	Protein tubes that form a skeletal-like framework that shapes and supports the cell	0.025 (diameter)
Filaments	Protein rods that are elements of the cytoskeleton and involved with cell movement	0.006-0.011 (diameter)
Flagellum	Long, hair-like, surface projection of microtubules that moves with a whip-like motion	$0.3 \times >10$ (up to 200)
Cilia	Short, hair-like, surface projections of microtubules that move together with a wave-like motion	$0.3 \times <10$
Centrioles	Tubular arrangements of microtubules that organize structures for cell division in some cells	0.15 (diameter)
Cell wall	Layers of tough material outside of some cells that protect and provide rigidity	1-5
Cell coat	A layer of complex molecules attached to outer surface of cell that monitors cell's environment	1-15
Vacuole	Membrane-enclosed, fluid-filled space that stores water and wastes in some cells	Variable

If not, there are at least two possible reasons: (1) the cells you observed did not have these particular organelles or (2) the organelles were present and your aided eye did not distinguish them.

Consider the first possibility. Compare a *Spirogyra* cell to a cheek lining cell. Also compare *Hematococcus* to *Paramecium*. List the obvious differences between the cells in these two comparisons.

13 *Spirogyra* to cheek lining

Explain why you agree or disagree with the statement, "Some organelles are found only in some cells?"

14

CONCEPTS/TERMS

Distribution of organelles

Biologists find that some cells lack a few organelles that are present in other cells. Cells of plants, fungi, plant-like simple organisms, and bacteria have a nonliving cell wall. Animal cells and animal-like simple organisms lack cell walls. Plant cells and simple green organisms have chloroplasts; fungi, animals, and some simple organisms lack chloroplasts. Animal cells have centrioles; plant cells do not. A few cells have flagella or cilia; most do not.

However, there are some organelles that are present in all cells that you have observed during this laboratory period. List these organelles.

15

EXPLORING

Can you see all resolvable organelles?

You may not see all possible organelles in the cells you observe because these cells may lack them. There is also a second reason. Some organelles may be present but you cannot distinguish them from other parts of the cell.

Compare the cells you observed both with and without stain, such as *Spirogyra* and cheek lining cells. What structures are visible both with and without the stain?

16

What structures are visible only in the stained wet mount or that look considerably different in the stained wet mount?

17

Does the malachite green stain increase, have no effect, or decrease your ability to see the structure of these cells?

18

CONCEPTS/TERMS

Differential staining techniques

Most parts of living cells are colorless and transparent to humans. When the parts cannot be distinguished from each other even though they are resolving power size, they are said to be lacking contrast. What is one organelle in *Spirogyra* that has a color contrast with neighboring organelles?

19

Cells are treated with special dyes to increase the contrast among organelles that lack visible differences. The process of using these dyes is called **differential staining.**

Dyes, or **stains,** are complex, organic, colored chemicals that react differently with one organelle or type of chemical in the cell than with other organelles or chemicals. This increases the contrast of the stained part compared to the neighboring parts of the cell. Some stains bind to acidic components; other stains bind to basic ones. Some bind to polar materials; others bind to nonpolar materials. Stains can be used to identify and increase the contrasts of specific chemical substances at their normal site in the cell. An example of this application is the identification of fats by their uptake of nonpolar, fat-soluble stains, such as sudan IV. Stains are chosen to increase contrast among parts of cells that are under study.

SCIENCE NOTE

It is important to observe without distortion to determine details of our living world. Scientists sometimes reject new techniques because they believe the techniques produce artifacts or changes that fail to give a true picture of living material.

Name two organelles stained by malachite green.

20 ______________________________

EXPLORING

Widening your exploration of cells

Observe two other algae, *Anabena* (pronounced AN-a-BEAN-a) and *Gleocapsa* (GLEE-o-CAP-sa). Also observe *Gleocapsa* with stain. Record your findings in Table 4-3.

Observe the commercially stained microscope slide of bacteria on the demonstration table. (This scope has the oil-immersion objective in operating position, which provides a total magnification of 970 or 1000 times.) Add information about these organisms to Table 4-3.

Table 4-3 Building blocks of life

Organism	Size (μm)	Diagram of unit(s)	Description
Anabena without stain			
Gleocapsa without stain			
Gleocapsa with stain			
Bacteria with stain			

Review the information in Tables 4-1 and 4-3. Specifically compare *Spirogyra* and *Anabena*. List their similarities and differences in Table 4-4.

CONCEPTS/TERMS

Prokaryotic and eukaryotic cells

Biologists have found differences in the number of membrane-bound organelles in cells. Some cells contain a large number of membrane-bound organelles and are called **eukaryotic** (U-carry-AH-tic) **cells** (*eu* [G], "good"; *kary*, "a nut" or "nucleus"). One membrane-bound organelle in eukaryotic cells is a sac of genetic material, a nucleus. Other cells lack or have very few internal membrane-bound structures. They are called **prokaryotic** (PRO-carry-AH-tic) **cells** (*pro* [G] meaning "before"; *kary* [G], "the nucleus"). Their genetic material lacks a membrane boundary; therefore prokaryotic cells do not have nuclei.

The prokaryotic cells are the older type of cell. Do they have a more, the same, or a less complex internal structure than do eukaryotic cells?

21 ____________________

In general, what is the size of the more advanced eukaryotic cells compared to prokaryotic cells?

22 ____________________

____

In classifying organisms into the large groups called **kingdoms,** biologists put together all organisms consisting of prokaryotic cells. One of the five kingdoms, kingdom Monera, contains all prokaryotic organisms. All other organisms, those composed of one or more eukaryotic cells, are distributed into four other kingdoms: Protista, Plants, Fungi, and Animals.

APPLYING

Which organisms have eukaryotic cells?

Which of the two organisms, *Anabena* or *Spirogyra,* is a eukaryote?

23 ____________________

Compare cells of all the organisms you observed. Make a list of organisms that are prokaryotes and those that are eukaryotes.

24 **Prokaryotes** ____________________

____

Eukaryotes ____________________

____

Table 4-4 Comparison of *Spirogyra* and *Anabena*

Organism	Common characteristics	Differing characteristics
Spirogyra		
Anabena		

Linking electron micrographs to observed organelles

Figure 4-2 shows three photographs taken through a compound light microscope. The photographs include prokaryotic and eukaryotic cells; they also include a plant, an animal, and a bacteria cell. (The magnification of the material is provided for each photograph.) Figure 4-3 has photographs of cell organelles taken through the electron microscope, which magnifies and resolves much more than a compound light microscope. To Figure 4-2, add the type of cell (prokaryotic or eukaryotic) and the kingdom (plant, animal, or moneran) to form a title for each photograph. Also match the letter of the organelle in each electron micrograph of Figure 4-3 to the corresponding organelle in the photographs of Figure 4-2 or to its location if it is too small to see.

A B C

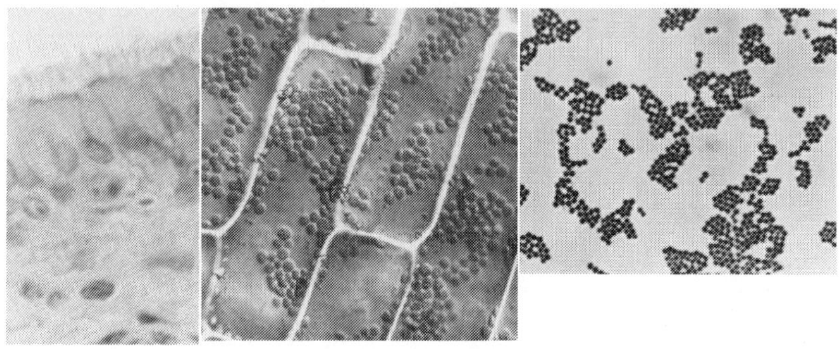

Figure 4-2
Compound light micrographs of cells.
A, ____________ (×400).
B, ____________ (×100).
C, ____________ (×970).

A B C

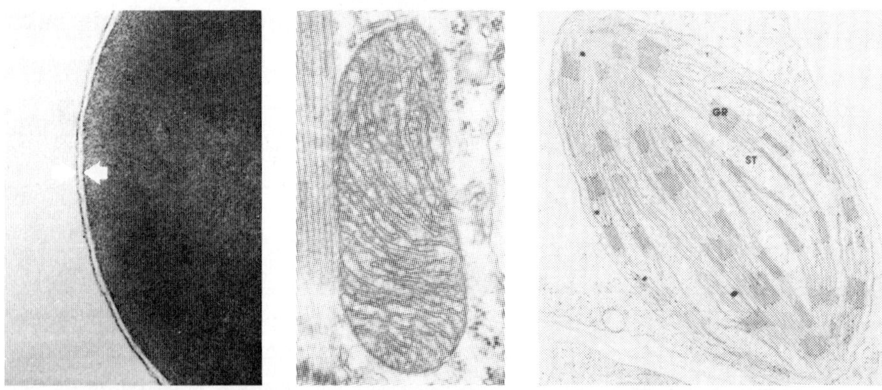

Figure 4-3
Electron micrographs of organelles.
A, ____________
B, ____________
C, ____________
D, ____________
E, ____________

D

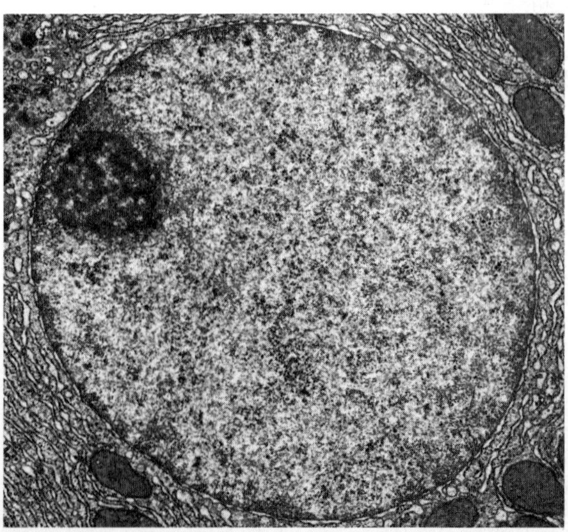

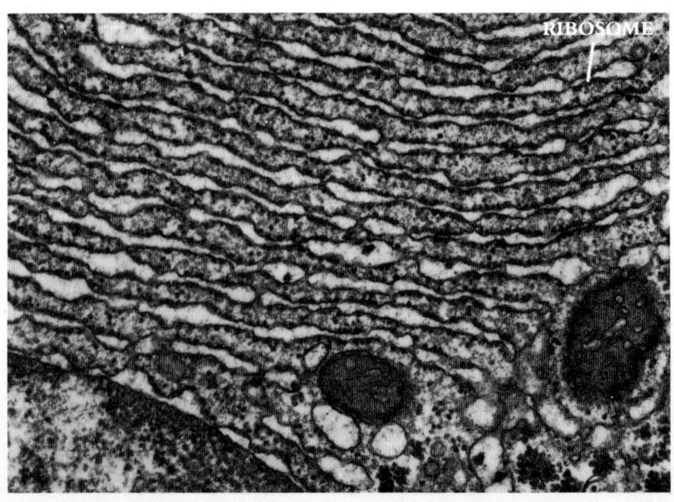

E

LABORATORY

5

Mitosis

Cell division for growth and repair

Cell division and genetic material

Organisms need cells for growth and for replacement or repair of old and damaged structures. In most eukaryotic (cells with nuclei) organisms, the process that provides new cells for these tasks is **mitosis.** This divisional process is one of the forms of cell reproduction called **asexual cell division.** The term *asexual* (the prefix *a* means "without") separates mitosis from another type of cell division that is involved in production of sex cells, such as eggs and sperm. This second type of cell division is called **meiosis.**

In the process of cell division, the parent cell passes instructions for life's activities to the resulting daughter cells. The molecules of deoxyribonucleic acid (DNA) in the parent cell contain the hereditary instructions. Located in the nucleus of eukaryotic cells, each DNA molecule consists of thousands of pairs of nucleotides and is associated with a large number of proteins. (Each nucleotide has a nitrogen base, a sugar, and a phosphate group.) Each cell in the body of an organism contains the same number of DNA molecules. In fact, each species of organism has its own characteristic number of DNA molecules in each cell. Among different organisms, the number of DNA molecules in nuclei varies from a few to several hundred.

EXPLORING

A separating problem

For a moment, explore a problem that cells may encounter as they divide into new cells. Obtain a container with two pieces each of several colors of sewing thread. For one minute, try to untangle these threads and sort them into two piles, each consisting of only one of the two threads of the same color. How many threads did you untangle and sort?

1 ______________________________

The DNA-protein complex is many times longer than it is wide, just as the threads are. In fact, the long DNA molecules, even with associated proteins, are too narrow (about 0.002 μm) to be seen in most types of microscopes. What is a possible problem that cells may have as they divide into new cells?

2 ______________________________

____

Describe how you think cells could overcome this problem.

3 ______________________________

____

CONCEPTS/TERMS

Packaging the genetic material

In the dividing cell, each DNA molecule coils many times into a length less than one one-thousandth (1/1000) or even one seven-thousandth (1/7000) its actual length. It is this intricately coiled (or **condensed**) DNA-protein structure, not the simpler double helix, that you see with the microscopes in the laboratory.

The slightly coiled DNA with its associated proteins is **chromatin.** The structure with the condensed genetic material is a **chromosome.**

EXPLORING

Mitotic cell division in plants

View the film of mitotic division in plant cells. Using time-lapse photography and several different kinds of microscopes, this film records the events occurring in plant nuclei during mitotic cell division. At times the coiled, condensed genetic material appears similar to strands of colored spaghetti. At other times it may appear as shadowy strands. In addition to the packaged DNA strands, you will see circle-shaped vacuoles and small, moving mitochondria. Follow the film through one time, concentrating on the condensed strands of DNA, particularly when these strands appear to move apart. Then run it a second time; if your projection equipment allows you to freeze the action, stop the film as needed for a more thorough inspection of the events. Notice that different nuclei were initially photographed at different points in the division sequence and photographed for different lengths of time. Several nuclei may be visible simultaneously, each at a different point in the process.

Specifically describe this sequence of events occurring in the nuclei of these cells. Begin your detailed description at the point when the chromatin begins condensing into chromosomes. This occurs after the nuclei have been through interphase.

4 ____________________

____

____

____

In Figure 5-1, draw the structure of a chromosome when it has structures identified in the film as sister chromatids. What happens to the sister chromatids of each chromosome as division occurs?

5 ____________________

____

Figure 5-1
Structure of a chromosome with sister chromatids.

Results of mitotic cell division

If each sister chromatid is formed from a DNA molecule exactly like the DNA forming the companion sister chromatid, describe the end result of mitotic cell division in regard to the genetic instructions in each of the new cells.

6 ______________________________

____

Mitotic cell division in animals

What differences, if any, would you expect in the mitotic process of plant and animal cells?

7 ______________________________

Explain your answer.

____

Observe the film showing mitotic division in animal cells. Compare the details of this sequence of nuclear events with the sequence in plant nuclei. Describe all similarities and differences in mitosis between these two types of organisms.

8 **Similarities** ______________________________

Differences ______________________________

CONCEPTS/TERMS

Terminology associated with mitotic cell division

Cell cycle

The sequence of events in a eukaryotic cell from the time it is formed by the mitotic division of its parent cell until it divides to form two new cells is the **cell cycle.** As shown in Figure 5-2, during most of the cell cycle, the cell is not dividing. This nondividing phase of the cell is **interphase.** The activities of the cell when it is forming two daughter cells like itself are **mitosis** and **cytokinesis.**

Interphase

During the lengthy interphase, DNA molecules are not coiled into chromosomes. They are "strung out" as chromatin. In this part of the cell cycle, chromatin carries on its usual activities; it provides instructions for making proteins and duplicates of each DNA molecule. Contained within the membrane-

Figure 5-2
Cell cycle.

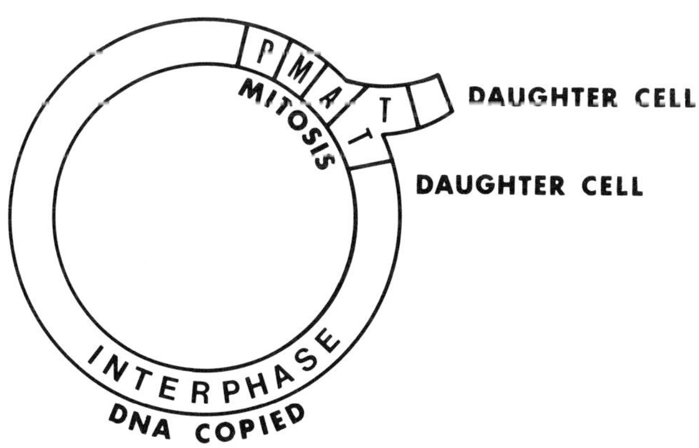

bound nucleus, the chromatin is easily stained. This form of DNA gives a granular appearance to the nucleus when viewed with the high power objective of laboratory microscopes. Circular nucleoli are also frequently visible within the nucleus.

Asexual cell division

One of the two processes that occur as a eukaryotic cell divides into two new cells is mitosis. It distributes one of the two DNA copies of each kind of parental DNA molecule to each of the two new nuclei. For easy reference, biologists break the continuous sequence of mitotic events into four phases: **prophase, metaphase, anaphase,** and **telophase.**

The other process in asexual reproduction is the division of the cytoplasm or cytokinesis. This process separates the parent cell's cytoplasm into two parts, one for each of the new nuclei.

Mitosis

Prophase. As the cell moves from interphase into prophase of the division sequence, the granular appearance of the nucleus becomes more evident until short, dense, strand-like chromosomes are visible under high magnification with the light microscope. At this time, each chromosome consists of two strands. Each strand consists of one condensed copy of a DNA molecule that was duplicated during interphase. These strands are attached to each other at a narrow section of the chromosome, the **centromere.** Each strand is a **chromatid.** Because each of the chromatids of a chromosome has a "kinship" with the other (each contains one copy of the same DNA molecule), these two chromatids are called **sister chromatids.** Other events occurring in prophase are the breakdown of the nuclear membrane, the disappearance of the nucleoli, and the formation of microtubule fibers that form an elongated, football-shaped structure, the **spindle.** Two spindle fibers, one extending from each end of the spindle, attach to each chromosome's centromere.

Metaphase. As the spindle fibers attach to each two-chromatid chromosome, the chromosomes move to a region in the middle of the spindle, the metaphase plate. Then each chromosome's centromere divides, providing a centromere for each of the chromatids. This marks the end of metaphase and the beginning of anaphase.

Anaphase. In anaphase, each centromere and its chromatid is pulled away from its sister chromatid as the fiber attached to each centromere shortens. At this point, each separate centromere and attached "chromatid" is called a *chromosome.* In anaphase, each chromosome has only one DNA molecule, which is coiled into the one strand of the chromosome. As the two chromosomes separate from each other, they move to opposite ends of the spindle. The regions at the end of the spindles are called *poles.*

Telophase. When all of the single-stranded chromosomes reach the two poles, the DNA partially uncoils and the nuclear contents re-form all of the structures of the interphase nucleus, such as the nucleoli and the nuclear membrane; the mitotic spindle breaks down. This process is telophase.

The end result of mitosis is the equal distribution of one of each kind of DNA molecule to each of the two resulting nuclei. Thus two nuclei are produced, each like the other and both with one copy of each kind of DNA molecule that was in the original parent nucleus.

Cytokinesis

After the nuclear division or, frequently, at the same time as telophase, the cytoplasm is divided into two parts. One of these parts surrounds each nucleus. There seems to be no mechanism to divide the cytoplasm exactly. This contrasts with the precise distribution of the nuclear material.

In the two films, you observed that the cytoplasm is divided differently in plants cells than in animal cells. In plants, what structure separates the cytoplasm in the vicinity of one nucleus from the cytoplasm near the other nucleus?

9 ______________________________

Describe how the cytoplasm is divided in animal cells.

10 ______________________________

Results of asexual cell division

What is the end result of asexual cell division in regard to hereditary information in the two new cells?

11 ______________________________

Frequently, the original cell is called the **parent cell.** The two products of mitosis and cytokinesis, which have the same hereditary instructions as the parent cell, are daughter cells. Each daughter cell now begins its own cell cycle.

EXPLORING

The location of mitotic cell division

If mitosis provides cells for growth, replacement, and repair, where do you think mitosis is occurring in your body?

12 ______________________________

Do plants and animals have different growth patterns?

13 ______________________________

Use whatever information you have about plants to answer the question, Where does mitotic activity occur in plants?

14 ______________________________

Hint: If you hammered a nail into a tree and returned 5 years later, you would observe that (1) the tree was taller, (2) the nail was embedded deeper into the trunk, and (3) the nail was still at the same height as 5 years earlier.

CONCEPTS/TERMS

Growth patterns

Plants and animals have differences in the location of mitotic cell division. This difference results in different growth patterns. Mitotic activities in young animals occur almost everywhere, and therefore animals increase in size throughout their bodies. Mitotic activity decreases as animals age. In contrast, mitotic cell division occurs in specific areas at the tips of plants (including trunk apex—its top, branches, and roots). The tips of plants grow. Some plants also have division in a layer near the surface of their stem. These plants also increase in circumference. Most plants continue growing throughout their life. In plants, the areas of cell division are called *meristem regions.*

APPLYING

Mitosis in dividing onion root cells

Materials and technique

Observe the events of mitosis occurring in the meristem of the onion root tip extending from a germinating (sprouting) onion seed or growing from an onion bulb. To do so, you need to prepare the cells for viewing. An easy technique to use is the **squash preparation.** This technique includes squashing the dividing tissue to obtain one layer of separated cells, which permits easy viewing of nuclear events through the microscope. Do you think that this technique is used to study the relationship between the nucleus and the plasma membrane of the cell or the relationship between cells?

15 ______________________________

Obtain the following materials:

- 1 150-ml beaker filled one-third with water and a few boiling beads
- 1 watch glass (a curved glass disk)
- 1 dropper bottle of aceto-orcein stain
- 1 dropper bottle of HCl acid
- 2 glass slides and cover slips
- Paper towels
- 2 pair of goggles (one for each student)
- 1 dissecting needle
- 1 pair of forceps
- 1 scalpel or single-edge razor blade
- 1 hot plate
- 1 oven mitt or clothespin
- 2 compound light microscopes

Procedure for the squash preparation

1. Wearing goggles, boil water in a 150-ml beaker on the hot plate.
2. Add nine drops of aceto-orcein (a dye that stains DNA) and one drop of HCl acid to the center of a watch glass.
3. Observe the onion root. Particularly notice the shape of the tip. In Figure 5-3, which shows the longitudinal section of a stained onion root, note the location of the meristem region.

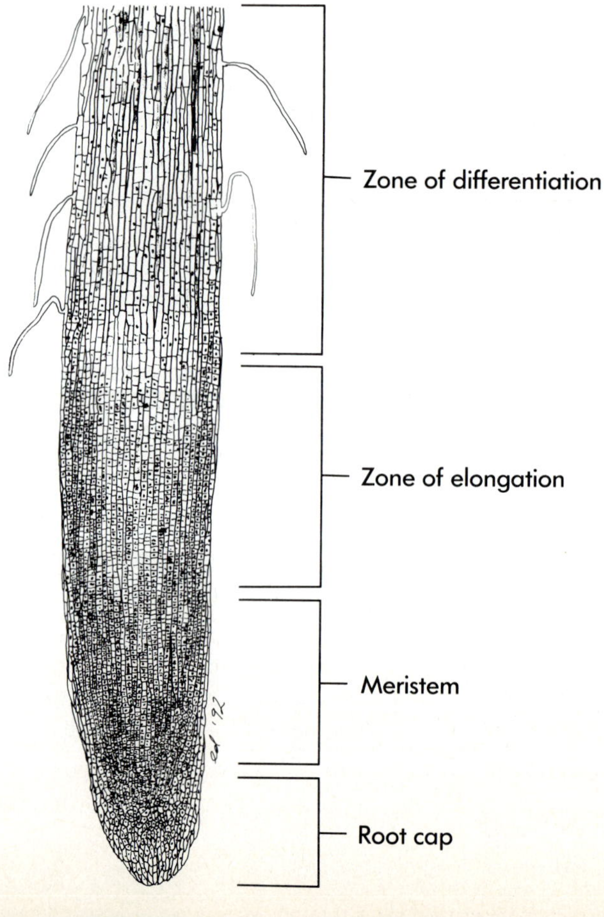

Figure 5-3
Onion root tip.

4. Cut off an entire root. Cut directly across the root so that you can easily distinguish the growing tip from the cut end.
5. Immediately submerge the tip of the root in the aceto-orcein and HCl puddle on the watch glass. Keep the root tip immersed in the dye solution. Two to four roots may be processed at one time. Considering the structure of plant cells, what do you think is the function of the HCl in preparing this squash?

16 ______________________________

6. Set the watch glass with the root (or roots) on the beaker of boiling water. Heat for 1½ minutes. If a small puff of smoke rises from the stain before 1½ minutes have elapsed, immediately remove the watch glass from the heat.
7. Holding the *cut* end of the root with forceps, move the stained root to a clean glass slide that has a drop of fresh aceto-orcein stain and *no* HCl. Cut *only* 1 mm of the tip from the root and move the tip into the drop of stain (Figure 5-4, *A*). Set aside the remainder of the root. It contains specialized, nondividing cells with thick cell walls too tough to be dissociated into one layer of cells. (If these cells remain attached to the tip, the squash will be useless.)
8. With a clean glass cover slip, cover the stain puddle with the 1-mm tip. Place the slide between paper towels on a flat surface. Gently apply downward pressure on the cover slip with your thumb (Figure 5-4, *B*). Why should you avoid lateral movement when squashing?

17 ______________________________

Observation of the squash preparation

Observe the slide under both low and high power of the microscope. Many nuclei will be in interphase, identified by their distinct membrane-bound structure and nucleoli. Search for an area of square-shaped meristem cells with groups of chromosomes. (These groups look similar to plates of red spaghet-

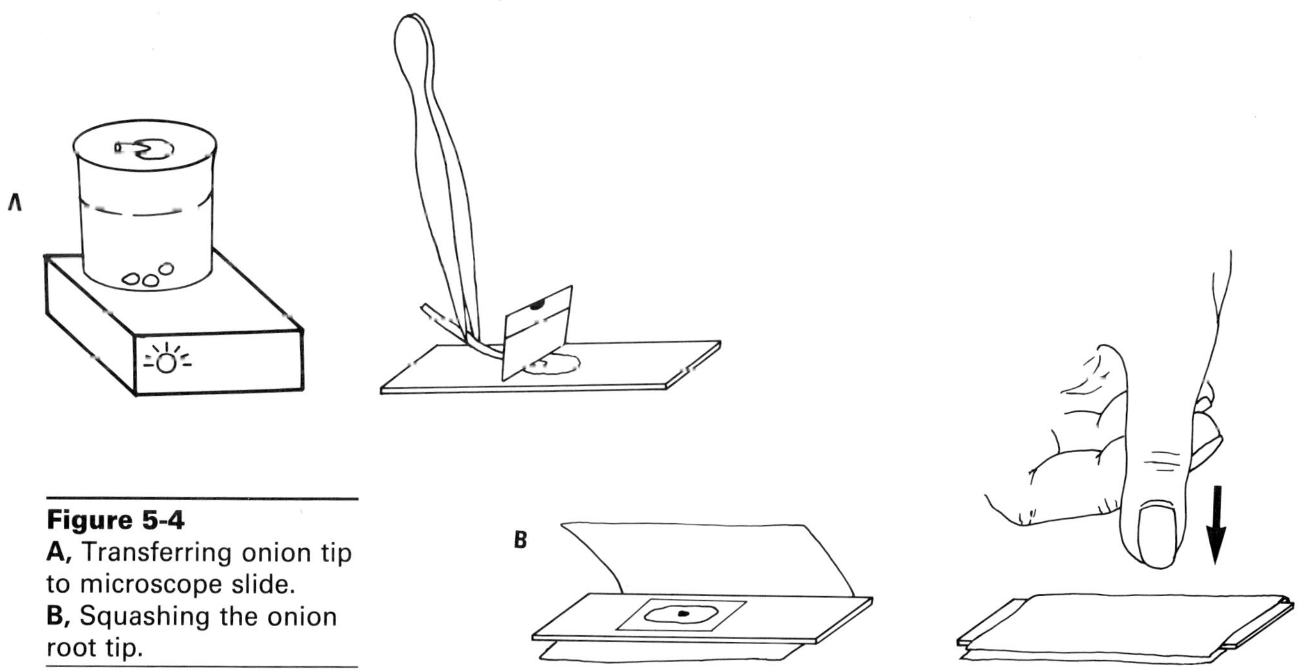

Figure 5-4
A, Transferring onion tip to microscope slide.
B, Squashing the onion root tip.

Figure 5-5
Mitotic cell division in whitefish.

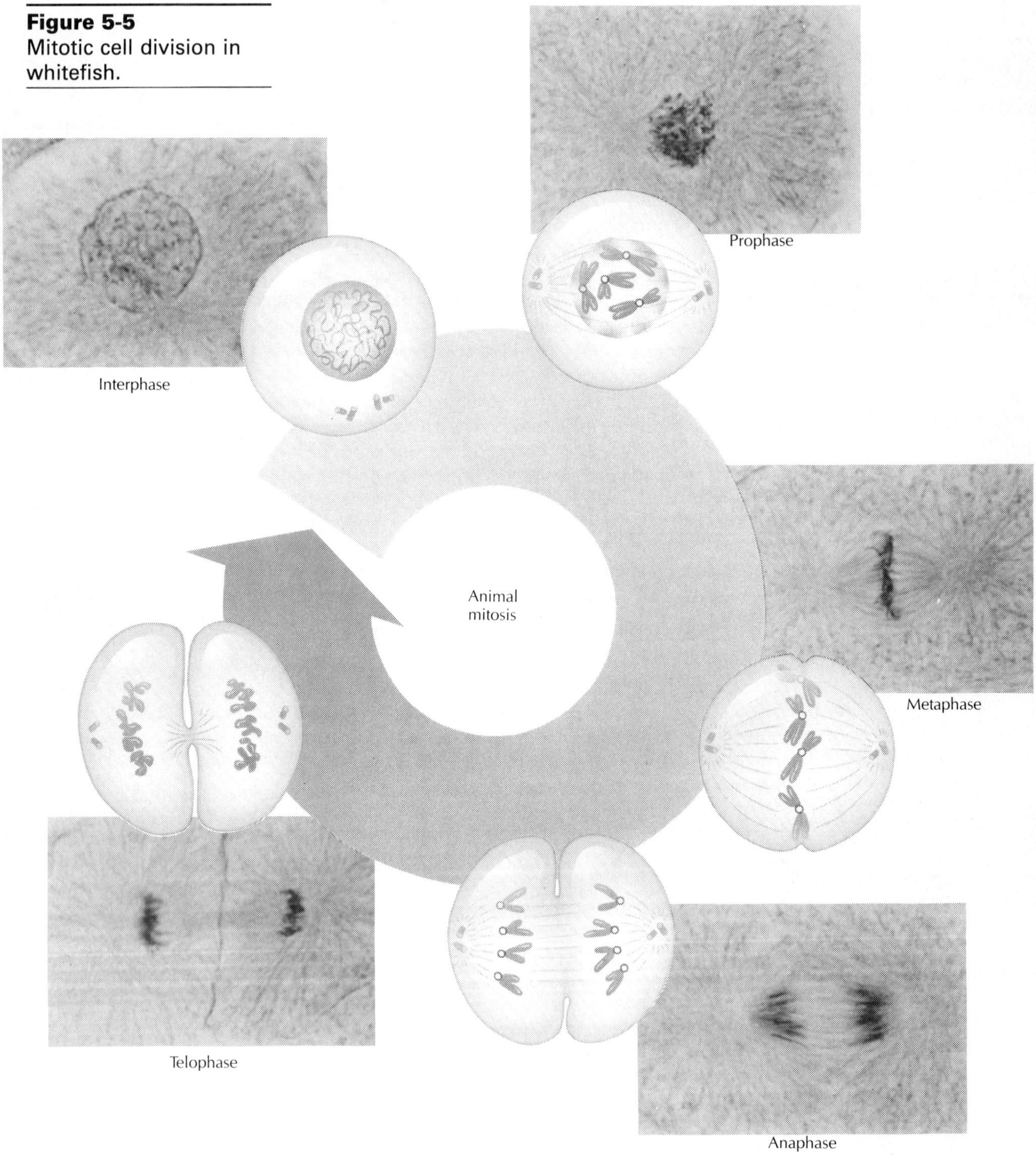

ti.) Compare the arrangements of chromosomes in the onion cells to the photographs of mitosis in whitefish (Figure 5-5 and Plate 3). Compare them also to the phases of mitosis you saw in the films of plant and animal cell division. Identify all four phases of mitosis and the cells that appear to be between phases. Compare cells with your neighbor. Using a cell in metaphase or early anaphase, try to determine the number of chromosomes in the onion. How many chromosomes does the onion have?

18 ______________________________

How closely do you and your neighbor agree?

19 ______________________________

Modeling the mitotic process

Obtain some paper clips and eight strands of yarn—two long pieces and two short pieces of one color, and two long pieces and two short pieces of a second color. Pair yarn strands to model the structure of chromosomes composed of sister chromatids. Use the paper clips as centromeres. In Figure 5-6, diagram your complete set of chromosome models. Identify the yarn color by outlining with colored pencils or shading in different patterns such as stippling or crosshatching. How many chromosomes are in your set?

20 ______________________________

How many chromatids?

21 ______________________________

How many centromeres?

22 ______________________________

How many DNA molecules?

23 ______________________________

Figure 5-6
Yarn chromosome models.

On a flat surface, move these yarn chromosomes through the mitotic process. When you are convinced that you understand this process, demonstrate it to your neighbor and model it for your instructor.

How many chromosomes are in each daughter cell?

24 ____________________

How many centromeres?

25 ____________________

How many DNA molecules?

26 ____________________

Experiment: effects of a chemical on mitosis (group experiment)

Do various chemicals such as nicotine, alcohol, caffeine, and colchicine affect mitosis? Seek information about this question using the scientific inquiry process. If you would like to review some aspects of the scientific process (see Laboratories 1 and 3), do so before initiating this project. Your instructor may ask that you design the experiment in groups or as a class.

Here are a few suggestions. Before you plan an experiment to test the effects of a chemical, you must decide which chemical (the independent variable) you want to investigate. Next, you need to form a hypothesis about the effects of that chemical on mitosis. In the experiment to test a prediction from the hypothesis, you can use squash preparations of roots to compare the percent of cells in each phase in one set of roots to the percent in each phase in another set of roots.

After designing your experiment, discuss it with your teacher. Be prepared to provide the following:

1. Hypothesis and prediction
2. Independent and dependent variables
3. A detailed procedure
4. A list of required equipment and supplies with the exact quantities needed

LABORATORY **6**

Meiosis and life cycles of organisms

PART ONE Meiosis: cell division for sexual reproduction

EXPLORING

Is there a use for a special division?

In animals as well as plants and many simpler organisms, two cells unite, initiating the formation of a new organism. The nuclei of these two cells contributed by the parental organisms fuse in the process of **fertilization.** The cells that unite are called **gametes.**

Let's consider a problem that may occur in the formation of these gametes and the resulting generations by examining humans. In the nuclei of your body there are 46 different chromosomes, each with a different molecule of DNA. How many chromosomes will be in the cells of your offspring if you and your mate produce gametes by the process of mitosis?

1 ______________________________

In Figure 6-1, write the number of chromosomes in the nuclei for each of the four generations shown. Assume that the gametes of both partners are always formed by mitosis.

What likely effect will the number of chromosomes in each cell of your great-grandchildren have on the size of the cells?

2 ______________________________

What problems might these cells have with importing sufficient nutrients from the environment and exporting their wastes?

3 ______________________________

____

Is there a use for a kind of division that is different from mitosis?

4 ______________________________

The increase in chromosome number in each cell from 46 to 368 in only four generations certainly presents a problem for the cell. The nuclei in your great-grandchild's cells will have to increase in volume about eight times just to fit all the additional chromosomes into each nucleus. You can imagine how this problem increases as the chromosomes double in number with each succeeding generation.

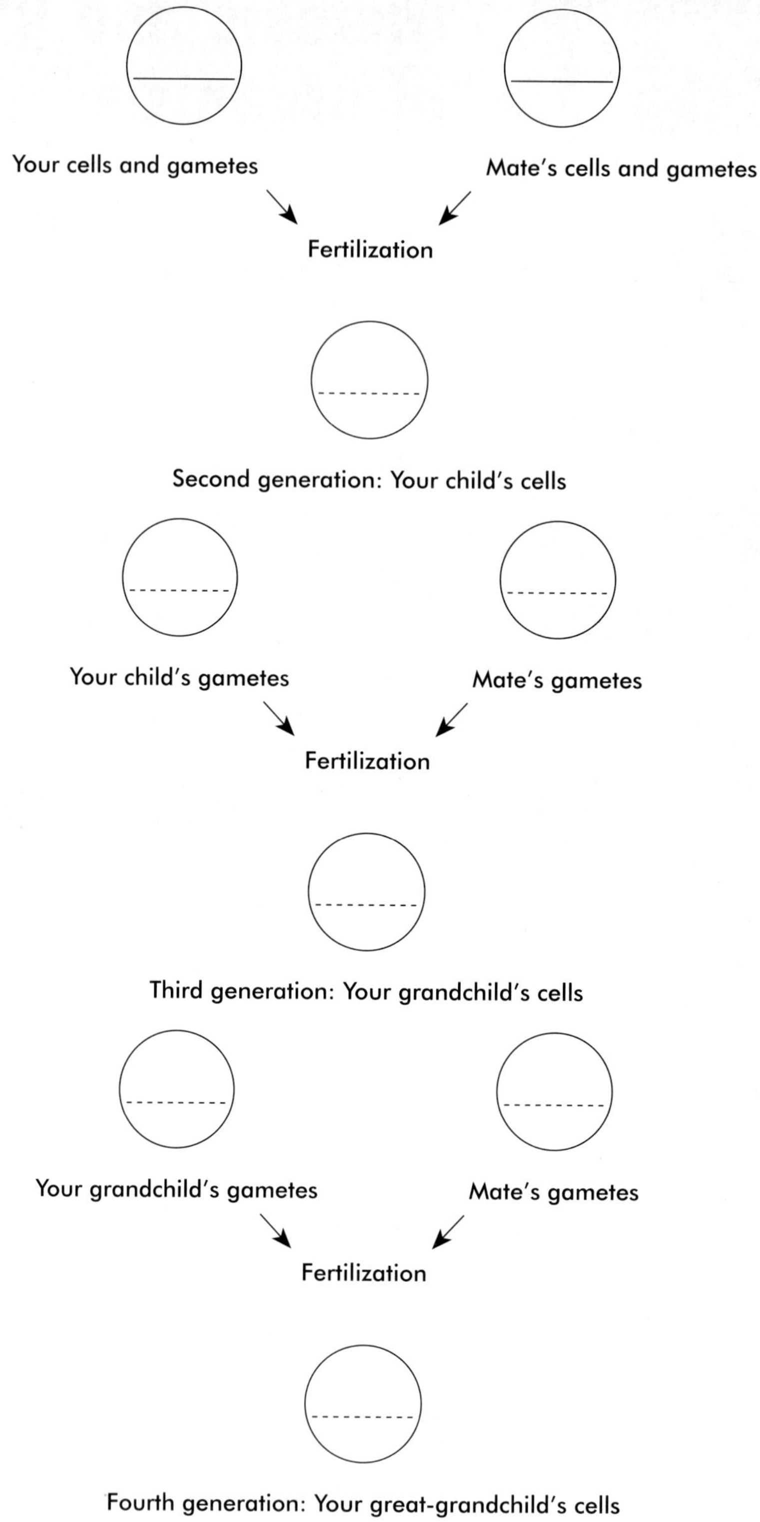

Figure 6-1
Number of DNA molecules in cells over four generations.

CONCEPTS/TERMS

Meiosis: a solution for the problem

Living organisms solved this problem when a different type of cell division evolved. This division reduces the chromosomes in the daughter cells to one half of the parent's chromosomes. It occurs in cells that produce gametes. Now when two gamete nuclei (each with half of the parental chromosomes) fuse in fertilization, the chromosome number in each offspring equals the number of chromosomes in the parents' cells. The fusion of gametes also gives rise to an

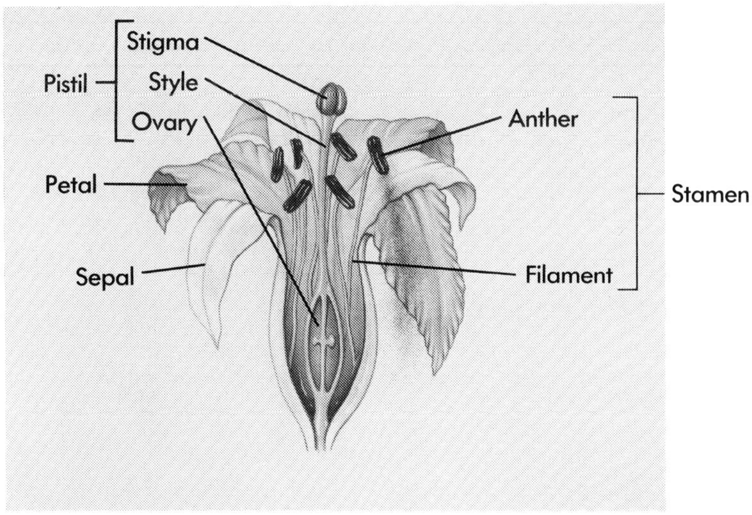

Figure 6-2
Lily flower.

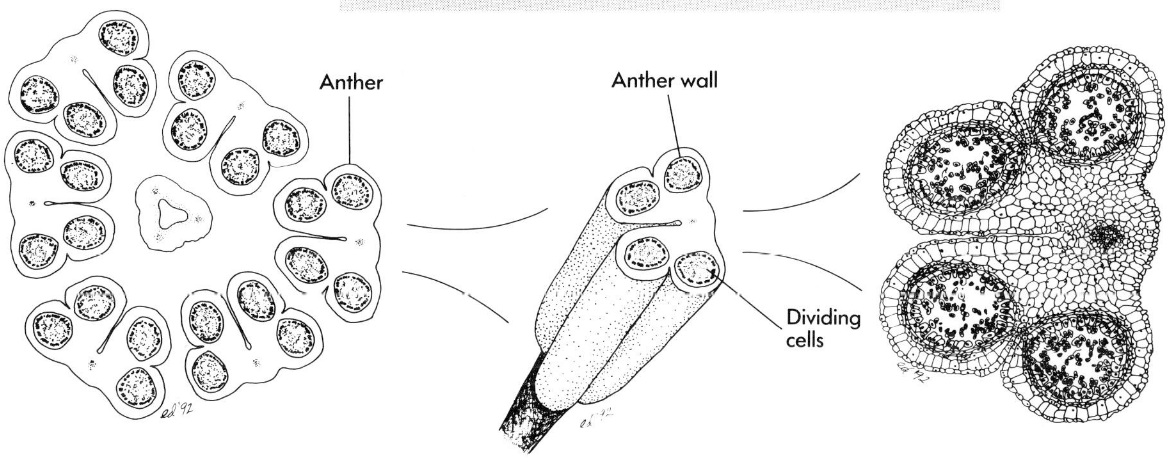

Figure 6-3
Lily anthers cross section.

offspring that has a unique combination of genetic information inherited from both parents. The reductional type of cell division is called **meiosis** (my-O-sis). It occurs in a few cells of all organisms that reproduce sexually. (Sometimes meiosis is called the *meiotic process* [my-AH-tic]; learn to spell and say this correctly because it is easy to confuse with the *mitotic process* [my-TAH-tic].)

EXPLORING

Meiosis in plants

Obtain your microscope and a series of slides prepared from sections of a lily anther. The **anther** is the male part of a flower (Figure 6-2 and Plate 4A). This series of slices, or sections, is taken from anthers as they aged over the life of the flower. The process of meiosis occurs within this structure as **pollen,** the source of future male gametes, is formed.

Using the "first prophase" slide of the series, locate four areas on each of the six anthers that contain dividing cells (Figure 6-3). Most species of lilies have 24 chromosomes. What are two or more reasons that you cannot count all 24 chromosomes in the cells in late prophase?

5 ____

____

Table 6-1 Events within the cavity of the lily anther

Slide name/phases	Diagram of cavity cells	Number of chromosomes	Description of events

Using each slide of the series in turn, follow the events that occur in these anther cells. As in the film on mitosis, all cells in the four dividing areas in the anther may not be in the same phase, but a large number will be synchronized. All phases of meiosis may not be represented in the sections on your slides. Your teacher will guide you through any missing phases. In Table 6-1, draw representatives of the dividing cells from each of the slides showing relative sizes of the cells and the condition of their nuclei. In the fourth column, describe the changes that occurred in the cells. Whenever possible, give the chromosome number in these representative cells.

Comparing observations: meiosis versus mitosis

Compare the sequence of events in the cells observed within the anther cavity with the events in mitosis (see Laboratory 5). Begin by comparing the number of divisional sequences for each original parent cell and the resulting number of daughter cells. Then compare the kinds of phases in mitosis and meiosis and provide the details of any differences in these phases.

6 ____

____

____

____

____

____

CONCEPTS/TERMS

Building concepts from your explorations

Terminology associated with meiosis

Biologists use certain terms to identify the cells of complex organisms that are in meiosis. In male animals, the parent cells that divide by meiosis are called **primary spermatocytes.** In females, they are **primary oocytes.** The final daughter cells in males are called **spermatids,** and they mature into **sperm.** In females, one final daughter cell is an **egg** (or **ovum**), and the others are **polar bodies.**

Meiosis

Meiosis consists of a series of two divisions. As you observed, the parent cell that initiates the meiotic process divides, producing two daughter cells. Each of these daughter cells then divides. This sequence of two divisions results in four daughter cells for each of the original parent cells. (In some ways, you can think of these four daughter cells as "four granddaughters" of each parental cell. Biologists, however, refer to them as daughter cells because they consider the two divisional parts of meiosis to be only one event.)

Before the first division of meiosis, the parental cells duplicate their DNA molecules just as cells dividing by mitosis do. However, the cells do not duplicate their DNA molecules between the first and second divisions.

In each of these two divisions, the phases of prophase, metaphase, anaphase, and telophase occur as they do in mitosis but differ in the details. The phases in the first meiotic division are identified with the Roman numeral "I." The second division's phases are identified with the Roman numeral II.

The primary differences between the phases of mitosis and the phases of meiosis occur in the first division of meiosis. Meiosis I can be summarized as follows:

Prophase I. In this phase (as in prophase of mitosis), the DNA molecules and their associated proteins condense into chromosomes, the nuclear membrane and nucleoli break down, and the fibrous spindle forms. In contrast to mitosis, however, prophase I chromosomes pair up and do not remain as noninteracting individual chromosomes. When two chromosomes pair, biologists say that the chromosomes are in **synapsis.** The members of each pair are identical in length and in the location of their centromeres. In fact, each carries genetic information for the same traits or characteristics of the organism. The two similar chromosomes are called **homologous** (home-OL-uh-gus) **chromosomes** or **homologues** (HOM-a-logs; *homo* meaning "same"). As the two homologous chromosomes intertwine lengthwise, they form a structure with four chromatids and two centromeres. Sometimes during prophase I, chromatids within these structures break and then reattach. If broken chromatids happen to reattach to broken ends of nonsister chromatids (those of the other homologue), the two homologues exchange parts of their genetic instructions.

Even though the homologues carry genetic information for the same traits of the organism, sometimes these two sets of instructions for a trait are slightly different. For example, one pair of homologues has a part of its DNA that provides instructions for determining your dominant hand; one of the homo-

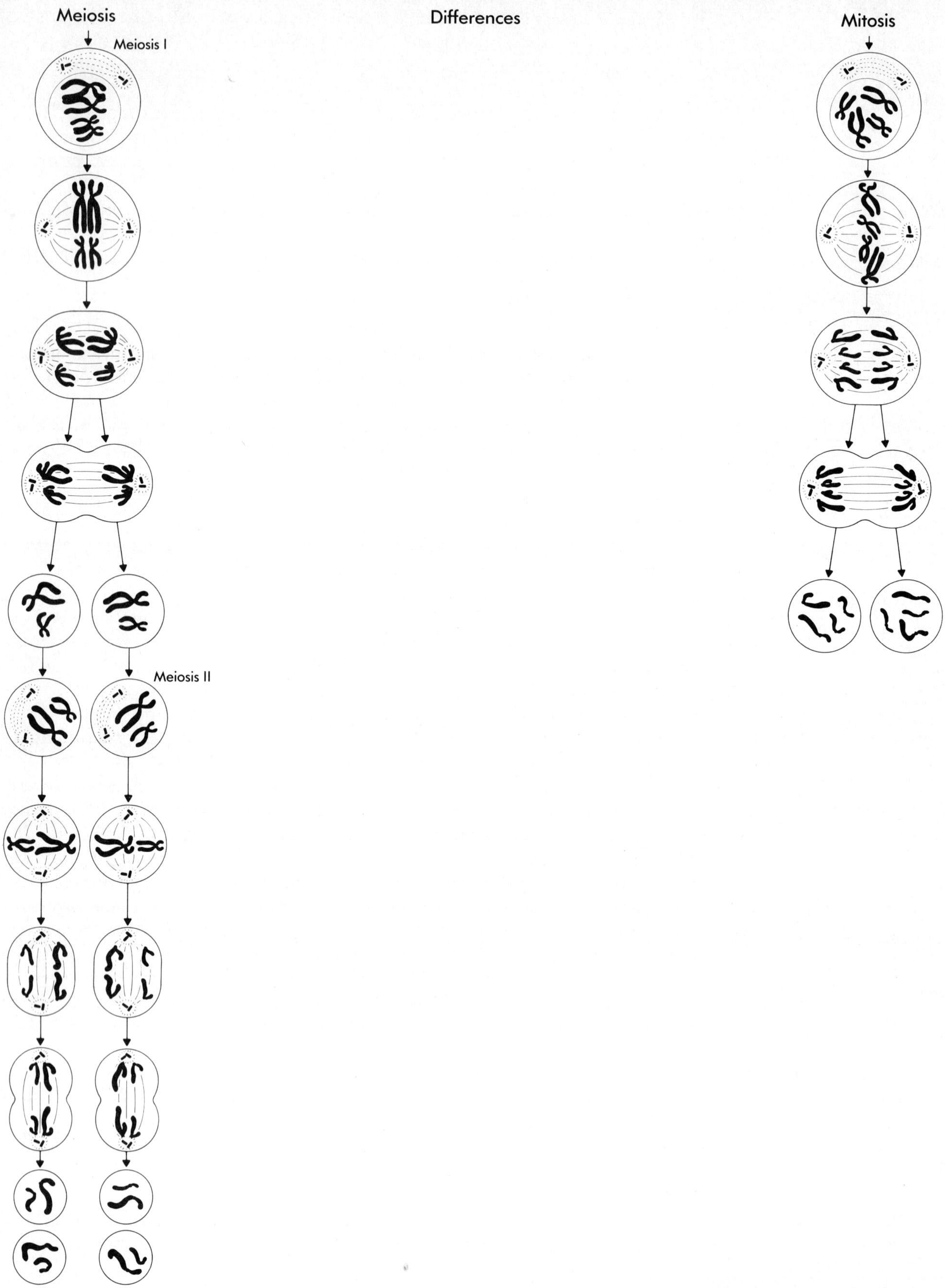

Figure 6-4
Comparison of meiosis and mitosis.

logues could have instructions for right-handedness, whereas the other homologue could have instructions for left-handedness. Each genetic instruction for a trait is an **allele.**

Metaphase I. The chromosomes pairs move to the metaphase plate with two spindle fibers attached to each pair. One fiber extends from one pole and is attached to the centromere of one homologous chromosome. The other spindle fiber extends from the opposite pole and is attached to the centromere of the other homologue.

Anaphase I. The attached spindle fibers shorten, pulling the two homologues toward opposite poles. In anaphase I, the two homologous chromosomes are separated from each other; in contrast, the sister chromatids are separated from each other in anaphase of mitosis.

Telophase I. When one member of each homologous pair of chromosomes reaches each pole, the nuclear membranes may re-form, giving the cell two nuclei. Each nucleus has only one half of the chromosomes of the parent nucleus. Importantly, the genetic instructions at each of the poles has only one version of the instructions for each trait of the organism.

Cytokinesis occurs, separating the cytoplasm between the two nuclei. This first divisional sequence results in two daughter cells. Each daughter cell has one complete set of genetic instructions distributed in one half of the original chromosomes. Because of this reduction in chromosome number, this first divisional sequence in meiosis is called the **reductional division.**

The structure of each chromosome in the daughter nuclei still consists of two chromatids attached by a centromere. Unless chromatids broke and reattached to the chromatids of the other homologous chromosome during synapsis, the chromosome has identical copies of the DNA molecules in its sister chromatids. In this structural state the chromosomes of each nucleus begin the second divisional sequence, **meiosis II.**

All four phases of the second division are similar to the four phases of mitosis. Two daughter cells form from each of the daughter cells of the reductional division. In meiosis II (as in mitosis), the **prophase II** chromosomes remain unpaired, centromeres divide in **metaphase II,** and the two sister chromatids are pulled apart in **anaphase II.** In **telophase II,** four nuclei form as the nuclear membranes are rebuilt. Each nucleus has only one member of each homologous pair of chromosomes in the parent cell, and each chromosome consists of only one strand of DNA.

Return to the chromosomes of the original parent cell. In prophase I, pairing of homologous chromosomes occurs. This pairing requires that each chromosome in the parent cell have a homologue. Having two chromosomes carrying information for the same inherited traits can be thought of as having two sets of genetic instructions. In fact, one set of genetic instructions carried by one member of each homologous pair comes from the male parent of the organism, and the set on the other homologous chromosome of each pair comes from the female parent. Having two sets of chromosomes in a cell is called **diploid** (DIP-loid; meaning "two sets").

The result of meiosis is the reduction of a two-set state to a one-set state. The one-set state is called **haploid,** meaning "half." The chromosomes in a haploid nucleus contain one version of the genetic instructions for all traits of the organism. This nuclear state is also called **monoploid** (meaning "one set").

In Figure 6-4, describe the differences between the phases of mitosis and meiosis that are diagrammed in cells containing four chromosomes.

APPLYING

Concepts of meiosis

Numbers to consider

How many chromosomes should be in each of the four daughter cells of meiosis in the lily anther?

7

How many centromeres?

8

How many homologous pairs?

9

How many DNA molecules?

10

If meiosis also occurs in the lily ovary as it produces an egg, how many chromosomes would be in the egg?

11

Modeling meiosis

Obtain paper clips and a set of 12 pieces of yarn, 4 long, 4 short, and 4 of medium length. In each of the subsets of 4, 2 of the yarn pieces should be of one color and the other 2 pieces should be of a second color. You may assume that one yarn color represents genetic information that an organism received from its male parent and the other color represents information from its female parent. From the yarn and paper clips, form chromosomes as they may look in cells that are in late prophase I of meiosis. How many chromosomes are in your set of chromosomes?

12

How many homologous pairs of chromosomes?

13

Model the meiotic process with your entire set of chromosomes. Consult with your neighbor. When you are convinced that you have the process accurately modeled, demonstrate it to your teacher.

Do all the chromosomes of one color (from one parent of this organism) necessarily end up in the same daughter cell when meiosis is completed?

14

How many different combinations of chromosomes do you think can be made with this set of yarn as daughter cells are formed by meiosis?

15

Observing meiosis in the uterus of *Ascaris*

At the demonstration table, observe the roundworm *Ascaris* (ASS-car-is). This worm is an intestinal parasite of horses. It is related to other roundworms that cause diseases in humans. You may have heard of such diseases as trichinosis, which results from eating encysted worms in undercooked pork, and elephantiasis, which is a swelling of the body resulting from worms plugging up the lymphatic system.

The adult female *Ascaris* lays up to 200,000 eggs a day. The cells that will become eggs are produced in the *Ascaris* ovary and are released in an immature state to the *Ascaris* uteri (plural of uterus). The immature eggs are fertilized and undergo meiosis within the uteri. (In contrast to *Ascaris,* the eggs of many organisms initiate the meiotic process before their release from the ovaries and before fertilization.) The fertilized eggs develop into new worms. When released from the female *Ascaris,* these developing worms exit with the host's feces and will infect new hosts.

Obtain a series of longitudinal sections from an *Ascaris* uterus. These sections begin at the part of the uterus closest to the ovaries and end at the portion of the uterus closest to the tube through which the young worms are released. After cutting, these sections were mounted on microscopic slides. (See Appendix E, which will help you understand the two-dimensional microscopic sections of three-dimensional structures, such as uteri and eggs.)

Using both the low and high power objectives of your microscope, observe the first slide in the series that your teacher designates. You may find a section cut directly along the tube-like uterus or one cut somewhat diagonally. (What would be the shape of the section if it were cut diagonally across the tube?

16 ______________________________

Many immature eggs are packed within this uterine section. These immature eggs are the primary objects of interest, but you may also note the triangular-shaped heads of the male sperm that have not entered one of the eggs.

On the slide, locate a large, round midsection through a ball-shaped egg with two nuclei. You may locate an egg section in which the sperm head is penetrating the egg covering or you may find sections that show the large sperm nucleus already inside the egg. As the sperm enters, the egg covering expands and raises from the surface of the egg. The space between the cover and the egg surface is a characteristic of a fertilized egg. In other sections, you may find ones that include either the egg or sperm nucleus but not both. You can tell the sperm and egg nuclei from each other by the sperm's slightly larger size and the more granular appearance of its interphase state. Most egg nuclei are beginning meiosis I, which can be identified by deeply stained condensed chromosomes. Many of the egg sections are cut through edges of eggs and are therefore smaller than the midsections. These edge sections probably lack one or both nuclei.

Using your high power magnification system on the slides, follow the events in the egg nuclei to the point where both the egg and sperm nuclei are large in size, in the interphase state, and have moved toward each other. Consider the following as you explore these slides:

How many chromosomes are in the egg nucleus as it begins meiosis?

17 ______________________________

How many chromosomes do you think are in the body cells of *Ascaris?* (Check your number with your neighbors. If you do not agree, seek guidance from your teacher.)

18 ______________________________

In prophase I, what is the structure of the chromosomes?

19 ______________________________

In metaphase I, how many homologous pairs of chromosomes are in the egg nucleus?

20 ______________________________

During anaphase I, what structures are separated?

21 ______________________________

At the end of telophase I, how equally is the cytoplasm divided between the two nuclei?

22 ______________________________

Can you identify prophase II by the chromosome number?

23 ______________________________

If so, how many chromosomes are in the prophase II egg cell (secondary oocyte)?

24 ______________________________

In telophase II, how many chromosomes are in each nucleus and what is their structure?

25 ______________________________

How equally is the cytoplasm divided at the end of meiosis?

26 ______________________________

Now compare cytokinesis after meiotic division in egg production to the cytokinesis observed in the lily anther. Describe the differences.

27 ______________________________

In egg production, cytokinesis divides the cytoplasm of the parent cell (primary oocyte) unequally. In both meiosis I and II, most of the cytoplasm remains with one of the daughter nuclei, and the other nuclei receive very little cytoplasm. The nuclei that receive almost no cytoplasm are **polar bodies.** They are ejected from the cell and disintegrate. In contrast, no polar bodies are produced in cytokinesis during sperm production because all four daughter cells receive approximately equal amounts of cytoplasm and each can mature into sperm.

Meiosis results in a different number of functional sperm and eggs. How many eggs are produced from each primary oocyte (parent cell) initiating meiosis?

28 ______________________________

How many sperm are produced from each primary spermatocyte? Thinking back to the lily, how many pollen grains (future sperm) will be produced from 10 of the primary spermatocytes (microsporocytes in plants) in the anther?

29 ______________________________

If cytokinesis during meiosis within the ovary of the lily is similar to that in the *Ascaris* uterus, how many eggs are produced from 10 primary oocytes (megasporocytes in plants) in the lily?

30 ______________________________

Life cycles

Now shift your thoughts from cells during their cell cycles to organisms during their life cycles. The term **life cycle** refers to the sequence of events in the life of an organism from its beginning when an egg and a sperm unite until it produces gametes for the next generation of organisms. Two major events occur in the life cycle of all sexually reproducing organisms. These two events are meiosis and fertilization.

EXPLORING

Constructing the human life cycle

Using information you may have from previous sources, give the human structures in which fertilization and meiosis occur.

31 ____

In Figure 6-5, organize the following terms in a circle to form the life cycle of humans: *adult male, adult female, baby, egg, fertilization, fertilized egg, gametes, meiosis, ovary, primary oocyte, primary spermatocyte, sperm,* and *testis.* Connect the appropriate terms with arrows to indicate the proper sequence in humans. (You may want to draw two life cycles next to each other, one for a male and one for a female.)

Now go one step further in thinking about our life cycle. What part of the human life cycle has haploid cells?

32 ____

What parts have diploid cells?

33 ____

Figure 6-5
Human life cycle.

Add the terms *diploid* and *haploid* to the appropriate locations in Figure 6-5. You may want to use 2n, which refers to the diploid state of chromosomes, and n, which refers to the haploid state of chromosomes.

Check your Figure 6-5 with your neighbors'. Do each of you have meiosis occurring in the primary oocytes within the female ovaries and the primary spermatocytes within the male testes? Is the term *fertilization* at the point where the egg and sperm unite? Do you have the term *diploid* associated with all stages of the organism from fertilization through the adult stage and the adult's production of the parent cells within the diploid sex organs? If not, seek help from your teacher.

Now add one more term to your human life cycle(s). At every point where mitosis occurs in forming the structure in the life cycle, add the term *mitosis*.

CONCEPTS/TERMS

Alternation of generations in sexual reproduction

The only haploid cells in the life cycle of humans and most other animals are the gametes and their precursor cells in meiosis II. These are located in the ovaries and testes. All other animal cells are diploid. Mitosis occurs only in our diploid cells. (Figure 6-6 shows a general diagram of the animal type of life cycle.)

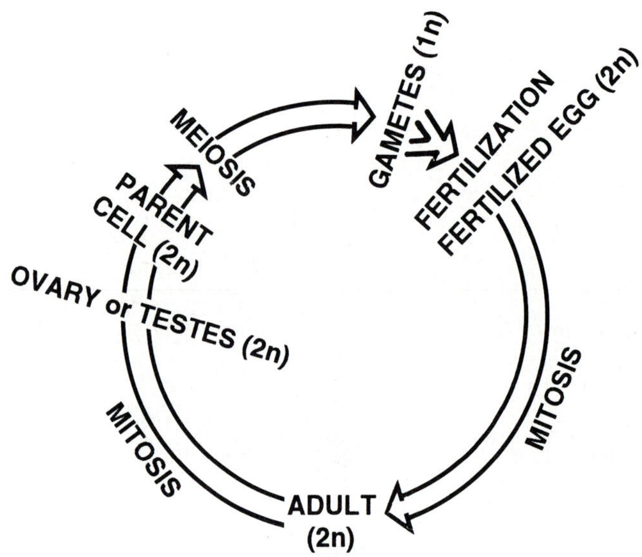

Figure 6-6
Life cycle of animals.

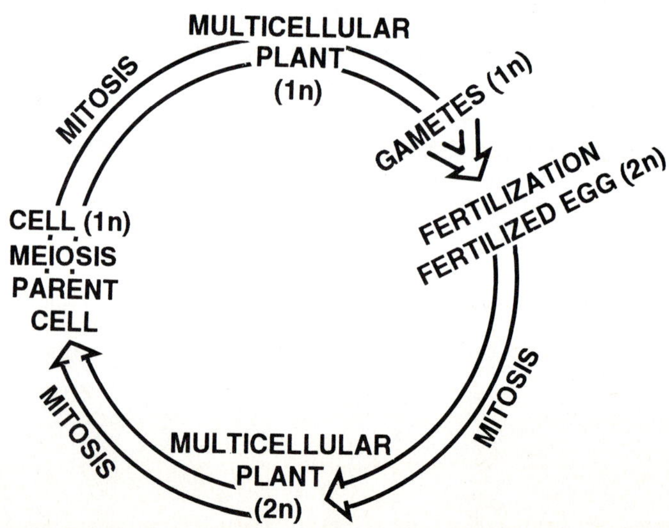

Figure 6-7
Alternation of generations life cycle.

In plants and some simpler green organisms, mitotic cell division occurs in both diploid and haploid cells. This mitotic division of the haploid cells results in multicellular structures of haploid cells. Basically, these organisms alternate between a multicellular haploid generation and a multicellular diploid generation within one life cycle (Figure 6-7). This is called the **alternation of generations** life cycle.

The key to understanding the basis of all varieties of life cycles is to recognize when and where meiosis and fertilization occur in the organism.

APPLYING

Alternation of generations life cycle in mosses

Using the following short description of the events in the life cycle of a moss, place the terms *diploid, fertilization, haploid, meiosis,* and *mitosis* on the diagram in Figure 6-8 and Plate 4B:

> The short green tuft that you probably recognize as moss is the most prominent part in the moss life cycle. These low-growing, soft, green plants frequently cover the soil in moist environments. Egg and sperm are produced in the tips of these tufts. During rain storms and after the gametes are mature, a sperm swims or is splashed over to the tip where the egg is located. The two gametes unite and a capsule-like structure forms from the resulting fertilized egg. Parent cells that divide meiotically are inside this capsule. The cells (spores) from this meiotic process are popped out of the capsule, fall to the soil, and form the multicellular green tuft.

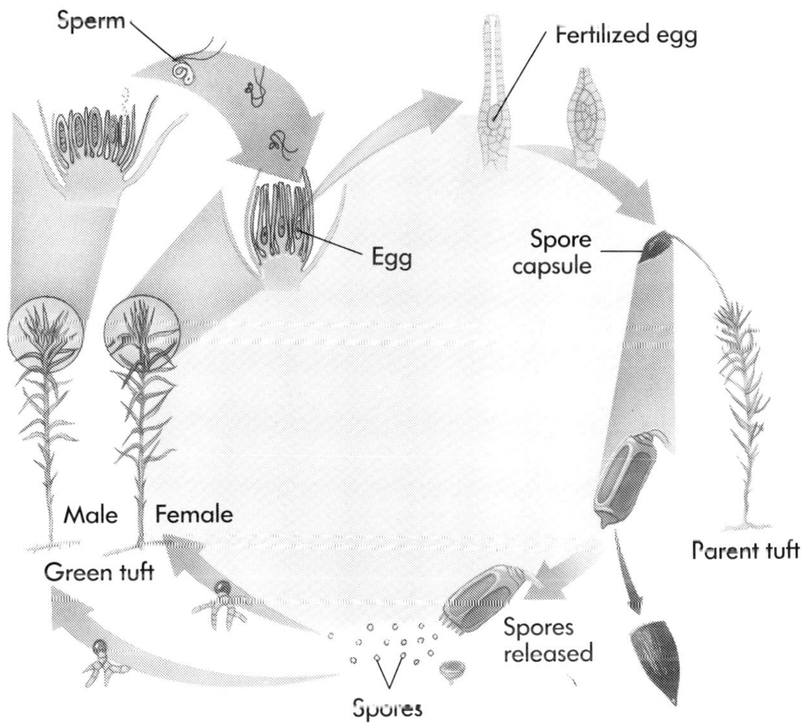

Figure 6-8
Moss life cycle.

LABORATORY

7

Enzymes

Catalysts of life

What is so important about enzymes? Biologists define **enzymes** as the catalysts of life, but what does that really mean to us and to other organisms? To begin to answer this question, let's examine the breakdown of starch molecules.

Starch is a large, spiral, carbohydrate polymer ("many units") of glucose. Starch is the primary energy storage molecule in plants. The starch stored in plants is a basic ingredient in our diets and in all organisms that feed on plants. Whether an organism is a human, grasshopper, tree, cow, mushroom, or *Spirogyra,* it must convert starch to its glucose monomer ("one unit") before it can use starch for energy. Some organisms, including humans, break down starch during the digestive process. Starch is broken down to sweet-tasting maltose (two attached glucose molecules) and then to glucose (Figure 7-1). These hydrolytic reactions (breaking bonds using water) provide the glucose that enters cells and is used for energy.

EXPLORING

Converting starch to glucose

Nonenzymatic breakdown

First, try breaking starch into sugars manually, with heat, or some other physical method. The following materials are available for your use:

- Starch, two forms:
  - Powdered
  - 200 ml of starch solution, 0.5%
- 4-5 150-ml or 250-ml beakers
- 2-3 50-ml beakers
- 1 10-ml graduated cylinder
- 3 test tubes
- 4-5 aluminum foil squares (form into containers)
- 4-5 droppers or Pasteur pipets
- 1 mortar and pestle (for grinding)

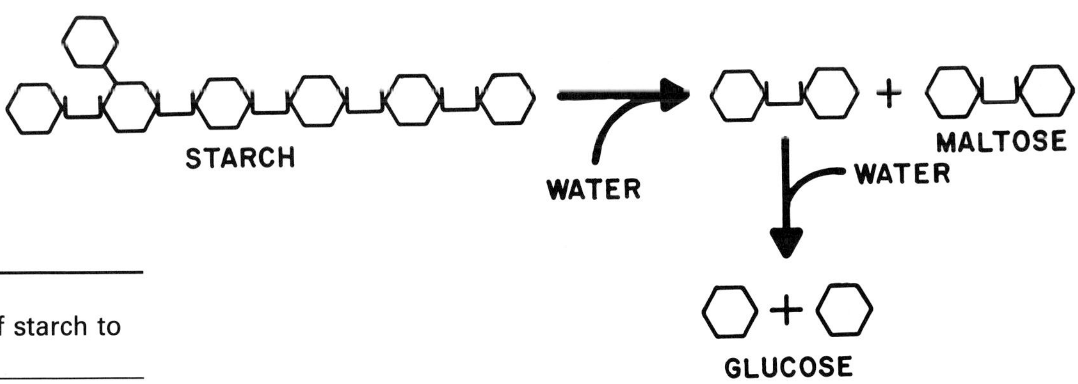

Figure 7-1
Conversion of starch to glucose.

CAUTION

Please be cautious and sensible in your method of trying to break starch into sugar. Wear goggles to protect your eyes. If you heat any liquid, add small glass boiling beads to the container to decrease the possibility of boiling solutions exploding out of their containers.

1 hot plate (for breakage of chemical bonds with heat)
Goggles
Glass marker
3-4 boiling beads

Describe the method of breakdown you want to try.

1 ____

After deciding on a method to break starch bonds, obtain approval from your teacher before beginning. Whatever your method is, your object will be to convert starch to sugar.

After applying your breakdown methods, you may want to test for the presence of starch and sugars to check your progress in converting starch to sugar. The following are quick, easy tests for starch and sugars:

Starch test. Obtain the following additional materials:
1 dropper bottle of I-KI (iodine-potassium iodide) solution
1 ceramic plate with small depressions (spot plate)

To test for starch, put several drops of the starch solution into one depression of the spot plate. Also put several drops of water in a second depression. Add one drop of I-KI solution to each depression. Describe the results from adding I-KI to starch and to water.

2 ____

If you get the same color reaction on adding I-KI to another solution as you did on adding I-KI to the starch solution, it indicates that starch is present (positive starch test). The I-KI solution by itself or in water without starch is golden-yellow.

Sugar tests. Two tests are described. Use the one suggested by your teacher.

Reagent strip test. Obtain the following additional materials:
1 dropper bottle of glucose solution
1 dropper bottle of maltose solution
2 narrow strips of paper with special chemicals absorbed at one end (glucose reagent strips)

The reagent strips are frequently used in doctors' offices to test for sugar in urine. The reagents react with glucose and maltose, causing a color change on the strip. To use, just add one drop of a solution to the chemical end of the reagent strip. Wait for the time indicated on the container's label. Compare the color of the test area to the label's color chart. This gives you information about the sugar concentration.

Benedict's test for sugars. Obtain the following:
1 container of Benedict's reagent
1 dropper bottle of glucose solution
1 dropper bottle of maltose solution
1 test tube clamp
1 hot plate

Add 2 ml of Benedict's reagent to 1 ml (20 drops) of each of the following liquids in separate test tubes: glucose solution, maltose solution, and water. Place the tubes in a beaker of boiling water (with two to three boiling beads), heat for 3 minutes, and then cool. Describe the results.

3 **Glucose** ____

Maltose ____

Water ____

Table 7-1 Converting starch to glucose

Method	Starch test	Glucose test

The results in the first two tubes indicates the presence of the sugars glucose and maltose, respectively. Why was the tube of water and Benedict's solution prepared?

4 ______________________________

Dispose of the used Benedict's reagent as directed by your teacher.

After completing your efforts to break starch into sugars, share your results with the class. You may record class efforts in Table 7-1.

Because you are interested in the starch breakdown process that occurs in living organisms, think about these breakdown methods in relation to the well-being of living organisms. If you used heat, think about the effect of high temperatures on living organisms. If you used grinding, think about the effect of grinding on the structure of living organisms. If you used another method, think about its effects on living organisms. Which of the methods in Table 7-1 are compatible with maintaining healthy, living organisms?

5 ______________________________

Enzymatic breakdown

Now try another method of breaking starch into sugars. Obtain the following materials:

1 dropper bottle of nonboiled α-amylase (a chemical obtained from a fungus)

1 dropper bottle of boiled α-amylase

Also keep the I-KI solution, starch solution, graduated cylinder, small beakers, spot plate, and sugar-testing materials used previously.

At specific time intervals in this procedure, quick action is needed. Read the entire instructions given below *before* beginning.

To each of two small beakers, add 10 ml of starch solution and 10 ml of water. When you are ready for starch testing (time zero), add 10 drops of nonboiled α-amylase fungal solution to one beaker. Working quickly, gently swirl the beaker to mix the contents and immediately transfer several drops of the mixture to a spot plate depression and test for starch. Repeat the same procedure for the second beaker, *but* add 10 drops of the boiled fungal solution in place of the nonboiled amylase solution. For each beaker, repeat the test for starch at 1 minute, 2 minutes, and every 2-minute interval thereafter (or at time intervals indicated by your teacher). Stop testing for starch whenever the starch solution in either one of the beakers fails to give a positive results for starch. At this point, conduct a test for sugar on the solution in each beaker. Record your results in Table 7-2.

Examine your data. Compare the activity of the nonboiled fungal α-amylase to the previous methods used to break down starch.

6 ______________________________

____

Of all the methods you and your classmates explored, which one is the most compatible with living organisms?

7 ______________________________

Explain your answer.

8 ______________________________

____

Compare the action of boiled fungal solution to the action of the nonboiled solution.

9 ______________________________

____

At this point, what explanation do you have for the differences in the action of these two fungal solutions?

10 ______________________________

CONCEPTS/TERMS

Activity of enzymes

The chemical α-amylase is an enzyme produced by cells of the fungus. As you have observed, enzymes help chemicals react without requiring drastic means that damage living things. As catalysts, they increase the speed of chemical reactions (such as starch conversion to maltose) without being consumed in the reaction. All of life's activities depend on thousands of different chemical reactions and thousands of enzymes. Each reaction is catalyzed by a particular enzyme.

Table 7-2 Time course for boiled and nonboiled fungal α-amylase activity

Clock time	Elapsed time (min)	Starch test	
		Nonboiled	Boiled
	0		
Glucose test			

Enzymes speed up chemical reactions by providing a specific area on their surfaces for the reacting materials called **substrates (reactants)** to attach. This attachment area contains particular atoms organized in a specific arrangement. This area is called the **active site.** When the substrates are orientated in particular positions as they attach to the enzyme's active site, stress is placed on bonds among the atoms within the substrate molecules. These stressed bonds

Figure 7-2
Enzyme activity.

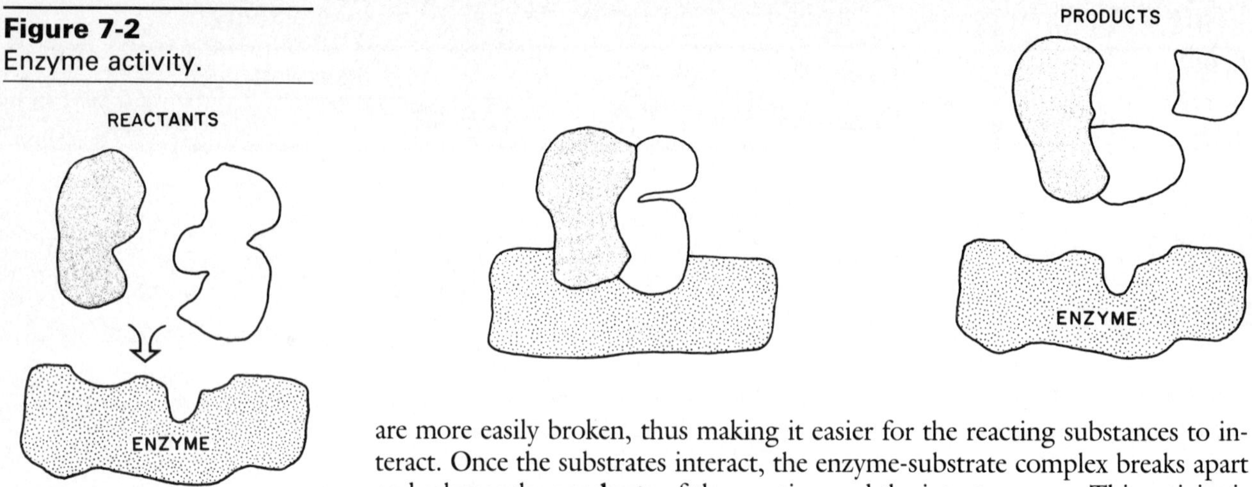

are more easily broken, thus making it easier for the reacting substances to interact. Once the substrates interact, the enzyme-substrate complex breaks apart and releases the **products** of the reaction and the intact enzyme. This activity is shown in Figure 7-2.

EXPLORING

The composition of enzymes

It is important to find out about the composition of enzymes because this information may give you some insight into the different results obtained with the boiled and nonboiled fungal solutions.

Today, you have used methods to test for starch (a complex carbohydrate) and sugars. Other types of molecules that make up living things are proteins and lipids. Tests can be used to identify proteins and lipids. Some of these tests are described later, and your instructor may suggest others.

Lipids

Obtain a dropper bottle of the nonpolar stain Sudan IV. It associates with lipids and is excluded from hydrophilic (water-loving) materials. Even though this reaction is not a good specific test for lipids, it will give you some indication of whether enzymes are lipids. For this test, add one drop of Sudan IV to the test material.

Proteins

In doctors' offices, urine is frequently tested for protein with a paper reagent strip. To test for protein in the α-amylase solution, use the protein reagent strip in the same manner as you used the sugar test strip.

Test for proteins, lipids, starch, and sugars using a pure α-amylase suspension provided by your teacher. Record the results in Table 7-3.

Table 7-3 Chemical composition of enzymes

	Possible molecular composition			
	Starch	**Sugar**	**Lipid**	**Protein**
Test				
Observed results				
Presence or absence in enzyme				

CONCEPTS/TERMS

Enzyme composition and factors that affect activity

Now that you have some information on the kind of molecule that composes α-amylase (as well as other enzymes), try to remember what you know about (1) the structure of this kind of molecule and (2) the type of bonds that hold atoms of this molecule in their proper positions. Summarize your information.

11 ______________________________

____

____

Using this information and the information about enzyme function, explain the difference in the activity of boiled and nonboiled α-amylase.

12 ______________________________

____

Compare your answer with your neighbor's. If you fail to agree, ask your teacher to listen to the explanations for each answer.

APPLYING

Some characteristics of enzymes

Experiments: factors affecting the activity of enzymes

You can see from Figure 7-2 that the shape of an enzyme is extremely important to its function. Considering the substances that enzymes consist of, hypothesize the effect that more moderate changes in temperature (less than boiling) will have on α-amylase activity.

13 ______________________________

What effects will pH changes have on the α-amylase activity?

14 ______________________________

Applying these hypotheses to specific temperatures and pHs, make a prediction from each hypothesis. For temperature, predict effects for 0° C, room temperature (__° C), and temperatures at 15° C intervals above room temperature (room + 15° C [__° C], room + 30° C [__° C], and room + 45° C [__° C]). For effects of pH, predict activity at pHs of 3, 5, 6, 7, and 9.

15 **Prediction I (temperature)** ______________________________

Prediction II (pH) ______________________________

In each of your predictions, identify the independent variable by underlining it; identify the dependent variable by circling it. The following are general suggestions for experiments to test a prediction from each of the two hypotheses. Working in groups, work out the details of experiments for one or both of these experiments (as directed by your teacher) and collect your data.

Effects of temperature on α-amylase activity. For the temperature experiment, you can modify the procedure used to explore the effect of α-amylase on starch breakdown over time. In this procedure you added 10 drops of enzyme solution to a beaker containing 10 ml each of starch solution and water at room temperature. Then you removed several drops of the enzyme-starch solution at specific time intervals to test for starch.

What kind of data will you be recording in this experiment?

16 ____

What are the standard conditions for this experiment?

17 ____

____

Be careful to have both the enzyme and starch solutions at the experimental temperature before mixing them. (Why is this necessary?) Also, be careful to keep each mixture at its experimental temperature throughout the experiment. Your teacher has the temperature environments ready for you.

When you have planned the details of your experiment and are ready to begin, describe it to your teacher and consider any last-minute questions or ad-

Table 7-4 Effects of temperature on a α-amylase activity using I-KI starch tests

	Temperature (° C)				
Time (min)	**0**				
0					

vice. Carry out your experiment, and record the data in Table 7-4 after adding your test temperatures and time intervals to the appropriate columns and rows.

To present your data, make a bar or line graph in Figure 7-3. Plot your independent variable on the horizontal axis and your dependent variable on the vertical axis.

At what temperature was enzyme activity at its fastest rate?

18 ____

The temperature at which the enzyme exhibits maximum activity is called the **optimum temperature.**

What do the data tell you about your prediction and hypothesis?

19 ____

Explain how you thought about this last question by citing specific observations and show how they support your statement.

20 ____

Consult with other groups that are conducting similar temperature experiments. Are their data similar to yours?

21 ____

Figure 7-3
Effects of temperature on fungal α-amylase activity.

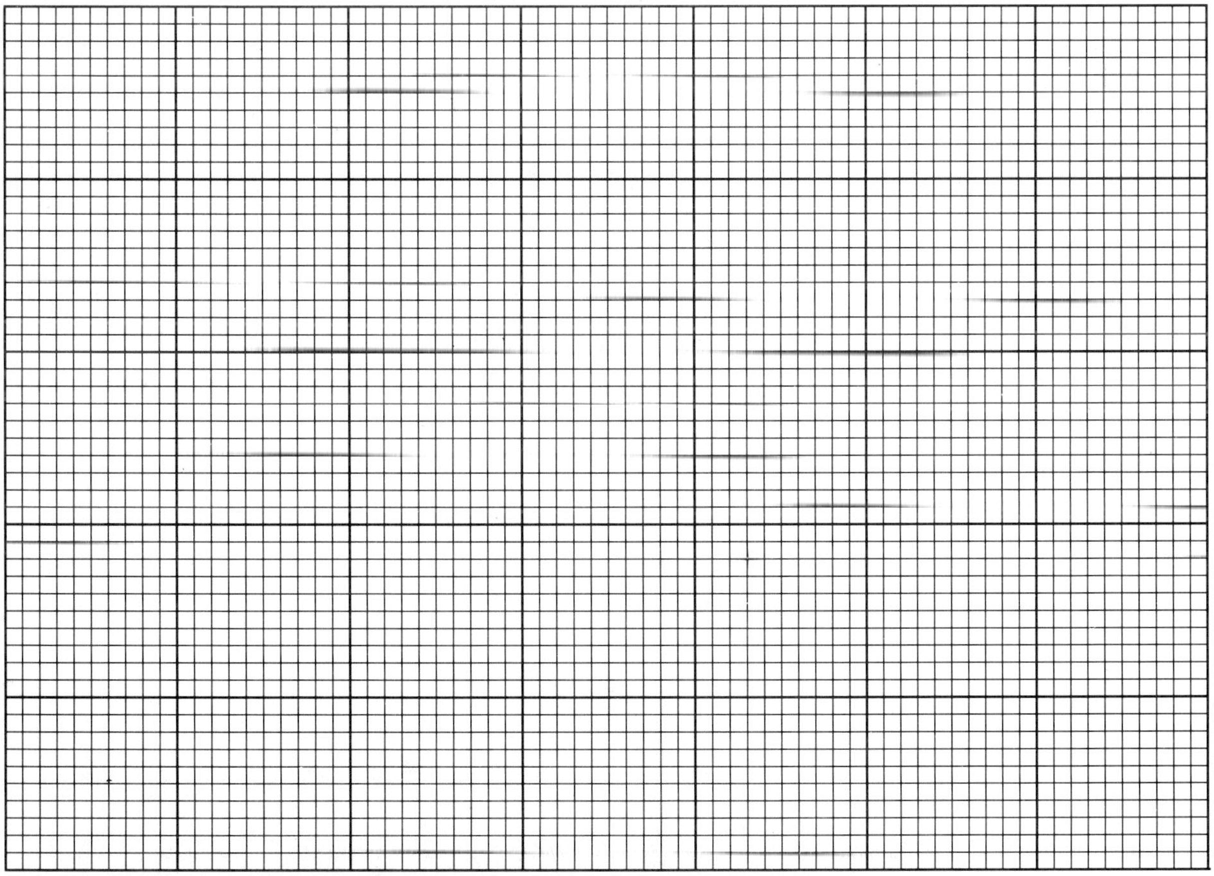

Do other classmates verify your conclusion?

22 ______________________________

If not, seek guidance from your teacher.

Effects of pH on α-amylase activity. For the pH experiment, you can also modify the procedures of your previous exploration of the effects of α-amylase on starch over a period of time. To make the starch solution of different pHs, you can dilute 10 ml of starch with an equal volume of the appropriate buffer solution.

What kind of data will you be recording in this experiment?

23 ______________________________

What are the standard conditions for this experiment?

Table 7-5 Effects of pH on α-amylase activity using I-KI starch tests

	pH				
Time (min)	3	5	6	7	9
0					

24

When your experiment is designed, describe it to your teacher and consider the questions and advice, run your experiment, and record data in Table 7-5.

Present your data by constructing a bar or line graph of the results in Figure 7-4. Plot the independent variable on the horizontal axis and the dependent variable on the vertical axis.

What is the optimum pH (pH at which the activity is highest) for this fungal enzyme?

25

What do the data tell you about the hypothesis?

26

Explain how you thought about this last question by citing specific observations, and tell how this evidence supports your statement.

27

Figure 7-4
Effects of pH on fungal α-amylase activity.

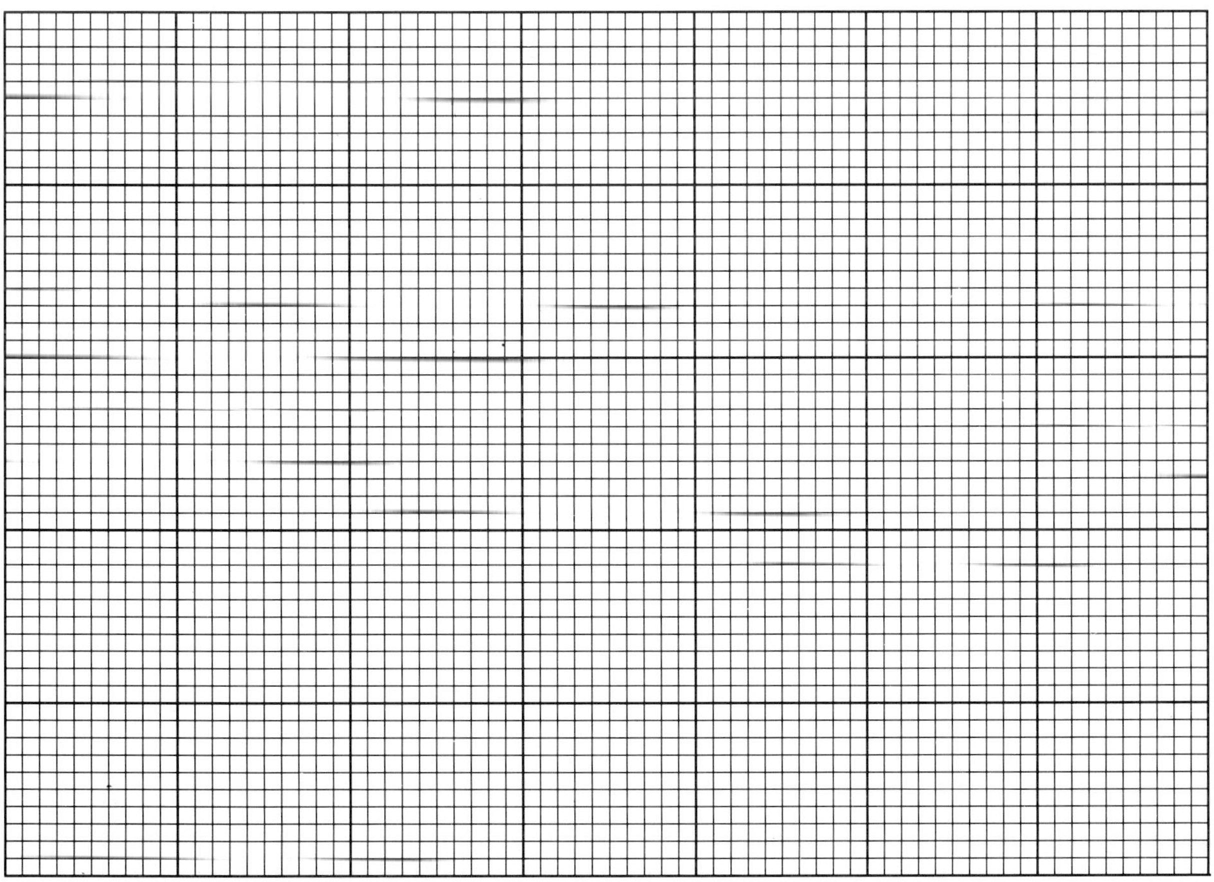

Do other classmates verify your conclusions? If not, discuss your results and conclusion with your teacher.

Experiment: do α-amylases differ among organisms?

Many organisms produce α-amylase, which allows them to use starch for energy. These organisms are found in widely varying temperature and pH environments. Do you think that α-amylases produced by organisms with different internal temperature and pH environments have different optimum pHs and temperatures?

28 ______________________________

Explain your answer.

29 ______________________________

____

Think for a moment about the source of the α-amylase you used and the pH and temperature of the environment in which its enzymes work. The α-amylases of many fungi are found in soils where they decompose plant material. The amylase will have to function at the temperature and pH of the soil or the fungi cannot use starch for energy. From your earlier explorations, what was the optimum temperature and pH for the fungal amylase?

30 ______________________________

What do you think are the optimum temperature and pH for amylase activity in a typical soil fungus?

31 ______________________________

CAUTION

As with blood, use precautions that prevent the contact of your skin or other body surfaces with another person's saliva. Although the chance of disease transmission is much lower than with blood, use gloves and goggles, wash hands and exposed areas, scrub the laboratory desk with detergent, and place all contaminated equipment in the container designated by your teacher for proper disposal.

The α-amylase enzyme is also found in human and pig mouths and intestines, functioning at the pH and temperature of these environments. Do you think that the amylase from fungi and the amylase from mammals have different optimum pHs and temperatures?

Form several hypotheses based on your expectations of optimum pH and temperature of the α-amylase in your own saliva or in the α-amylase from a pig. Design experiments to test a prediction from one of these hypotheses. After having your teacher check your proposed procedure, run your experiment. Record the data in a table and present it in a graph. Attach these records to this manual. Summarize the data.

32 ______________________________

____

How do the data relate to the prediction and its hypothesis?

33 ______________________________

____

____

What are your conclusions about the hypothesis?

34 ______________________________

LABORATORY

8

Cellular respiration

Providing energy for life

Energy and work

When you move any part of your body, you do work. As you construct molecules of protein, carbohydrates, or lipids, you do work. In addition, you work constantly just to maintain your living material in its highly organized state. Whenever you do any work, you require energy.

The concept of energy is difficult to understand. Frequently, we use the term *energy* in such statements as, "I need energy this morning to take the test, so I'll have two doughnuts for breakfast" and "It takes more energy to jog a mile than to walk a block." In the first statement, your brain cells will do the work, and in the second case, your muscles will do the work.

Energy is somewhat unsatisfyingly defined as the ability to do work. We come in contact with some forms of energy on a daily basis. Some forms of energy are heat, light, radiowaves, radioactivity, sound, electricity, and mechanical, nuclear, and chemical energy.

We can convert energy from one form to another, but we cannot create or destroy it. This concept is the **first law of thermodynamics.** The total amount of energy in the universe 4 billion years ago is the same as it is today and the same as it will be billions of years in the future. However, the proportion of the total energy in each form can vary.

The transformation of energy in living organisms involves only light, chemical, mechanical, and heat energy. Animals convert the chemical energy in molecular bonds to heat and the mechanical energy of movement. Photosynthesizing plants transform the energy of sunlight to the chemical energy stored in plant molecules; they also convert chemical energy to heat. Once energy is in the form of heat, living organisms lack the ability to use it for work. Therefore heat is wasted energy as far as life it concerned. Heat is continually increasing as other forms of energy slowly decrease.

As you may recall from the discussion on diffusion, heat is the energy in the random movement of molecules that make up the matter of the universe. Increasing heat increases this random movement and the resulting disorder of molecules. A measure of the constantly increasing disorder in the universe is called **entropy.** Imagine what happens to marbles if they are spilled from their container. The marble-organizing container holds the marbles within a limited space. The container can do this because the energy-rich bonds of its molecules keep it in a stable and organized arrangement. Without the container, marbles move in all directions to a state of greater disorder or greater entropy. The concept that disorder in the universe constantly increases is the **second law of thermodynamics.**

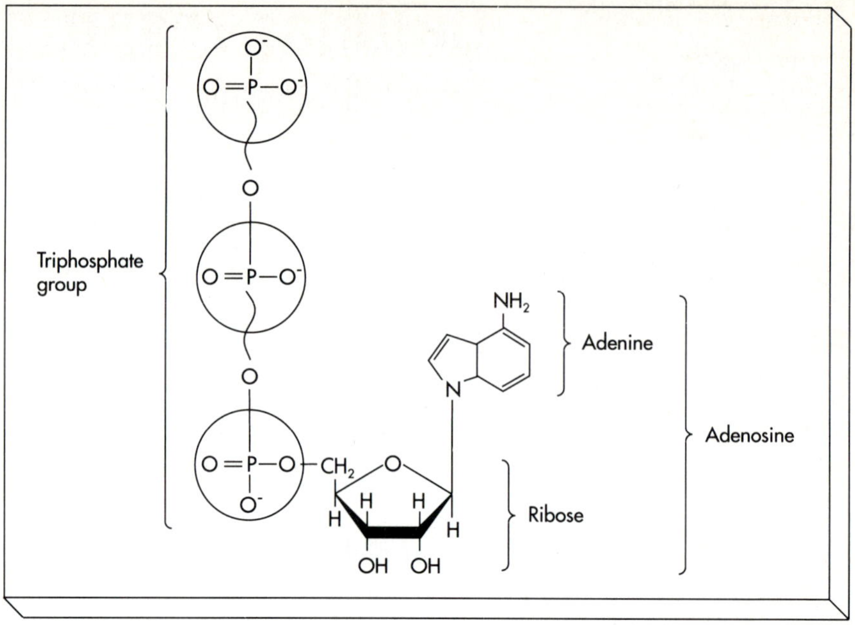

Figure 8-1
Adenosine triphosphate.

Living organisms and energy

Because living organisms lose a usable form of energy as it is converted to heat, they require a constant input of energy from other sources. The ultimate energy source for all life on Earth is the sun. The energy in sunlight enters living material when green plants transform it into chemical energy, storing it in the bonds of stable glucose molecules. Plants manufacture glucose in the energy-transforming process of photosynthesis and frequently store it as starch in their cells. This stored chemical energy is the source of energy for all of life's activities. Because animals, fungi, and most other organisms lack the necessary structures to transform light into chemical energy, they depend on plants for their energy.

As cells of organisms use glucose, the chemical energy stored in the bonds among atoms of the glucose molecule is released as the bonds are broken. Part of this energy is temporarily stored as chemical energy in **adenosine triphosphate (ATP)** molecules and part of the energy is lost from the cell as heat (Figure 8-1). The energy in the ATP molecule is used directly by the cell to power all its activities. In some ways, ATP is similar to the dollar bills that you use to pay for products and services, and glucose is your bank account from which the dollars (ATP molecules) are obtained.

If you heated a teaspoon of sugar sufficiently, it will begin to burn. When glucose burns, its energy is rapidly released as heat. If this quick transformation of glucose's chemical energy to heat occurred inside cells, the heat would damage the cells and provide no energy to run cell activities. In cells, glucose is "burned" but in a very different way than in a fire. In living cells, glucose is broken down in a sequence of chemical reactions that releases small amounts of energy at a time.

In this laboratory session, you will be exploring glucose breakdown.

EXPLORING

The breakdown of glucose

Products of glucose breakdown

Obtain the following materials:

1 container of barium hydroxide (BaOH) solution
1 small beaker
1 drinking straw
1 10-ml graduated cylinder

Transfer 3 or 4 ml of BaOH solution to the small beaker. Using the straw, bubble your lung air through the BaOH solution. After blowing for a minute or two, what happens to the barium water?

1 ______________________________

At the demonstration table, observe the beaker of barium water that has been enclosed within a darkened container with a green plant. Compare this barium water to a beaker of barium water that has been in a closed container without a plant. What do you observe?

2 ______________________________

Is there any evidence of excess moisture in the air of either container?

3 ______________________________

If so, describe this evidence.

4 ______________________________

____

Barium hydroxide reaction

Carbon dioxide (CO_2) and BaOH react chemically, producing barium carbonate. Barium carbonate is insoluble in water; it solidifies and settles out as a white precipitate. Barium carbonate is the chalky material you observed in some of the beakers.

Think about the possible sources of carbon dioxide that reacted with the BaOH in the beakers. In each case of barium carbonate formation, what is the source of carbon dioxide?

5 ______________________________

____

____

____

____

CONCEPTS/TERMS

Glucose breakdown

Products

When glucose is burned in cells, among the products are carbon dioxide and water (H_2O). (These products are also produced when glucose is burned in fire.) As you blew air from your lungs into the BaOH solution, carbon dioxide in your exhaled air reacted with the BaOH, forming the barium carbonate precipitate. The carbon dioxide, along with water, was produced as your body's cells used glucose. This gas was then transported to your lungs and diffused into their cavities.

Plants also burn glucose. They release the carbon dioxide and water into the air through leaf openings. The carbon dioxide that reacted with BaOH in the beaker enclosed with the covered plant came from the plant's breakdown of glucose. (Air contains a small amount of carbon dioxide, about 0.03%. If you observed some white precipitate in the beaker of barium water in the container without a plant, the carbon dioxide source was the small amount of this gas in the air.) The water from glucose breakdown may have condensed on the surface of the enclosing container.

$$1\ C_6H_{12}O_6 + 6\ O_2 + 36\ ADP + 36\ P_i \longrightarrow 6\ CO_2 + 6\ H_2O + 36\ ADP{\sim}P + \text{Calories}$$

Glucose (high energy)	Oxygen (low energy)		Carbon dioxide (low energy)	Water	ATP (high energy)	Heat

Figure 8-2
Summary of complete glucose breakdown.

Concepts of aerobic respiration

The chemical reactions occurring in the glucose breakdown process are summarized in the chemical equation of Figure 8-2. If glucose is broken down completely in living organisms, oxygen, **adenosine diphosphate (ADP),** and inorganic phosphate (P_i) are required and the final products are carbon dioxide, water, heat, and ATP molecules (ADP ~ P). This series of chemical reactions is called **aerobic cellular respiration.** The high-energy bond (denoted by the squiggly line) between the second and third phosphate group of the ATP molecule holds the chemical energy transferred from the glucose molecule. The organism then uses this unstable ATP molecule for cellular activities by breaking this high-energy bond, releasing its energy. (Heat is measured in calories. A calorie is the amount of energy needed to increase the temperature of 1 g of water 1° C.)

Incomplete glucose breakdown: fermentation

When oxygen is low or absent, some organisms such as yeast and some bacteria do not burn glucose completely. In fact, when you are exercising maximally, your muscles may lack sufficient oxygen to burn glucose completely. This incomplete burning results in the production of lactic acid in the muscles instead of carbon dioxide and water. This acid sometimes causes damage to the muscle, resulting in cramps or sore muscles. Incomplete burning of glucose is called fermentation. In fermentation, oxygen is not a reactant.

EXPLORING

Fermentation

Obtain the following materials:

1 container of boiled yeast suspended in water
1 container of nonboiled yeast suspended in water
1 container of glucose solution
2 fermentation assemblies
1 mm ruler (needed only if the fermentation assemblies lack graduated markings)
1 dropper bottle of methylene blue
1 10-ml graduated cylinder
2 1-ml pipets
1 pipeting bulb

When needed, get one small beaker with about 2 ml of BaOH.

Your teacher will demonstrate how to use a pipet. Practice pipeting different volumes of water. When you can pipet accurately, proceed with the remainder of the activities.

Set up two fermentations assemblies, one with nonboiled yeast and the other with boiled yeast. Your teacher may change the volume of the materials. Before adding yeast, thoroughly mix each suspension to distribute the yeast evenly. After completing the set-ups, place the fermentation systems into the warm water bath designated by your teacher. Note the time and the volume of air that may be in the collecting tubes.

Preparing fermentation assemblies

Fermentation tubes with upright arms

1. Put 20 ml of warm glucose solution and four drops of methylene blue into a fermentation tube with the upright arm.
2. Cover the opening and rotate the tube back and forth to mix; add 1 ml yeast solution, and mix thoroughly.
3. Rotate the tube to fill the closed "upright arm" with fluid.
4. Place the filled tube in the warm water bath.

Fermentation setup of two connected tubes

1. Completely fill the tube with ml marks with water and place it upside down in a beaker of water. This is the gas-collecting tube.
2. Fill the other tube to 2.5 cm of its top with warm glucose solution. This is the fermentation tube. Add four drops of methylene blue. Mix.
3. Add 1 ml of yeast suspension to the glucose and mix.
4. Immediately stopper. Attach one end of the flexible tubing to the glass tube in the hole of the stopper.
5. Place the free end of the tubing into the water-filled open end of the collecting tube. Put the entire assembly in the water bath.

Nested fermentation tubes

1. With warm glucose solution, fill the smaller, graduated tube to 2 cm of its top; add four drops of methylene blue and mix.
2. Add 1 ml of yeast suspension and mix.
3. If the tube is not completely filled, add additional glucose.
4. Invert the larger tube and lower it over the smaller one until the bottom of the larger tube is pushed against the top of the smaller tube.
5. Invert the nest of tubes quickly so that the fluid remains in the smaller tube. Set them in a rack in the water bath.

Observe the two fermentation assemblies every 15 minutes for the next 45 minutes or longer. In Table 8-1, describe the color of the fluid and the amount of gas produced. You may want to use a color recording system with five plusses (+++++) representing the most intense blue and fewer to zero plusses

Table 8-1 Nonboiled and boiled yeast fermentation of glucose with methylene blue

Time		Nonboiled yeast		Boiled yeast	
Clock	**Elapsed (min)**	**Gas volume**	**Color**	**Gas volume**	**Color**
	0				
	15				
	30				
	45				
	60				

for lesser intensities. If the gas collecting tube of the assembly lacks millimeter markings, measure the part of the collecting tube that is filled with gas.

After the run is complete, remove the gas with a Pasteur pipet and bubble it into a small amount of BaOH solution. If you have the fermentation tube with the gas-trapping arm, collect the gas by the following method:

1. Cover the tube's opening and tilt the tube to allow the gas accumulated in the upright arm to move into the bulb portion of the tube.
2. When ready to pipet, lift the cover and remove gas with the Pasteur pipet.

To collect gas from the two-tube assembly, do the following: Pinch the flexible tubing closed at the collecting tube and remove it from the beaker of water. When ready to pipet, release your pressure on the tubing and pipet the gas from it.

To collect gas from the nested tubes, invert the pair, immediately remove the outer tube, and cover the inner tube until you are ready to pipet out the gas.

What information have you obtained from bubbling the gas through the BaOH?

6 ____

On what evidence do you base your answer?

7 ____

What is your best explanation for the lack of gas production in the boiled yeast tube?

8 ____

Hold the tube of nonboiled yeast a few inches below and in front of your nose. With your free hand, wave some of the air near the tube up toward your nose. If you recognize this odor, what is it?

9 ____

CONCEPTS/TERMS

Fermentation

Summary of fermentation chemistry

The equation that summarizes the chemical reactions in the yeast fermentation of glucose is shown in Figure 8-3. Fermentation produces carbon dioxide and ATP as in aerobic respiration but no water. It also produces alcohol.

Comparing fermentation and aerobic respiration

The relationship between aerobic respiration and fermentation is shown in Figure 8-4. Both use a series of reactions called *glycolysis;* the remainder of their reactions differ.

In glycolysis, the 6-carbon glucose is broken to two 3-carbon molecules with lower energy—**pyruvic acid.** In this process, some electrons (and their energy) are transferred to an electron carrier molecule called **nicotinamide ad-**

Figure 8-3 Summary of fermentation.

$$1\ C_6H_{12}O_6 + 2\ ADP + 2\ P_i \longrightarrow 2\ C_2H_5OH + 2\ CO_2 + 2\ ADP{\sim}P + \text{Calories}$$

$1\ C_6H_{12}O_6$	$2\ ADP + 2\ P_i$	$2\ C_2H_5OH$	$2\ CO_2$	$2\ ADP{\sim}P$	Calories
Glucose (high energy)	(low energy)	Alcohol (high energy)	Carbon dioxide (low energy)	ATP (high energy)	Heat

enine dinucleotide (NAD^+). Hydrogen ions transfer with the negatively charged electrons. Therefore NAD^+ with its load of electrons and hydrogen is frequently written as *NADH.* Each time these electrons and hydrogens are removed from the former glucose molecule, the remaining molecule decreases in energy and the carrier molecule increases in energy.

If oxygen is present and the organism has the appropriate organelle and enzymes, aerobic respiration continues with two series of chemical reactions. One series, the **Kreb's cycle,** uses pyruvic acid from glycolysis. The other series, the **electron transport system (ETS),** uses the electrons and hydrogens on the carrier molecule NADH (and FADH).

In the Kreb's cycle, the electrons and hydrogen are transferred from pyruvic acid to additional carrier molecules (NAD^+ and FAD^+), and two molecules of ATP are made. The carbon and oxygen atoms of the pyruvic acid form carbon dioxide—a low-energy waste product of respiration. The Kreb's cycle is the source of your exhaled carbon dioxide.

In the ETS, the carrier molecules (from glycolysis and Kreb's cycle) pass their electrons to a row of other carrier molecules and finally to oxygen. The oxygen combines with the hydrogen, forming water. Each time an electron transfer is made, some energy is released. Part of this energy is used to build a gradient of hydrogen ions. The hydrogen ions flow from areas of higher concentration to areas of lower concentration. This flow of ions releases energy that chemically attaches the third phosphate group to ADP; thus ATP is produced.

If oxygen is not present, fermentation occurs. In the reactions of fermentation, the Kreb's cycle, ETS, and oxygen are *not* used, but the products of glycolysis are (Figure 8-4). After glycolysis in yeast fermentation, one of the three carbons and two oxygens of pyruvic acid are split off and form carbon dioxide. Next, electrons are passed to the remainder of the pyruvic acid molecule from NADH, producing ethyl alcohol. The carbon dioxide produced in the process of fermentation raises bread dough and puts bubbles in beer. Ethyl alcohol is the chemical responsible for changes in reactions of our nervous systems if beer, wine, or bourbon is consumed. Other organisms may produce different fermentation products.

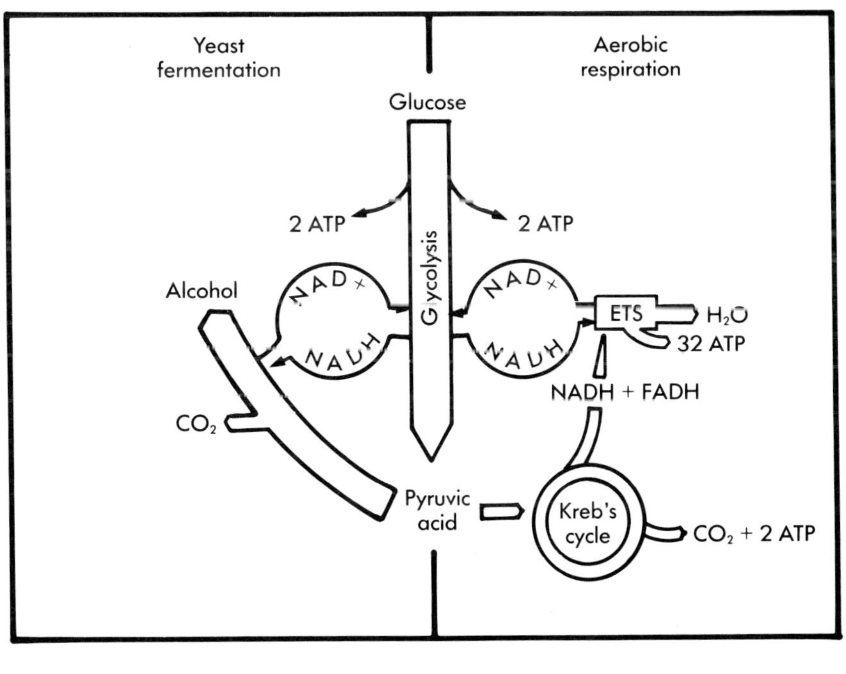

Figure 8-4
Relationship of fermentation and aerobic respiration.

The number of ATP molecules produced in fermentation is much less than the number produced in aerobic respiration. Scientists have determined that 2 ATPs are produced for each glucose molecule through fermentation, whereas 36 ATPs are produced through aerobic respiration. Alcohol, one of yeast's fermentation products, still contains a lot of chemical energy.

Consideration of the observed color change

Now consider the methylene blue color changes that occurred during yeast fermentation. In glycolysis, the normal living organism uses NAD^+ as the carrier molecule for the electrons removed from glucose. Methylene blue can also carry electrons. Scientists say that a compound, such as methylene blue, is **oxidized** when it does not have extra electrons. The oxidized form of methylene blue is blue in color. After a compound picks up electrons and hydrogen, scientists say that it is **reduced.** Reduced methylene blue is colorless.

With this information, give your explanation of the observed color change in the yeast-glucose-methylene blue mixture.

10 ____

____

____

APPLYING

Experiment: do fast-acting yeasts do a better job?

One product on grocery store shelves is a fast-acting yeast for bread baking. The directions on the package give the following instructions:

1. Mix the dry granular yeast with the dry ingredients (such as flour and sugar).
2. To this dry mixture, add water that is 125° to 130° F (55° C).

This water temperature is considerably hotter than the water temperature used with regular dry yeast. Regular baking yeast has package instructions that direct the user to dissolve the yeast in 110° to 115° F (45° C) water with a small amount of sugar and then to add the mixture to the dry ingredients.

Several questions arise about the use of the fast-acting yeast. A general question is, "Does it really raise the bread more rapidly?" For biologists, this question could be rephrased as "Does the fast-acting yeast produce carbon dioxide more rapidly?" If so, there are at least two questions about how carbon dioxide is produced more rapidly:

1. Is carbon dioxide production increased because the fast-acting yeast's enzymes are in a warmer environment (55° C) than the regular yeast's enzymes (45° C)?
2. Is carbon dioxide production increased because the fast-acting yeast has faster working enzymes with higher optimum temperatures than the regular baker's yeast?

If you had experimental data on the effects of temperature on the production of carbon dioxide by these two yeasts, how could you apply the information to the questions above to get complete or at least partial answers?

11 ____

____

____

When the answers to these questions are determined, you have information about whether the fast-acting yeast is a better product than regular yeast. In other words, you have answered the question "Do fast-acting yeasts do a better job?" In addition, if the fast-acting yeast is more expensive, you can judge whether its activity is worth the extra cost.

Design an experiment that gives data that will help answer questions such as the following:
What is the optimum temperature for regular yeast fermentation?

12 ____________________

What is the optimum temperature for fast-acting yeast fermentation?

13 ____________________

Which yeast is the faster fermenting organism?

14 ____________________

What temperature would you recommend that bakers use for raising bread dough?

15 ____________________

Do you recommend the purchase of fast-acting type yeasts for bread baking? (You should be able to explain your reasoning that resulted in this answer.)

16 ____________________

In designing your experiment, you need to consider the following questions:
What is your hypothesis and its prediction?

17 ____________________

What is the independent variable?

18 ____________________

What is the dependent variable?

19 ____________________

What procedure provides the most comprehensive data for answering the questions about the two yeasts?

20 ____________________

____

____

What factors should be controlled?

21 ____

____

When you have designed your experiment, check with your teacher, then run your test of the two yeast types. After you collect your data, construct a graph and answer the questions about these yeasts. Explain how your observations justify your conclusions.

LABORATORY 9

Photosynthesis

Light and pigments

Transformation and storage of energy: photosynthesis

In chloroplasts of plants and some simpler organisms, two series of reactions occur that are necessary for most organisms to exist. These reactions convert light energy from the sun to chemical energy stored in glucose. Glucose is the primary food molecule, either directly or indirectly, of organisms. The energy in glucose powers life's activities. These reactions that use light to produce glucose are called **photosynthesis** (*photo,* "light" and *synthesis,* "put together").

The first part of photosynthesis transforms light energy to chemical energy. Special molecules called **pigments** absorb light energy, which increases the energy of their electrons. Some pigment electrons are passed through a chain of carriers to the electron-hydrogen carrier molecule **nicotinamide adenine dinucleotide phosphate ($NADP^+$).** As electrons are passed to the $NADP^+$, NADPH is produced and some of the energy is used to attach a third phosphate group to ADP, forming ATP. In addition, water molecules split; this splitting supplies electrons, hydrogen ions, and oxygen. The oxygen is released as a waste product of photosynthesis. This series of reactions is called the **light reaction** (Figure 9-1).

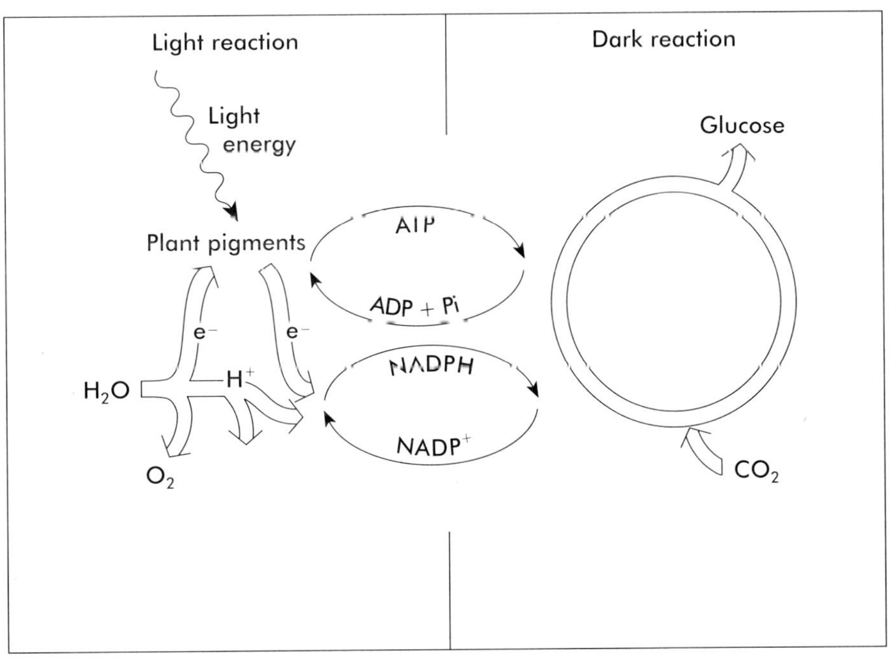

Figure 9-1
Summary of photosynthesis.

The second part of photosynthesis uses the chemical energy stored in the unstable products of the light reaction (ATP and NADPH) to synthesize the stable, high-energy molecule of glucose from low-energy carbon dioxide. Because this formation of glucose can occur in the absence of light, this part of photosynthesis is called the **dark reaction** or the **light-independent reaction.**

To understand photosynthesis, it is essential to know some of the properties of light and the pigment molecules that respond to it. These are the foci of the laboratory activities today.

EXPLORING

Visible light

Obtain a triangular column of acrylic or glass (a **prism**), a lamp, and a sheet of white paper. Shine a narrow beam of light on the prism. Hold the piece of white paper a foot or so away from the other side of the prism, or project the light from the prism on a nearby wall or screen. In Figure 9-2, record the color of the bands in their top to bottom order.

CONCEPTS/TERMS

Light

Light is a part of one form of energy that travels through space as radiating waves. This energy is called **electromagnetic energy, or radiation.** The waves, measured from crest to crest, are of various lengths. Some waves are many meters long, as in radiowaves; others are several billionths of a meter, as in x rays and gamma rays. The electromagnetic energy we detect through our eyes and brain, **visible light,** varies in wavelength from 760 nanometers (nm) of red light to 390 nm (or 0.760 to 0.390 μm) of purple light (see Plate 5A).

The visible light from an light bulb is composed of all the visible wavelengths. (This is sometimes called **white light.**) When rays (beams) of white light shine on a prism, the light of different wavelengths is dispersed into a rainbow-like spectrum. This sorting of the wavelengths is based on **refraction.** (Refraction is the bending of light as it passes from one transparent medium to one of different density, such as from air to glass.) Shorter wavelengths are refracted more than longer wavelengths. Hence purple light is bent more than any of the other colors and red is bent the least.

When sunlight penetrates raindrops, the water droplets in the air act as prisms. By refraction, they separate the different wavelengths in sunlight into the colors of the rainbow.

EXPLORING

Colors of objects

Obtain two tubes of liquid labeled A and B. With the light coming from behind you, look at the container of liquid A. What is the color of the fluid?

1 ______________________________

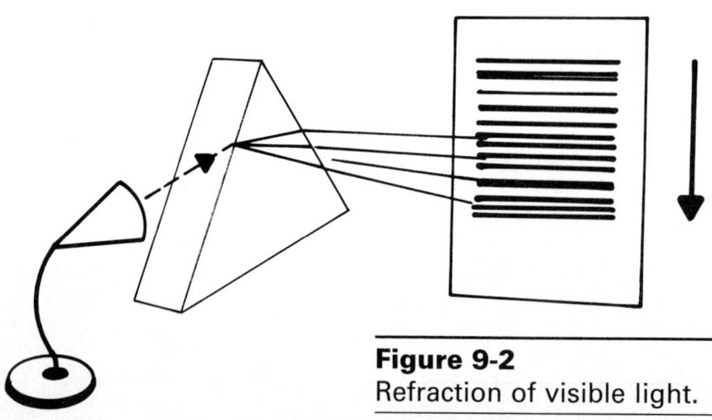

Figure 9-2
Refraction of visible light.

1. ____________________
2. ____________________
3. ____________________
4. ____________________
5. ____________________
6. ____________________

Hold the container of liquid A up to the light. What is the color of the fluid now?

2 ______________________________

Next, look at the container of liquid B. What is its color?

3 ______________________________

In this case, does the color appear the same if the lamp is on your side of the tube as when it is on the far side of the tube?

4 ______________________________

CONCEPTS/TERMS

Terminology associated with light rays

When white light from the lamp is on your side of the container, some of the light that reaches the liquid bounces off the molecules of the liquid and back to your eye. This is **reflected light.** If certain cells of the eye are sensitive to light of this particular wavelength, they send messages to the brain, which interprets the reflecting object as having a specific color.

Reflection is not the only possibility that can occur when rays of light hit material. Light rays can also pass through the material; this is **transmitted light.**

If the light rays have the right amount of energy, a third possibility is conversion of the light energy into the movement of electrons in the molecules. This is **absorbed light.** This is an energy-transforming step. It occurs in the pigment molecules of the chloroplast.

APPLYING

Terms and the tubes of fluid

As light rays hit liquid A, what color of light is reflected?

5 ______________________________

What color of light is transmitted?

6 ______________________________

Do you think that liquid A absorbs the blue wavelengths of light?

7 ______________________________

Explain the reasoning that led to your answer.

8 ______________________________

____

____

What colors of light are absorbed by liquid B?

9 ______________________________

The liquid in container B appears black because all the rays in white light are absorbed by this liquid and no light rays are transmitted or reflected to your eyes. Your brain interprets this absence of light from the object as black. In Figure 9-3, identify the types of light as white light shines on the test tube of material.

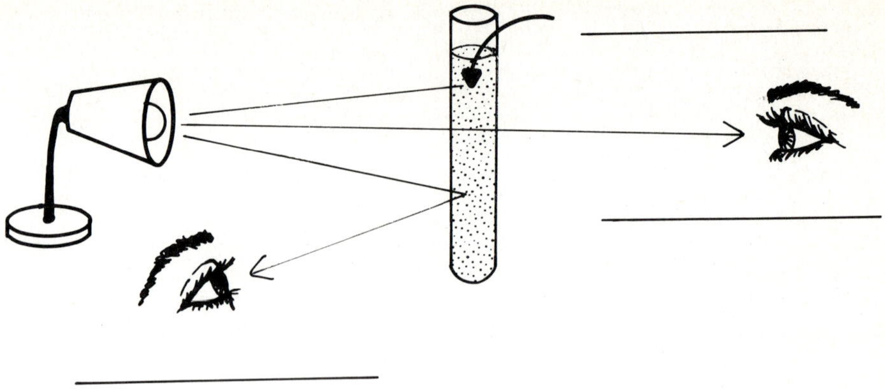

Figure 9-3
Visible light shining on a container of material.

EXPLORING

Light-absorbing molecules of photosynthesis

Extracting plant pigments

Working in pairs, obtain the following materials:

1 mortar and pestle
3-4 single-layers of cheesecloth, 10-cm square
1 funnel
2 beakers
1 container of ethyl alcohol
1 25-ml graduated cylinder
3-4 small beakers
Pinch of sand
Goggles

CAUTION
The chemicals you are using in this exploration of plant pigments evaporate easily, catch fire readily, and can damage body cells if inhaled or splashed in the eyes. Work safely. Wear goggles. Work in well-ventilated areas. Keep reagent bottles tightly closed. Work in areas that are free of fires, hot surfaces, and smoking.

Select a handful of leaves of the plant designated by your teacher. Remove big veins and "little stems" (petioles) of the leaves. Place leaves in the mortar, add a pinch of sand, and add about 15 ml of ethyl alcohol. Working under the laboratory fume hood or in a well-ventilated area, grind the leaves thoroughly to break open all cells and their organelles. If necessary, add small amounts of additional alcohol to obtain a smooth mixture about the consistency of very thin pudding.

Filter this mixture into a beaker through three to four layers of alcohol-dampened cheesecloth laid in the funnel. Discard the unbroken leaf cells remaining in the cloth. The fluid that passes through the cloth is an **extract.**

To remove remaining cell fragments, filter the extract through a coarse, alcohol-dampened filter paper. (Your teacher will demonstrate how to fold filter paper to fit the funnel.) The transparent extract should be free of particles. If the extract becomes cloudy, your teacher will suggest a method to clear it. This extract contains the plant's energy-transforming pigments. Using the small beakers, divide your extract into several volumes of 2 to 3 ml (label). Keeping one beaker of extract, give others to the laboratory groups that do not have extract from this plant. From other groups, obtain extracts from other types of plants.

Check each extract (as with liquids A and B) for reflected and transmitted light. Record the colors of the reflected and transmitted light of each extract in Table 9-1. In the fourth column, predict the colors of the absorbed wavelengths of light.

Separating plant pigments: chromatography of plant extracts

Continuing to work in pairs, obtain the following materials:

2-3 ml of four plant extracts
4 capillary tubes (very small–diameter tubes)
1 piece of chromatography paper (Keep the paper free of fingerprints, dust, oils, and other materials.)

Figure 9-4
Preparing a chromatogram.

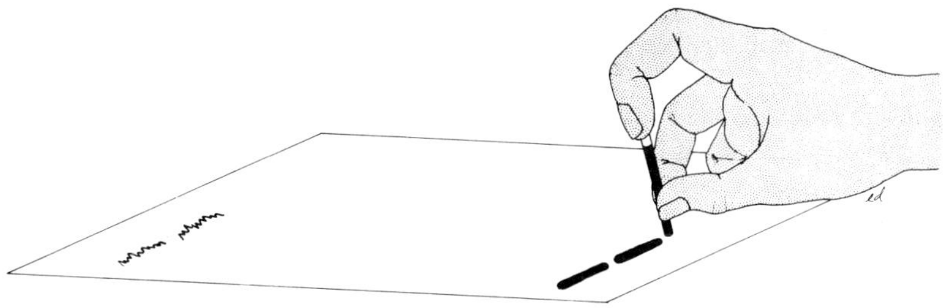

Table 9-1 Light and pigment extracts

Plant	Colors		
	Reflected	Transmitted	Absorbed
1.			
2.			
3.			
4.			

1 container of petroleum ether:acetone (9:1)
Ruler with 1-mm increments

Prepare a chromatogram, which is a technique used to separate and identify compounds. With a pencil and ruler, lightly mark a line along the bottom of the chromatography paper, 2 cm from the bottom edge. Divide this line into four segments with small spaces (about ¼ cm) between segments (Figure 9-4).

Apply a very narrow line of one of the extracts on the first line segment with a capillary tube. Control the width of the line by regulating forefinger pressure on the end of tube as demonstrated by your teacher. On each of the other three segments, apply one of the other extracts. Allowing each extract to dry between applications, repeat the applications at least 5 times. Try to get the most concentrated application of extract on the narrowest of lines. On the edge opposite the application line, identify the plant for each segment in pencil.

Under the hood or in a well ventilated area, pour the prepared solvent mixture of petroleum ether:acetone (9:1 ratio) into a dry chromatography chamber (a quart jar) to the depth of 1 cm. Immediately place the lid on the chamber to prevent air pollution in the laboratory and to let the solvent molecules saturate the air in the jar.

Roll the chromatography paper with the extract segments into a cylinder, with the extract application line at the bottom of the cylinder. Staple together the touching but not overlapping sides.

Without touching the sides of the jar, place the cylinder into the chamber, extract segments down. Quickly replace the lid of the chamber, and wait about

15 minutes for the organic solvent (petroleum ether: acetone) to move up the paper to approximately 1 cm of the top edge. Do *not* let the solvent reach the top edge of the cylinder.

Describe what happens as the organic solvent moves up the chromatogram.

10 ______________________________

____

____

Remove the chromatogram *before* the solvent reaches the top edge of the paper. With a pencil, immediately mark the top (leading) edge of the solvent, the **solvent front.** Allow the paper to dry standing up under the hood or area designated by your teacher. As soon as the solvent evaporates, remove the staples and outline the colored bands. Identify the colors also because these bands of color fade with time.

Obtaining information from the chromatogram

Each one of the colored bands from a plant's extract is a different pigment. How many pigments did each plant have?

11 ______________________________

____

How many plants have bands that are at the same height *and* of the same color as other plants?

12 ______________________________

Measure the distance the organic solvent moved past the extract application line and record it in Table 9-2. Also measure and record the distance each pigment band traveled in each of the plant extracts. Calculate the distance that each pigment traveled compared to the distance past the application line traveled by the solvent front. This relative distance is called the **Rf value.** Attach your chromatogram to Table 9-2.

Concepts of the technique of paper chromatography

Paper chromatography is a technique used to separate and identify substances. The basis of this technique is the use of two solvents in which the substances to be separated have different solubilities. One of these solvents is called the **stationary phase** because its molecules remain in place. The other solvent is called the **mobile phase** because it moves past the molecules of the stationary solvent. In this laboratory, the mobile solvent is the petroleum ether: acetone mixture and the stationary solvent is water. Water molecules from the air attach to the molecules of the chromatography paper by hydrogen bonds.

What do you think happens as the mobile phase moves up the paper if the pigment applied to the line segment is very soluble in the mobile phase and not in the stationary phase?

13 ______________________________

Because the distance traveled by each substance is a measure of its differences in solubility in the two solvents, you can partially identify unknown substances by comparing the distance they traveled to the distance traveled by a known substance. If these distance are the same, the two substances may be the same.

Table 9-2 Plant pigments: chromatographic results

Color of band	Distance traveled	Rf value*	Color of band	Distance traveled	Rf value*	Color of band	Distance traveled	Rf value*	Color of band	Distance traveled	Rf value*
EXTRACT OF ________			EXTRACT OF ________			EXTRACT OF ________			EXTRACT OF ________		
DISTANCE OF			DISTANCE OF			DISTANCE OF			DISTANCE OF		
SOLVENT FRONT			SOLVENT FRONT			SOLVENT FRONT			SOLVENT FRONT		

*To calculate Rf values, divide the distance each pigment moved from the extract line by the distance traveled by the solvent front. This gives a relative distance value (Rf) for each of the pigments compared to the distance the solvent traveled on the chromatography paper, so Rf = distance the pigment traveled ÷ distance the solvent front traveled.

CONCEPTS/TERMS

Using Rf values to identify substances

The Rf value is characteristic of the particular pigment in the band. The Rf values of specific pigments will always be in the same order whenever these pigments are chromatographed with the same two solvents. If chromatographed under identical environmental conditions with the same solvents, the Rf values for each pigment will be the same from one run to the next. If a different mobile phase or a different type of stationary phase is used, the distance traveled by each pigment will differ from the Rf value obtained in the water/petroleum ether:acetone system; even the pigments' order can change.

Unknown substances can be identified by comparing their Rf values to Rf values of known substances in a number of solvent systems. An additional help in identifying plant pigments is the color of the band.

Table 9-3 gives reference Rf values and colors of a few plant pigments. Compare the pigments from your chromatogram with this information to identify most (if not all) of the pigments in the plant extracts. If your environmental conditions were somewhat different than those when the Rf values in Table 9-3 were determined, your values may differ. In this case, match your values to the closest Rf values in Table 9-3, particularly observing the order of pigments. Add the names of the plants that have each pigment in the last column.

Relationships among your observations

Name the individual pigments you found in each of the four plant extracts.

14 1. ______________________________

2. ______________________________

Table 9-3 Known plant pigments

Pigment names	Band color	Rf value*	Molecular weight	Location in cell	Plants with pigment
CHLOROPHYLLS	Green				
a	Bluish-green	0.171	893	Chloroplast	
b	Yellow-green	0.108	907	Chloroplast	
CAROTENOIDS	Yellow, orange, red				
Carotenes					
α	Yellow	0.980	536	Chloroplast	
β	Yellow	0.996	536	Chloroplast	
Xanthophylls	Yellow, green-yellow				
Crytoxanthine	Yellow	0.712	552	Chloroplast	
Leutin	Yellow	0.451	568	Chloroplast	
Zeaxanthine	Yellow	0.274	568	Chloroplast	
FLAVINOIDS	Variable				
Anthocyanins	Red, blue, purple	0.001	423†	Vacuole	
Flavinols	Ivory to yellow	0.001	417†	Vacuole	

*In a solvent system consisting of petroleum ether:acetone in mobile phase; water in stationary phase.

†These pigments attach to other molecules. The actual molecular weight of the pigments depends on the specific molecule to which these pigments are attached.

3.

4.

Which of the pigments found in all four extracts are found in chloroplasts (use Table 9-3)?

15

Name the pigments that are not common to all plants.

16

Which of these pigments are found in chloroplasts?

17

Do you see all the pigments in a plant by looking at its leaves or its extract?

18

What evidence serves as the basis of your reasoning process?

Do you think that photosynthetic pigments that capture energy for building glucose are found in all plants?

19

Explain your answer.

20

APPLYING

Hidden pigments, photosynthesis, and shaded plants

Review the decisions you made for Table 9-1. Considering the information you gained from the chromatogram, make any appropriate changes in Table 9-1 by completing Table 9-4. Do you think you should include the wavelength colors of the "hidden" pigments in the reflected and transmitted columns? Considering that each of these four plants photosynthesize, which pigments do you think are involved directly in the process of photosynthesis?

21

Explain how you developed your answer.

22

With the knowledge you now have, give a possible explanation for the seasonal change in the color of fall leaves.

Table 9-4 Pigment extracts and reflected, transmitted, and absorbed light

Plant	Colors		
	Reflected	Transmitted	Absorbed
1.			
2.			
3.			
4.			

23 ______________________________

____

____

Some plants that grow well in the shade produce less red pigment (anthocyanins) than when they grow in a sunny area. Speculate about the effect of the decrease in anthocyanin on the rate of photosynthesis.

24 ______________________________

Explain your answer.

25 ______________________________

____

LABORATORY
10

Photosynthesis

Absorption and food storage

PART ONE Absorption: energy transformation

Organisms and energy

Energy is fundamental to the existence of living organisms. They require a constant input of energy because their chemical energy is continually converted to unusable heat energy. The energy in living organisms is replenished through the process of photosynthesis. In photosynthesis, plants transform the light energy of the sun into the chemical energy stored in the stable bonds of glucose molecules. Glucose is then available, either directly or indirectly, as food that supplies the chemical energy for all organisms. Today, the laboratory activities focus on the energy-transformation process and the storage of **starch,** a glucose polymer.

First, return to the consideration of the electromagnetic spectrum of which visible light is a part. Plate 5A shows the amount of energy associated with the different wavelengths of radiation. Each wavelength of radiation is characterized by special energy units, **photons.** In fact, radiant energy can be thought of as packets of specific amounts of energy as well as radiating waves. The particular nature of their interaction with matter distinguishes photons of the different energy levels.

EXPLORING

Energy of radiation

Examine the electromagnetic spectrum using Plate 5A. What is the range of energy levels that correspond to the wavelengths of visible light?

1 ______________________ **(purple)** to ______________________ **(red)**

What is the energy range of the photons in x-ray radiation?

2 ______________________

In infrared?

3 ______________________

Look at the entire spectrum. In general, what type of relationship exists between the lengths of the radiation waves and their photon energy levels?

4 ______________________

Recall what you know about the effects of overexposure of skin to sunlight (even on a cloudy day). Would you expect ultraviolet (UV) rays to be harmful to living material?

5 ______________________________

Or x rays?

6 ______________________________

Put together what you know about energy levels of these rays with your knowledge of what holds molecules together. Propose an explanation for the effects of UV and x-rays on organisms.

7 ______________________________

CONCEPTS/TERMS

Radiation energy and organisms

There is an inverse relationship between wavelength and energy; short wavelengths of radiation have high energy levels. High radiation energy can break molecular bonds, damaging living material. In fact, it can disrupt structures of molecules sufficiently to cause the processes of life to cease and the organism to die. The hereditary material, DNA, can be damaged by the energy in UV rays, whereas other molecules in organisms are damaged only with exposure to photons of higher energy levels.

Energy levels within the visible spectrum do not usually damage molecules of living organisms. Some molecules can convert these specific photon energy levels to chemical energy, the movement of the electrons. These molecules are changed by only *some* of the different energy levels found in the wavelengths between 380 and 780 nm. For example, one type of molecule may be readily affected by the specific photon energy only in the wavelengths close to 540 nm. A different molecule may be affected by other energy levels such as those in wavelengths of 435 and 650 nm and not 540 nm or other visible wavelengths. The effect of radiant energy levels on molecules depends on the structure of the particular molecule.

A substance's absorption of the radiant energy in different wavelengths can be measured and plotted; this is called an **absorption spectrum.**

EXPLORING

The absorption of spinach pigments

Prediction

In Table 9-4, you predicted that certain wavelengths of light would be absorbed by four plants, one of which was spinach. Now do some preliminary checking on your prediction.

At the demonstration table are the following materials:

1 container of spinach extract
1 prism
1 white sheet of paper or projection screen
1 narrow beam lamp

Place the extract container in the beam of white light between the lamp and the prism. Observe the effects of the extract on the visible light bands produced by the prism. In Table 10-1, record the presence (+) or absence (−) of colored bands of the spectrum with and without the spinach extract in the light path.

Using this information and the information on which you based your predictions recorded in Table 9-4, predict the wavelengths of visible light that this pigment extract absorbs. State your prediction.

8 ______________________________

Table 10-1 Colors of visible light wavelengths with and without plant extracts

Color band	Without extract	With extract
Violet		
Blue		
Green		
Yellow		
Orange		
Red		

Give the specific evidence on which you based your prediction. Also explain how you linked these observations together to develop your prediction.

9 ______________________________

____

____

____

Absorption spectrum of spinach extract

To measure the absorption of spinach extract at the different wavelengths of visible light, you need some information about an instrument called a *spectrophotometer* (see Appendix G). If you have questions about the basic concepts of spectrophotometry or how the particular model of spectrophotometer in your laboratory works after reading Appendix G, ask your teacher for clarification.

Work with a partner or as directed by your teacher. When you are ready to proceed, obtain the following materials:

2 optically matched spectrophotometer tubes
Access to a spectrophotometer
2-3 ml spinach extract (ethyl alcohol)

Slowly pour the spinach extract into one spectrophotometer tube to about a 1-cm depth. Dilute this extract by adding enough additional solvent (ethyl alcohol) to bring the volume to the label of the tube. Cover the tube and thoroughly mix the extract with the additional solvent by rotating the tube back and forth.

Following the directions for using your spectrophotometer, measure the transmittance of this diluted extract in light of 500 nm wavelength. What tube and substance should be used for the reference blank?

10 ______________________________

Table 10-2 Absorbance and transmittance of spinach pigment extract

	Wavelength (nm)												
	380	400	420	440	460	480	500	520	540	560	580	600	620
Absorbance													
Transmittance													

Explain your answer.

11 ______________________________

If the extract does not measure between 75% and 85% transmittance at 500 nm, adjust the dilution of the extract (add more diluting solvent or more concentrated extract) until the reading is within this range. This concentration of extract allows the absorption spectrum's readings to be within or close to the accuracy range of the spectrophotometer. (This accuracy range is between 20% and 80% transmittance.)

When you have adjusted the extract's concentration to give the appropriate range of readings, measure the transmittance and absorption of the extract at different wavelengths. Use 20-nm intervals ranging from 380 to 620 nm. If your spectrophotometer also produces light in the 640- to 700-nm range, include these in your data. (Remember to restandardize the machine every time you change wavelengths.) Record the readings in Table 10-2.

On Figure 10-1, prepare a graph of these readings using two vertical axes. Label one Y-axis *Percent transmittance.* Use this axis for plotting your transmittance readings at each of the measured wavelengths. Place 0% at the bottom of the percent transmittance axis and other appropriate numbers up to 100% transmittance at corresponding locations on the remainder of the axis. Label the other Y-axis *Optical density (OD)* or *Absorbance.* Use this axis for plotting the absorbance readings for each wavelength. The second Y-axis should extend from 0 at the bottom near the X-axis, to 2 near the top, close to the same level as 100% transmittance. (After 2, the absorbance quickly moves toward infinity.) What kind of relationship do you expect between the two plotted lines?

12 ______________________________

Did you base your answer on the relationship between transmittance and absorbance? If not, review Appendix G, particularly the section discussing the relationship between these two measures of light. Also, think back to what happens to light as it shines on a tube of material such as plant extract (see Laboratory 9). Correct your answer if necessary. Connect the points for transmittance readings and for absorbance readings. Identify each line. Does the graph support your prediction? If not, ask your instructor for clarification.

CONCEPTS/TERMS

Relation of spinach absorption spectrum to the absorption spectra of individual pigments

How many different pigments were contained in the spinach extract as determined by the chromatogram of Laboratory 9?

13 ______________________________

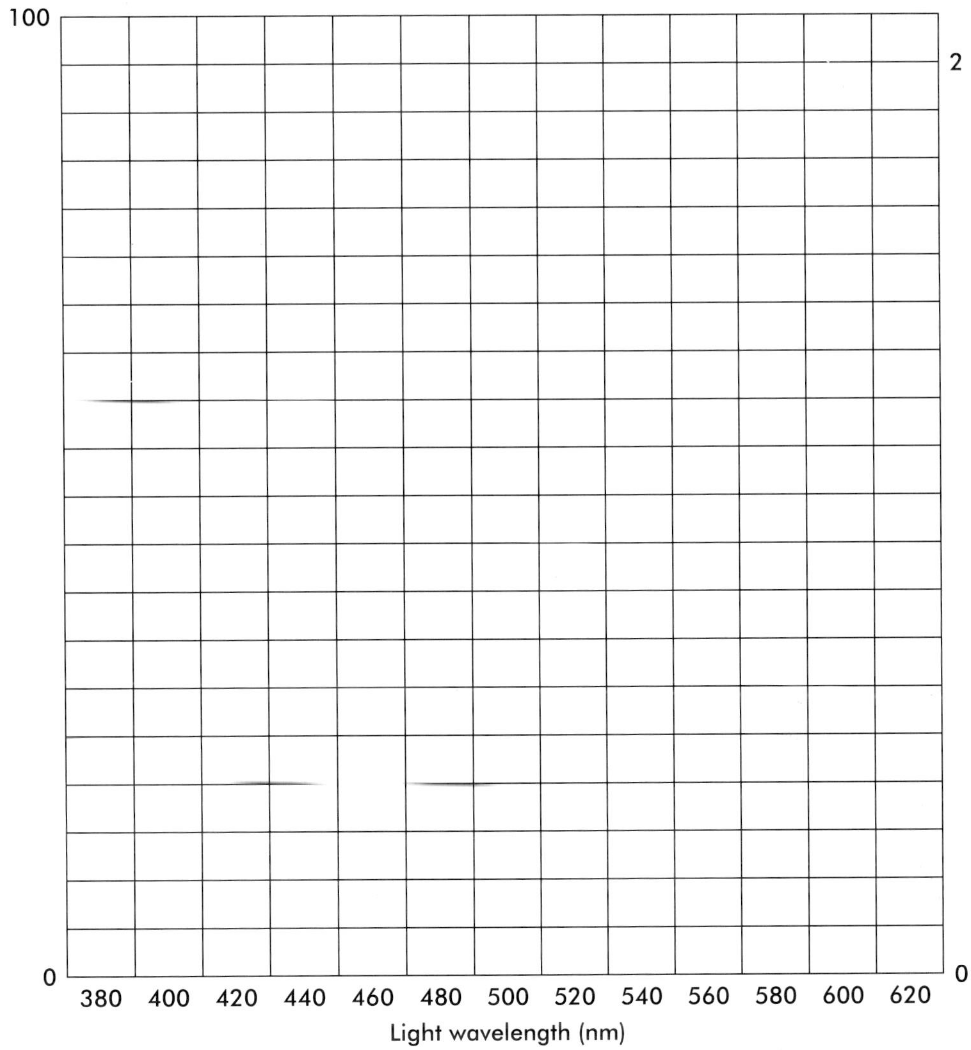

Figure 10-1
Absorbance and transmittance spectra of spinach extracts.

What colors of light were these pigments *not* absorbing?

14 ______________________________

Examine your spinach absorption spectrum. Do you have very low absorption where you expected your pigments would not be absorbing?

15 ______________________________

If not, ask your teacher for clarification.

Now, consider the wavelengths where you expected spinach pigments to absorb light energy. The absorption spectrum you constructed for spinach is the absorption activity of all pigments contained in the extract, not each pigment individually. Each pigment, including β-carotene, each xanthophyll, and chlorophylls *a* and *b,* has its own absorption pattern (Figure 10-2 and Plate 5B). Based on Figure 10-2, do the absorption spectra of all pigments in the spinach extract account for the absorption spectrum of the spinach extract?

16 ______________________________

If not, consult with your teacher.

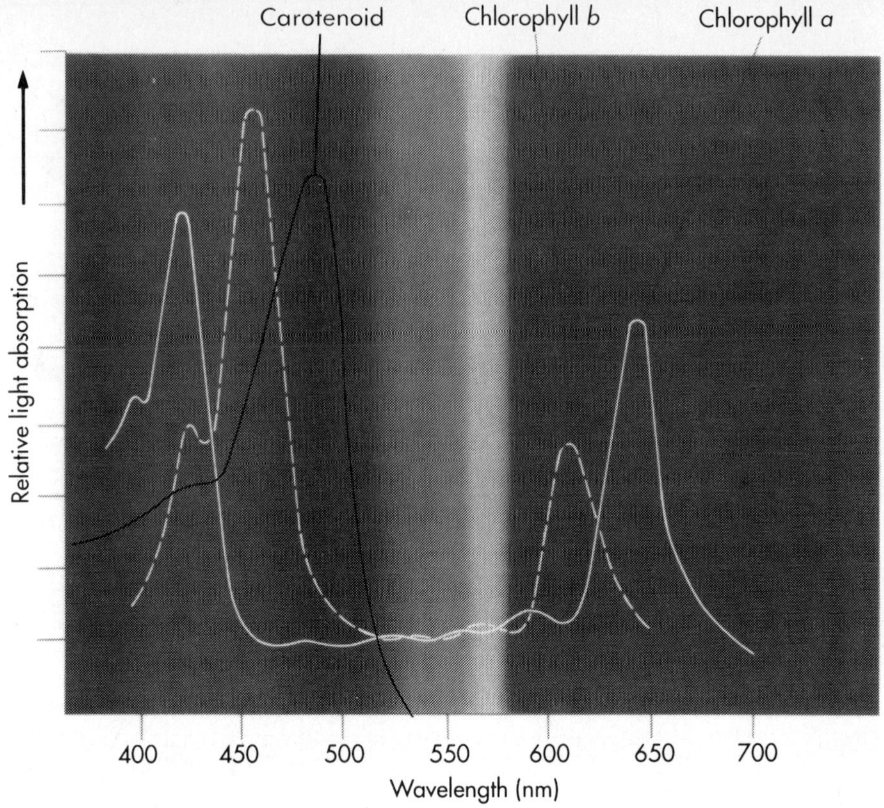

Figure 10-2
Absorbance spectra of some common pigments.

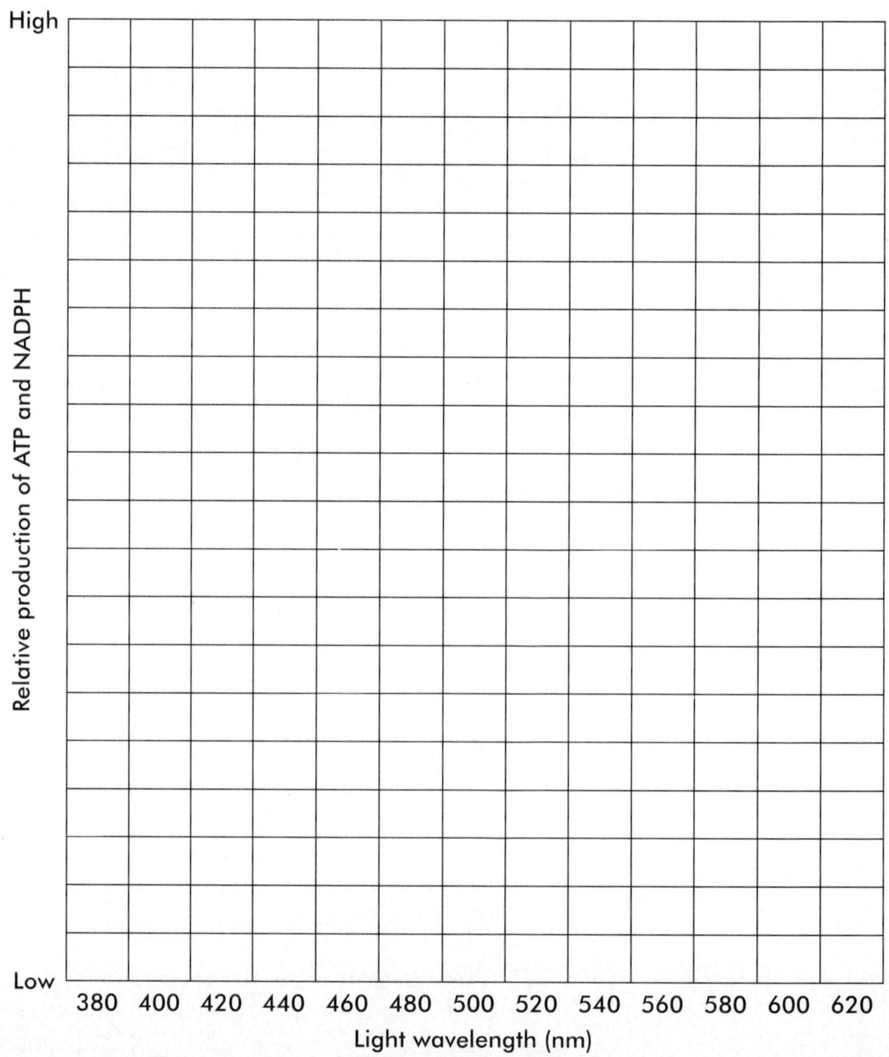

Figure 10-3
Predicted effects of light on the production of ATP and NADPH.

APPLYING

Concepts of light absorption to production of photosynthetic products

Recall what you now understand about the process of photosynthesis. Link the information you have about the absorption spectrum of plant pigments to the production of high-energy molecules. What do you think is the relationship between the absorption at different wavelengths and the production of glucose?

17 ______________________________

____

____

On Figure 10-3, plot what you think will be the relative amount of the ATP and NADPH molecules produced as the result of shining light of each different wavelength on a green plant such as spinach.

The amount of photosynthetic activity, as measured by photosynthetic products, plotted against the wavelength of the light is the **action spectrum** of photosynthesis.

Do your neighbors have an action spectrum similar to yours? If not, discuss how you each thought through the problem of predicting this action spectrum. Then, if there is disagreement, check with your teacher.

PART TWO Food production

EXPLORING

Effects of environmental light on stores of starch

When glucose production exceeds the glucose used for cellular respiration and other needs of the plant, the excess glucose is converted to starch, a glucose polymer, and stored in membrane-bound "bags" within the cells.

There are four plants at the supply table. Each one has been under a different environmental light condition for 3 or more days. These conditions are (1) the dark, (2) the sunlight or a plant light, (3) a green light, and (4) a red light. In the third column of Table 10-3, write your predictions regarding starch production in each of these plants. Explain the reasoning that led to your predictions.

18 1. ______________________________

2. ______________________________

3. ______________________________

4. ______________________________

Test a leaf from each plant for starch. Working with your partner or as designated by your teacher, obtain the following materials:

1 hot plate
2 beakers (one about 400 ml and the other about 150 ml)
2 petri dishes
4 pieces of thread or string, 12 to 15 cm long
1 dropper bottle of dilute iodide-potassium iodide (I-KI)
1 bottle ethyl alcohol, 85%
1 oven mitt or hot pad
Access to a bottle of India ink and a pen

Add water to the larger beaker until it is about one-third full. On the hot plate, heat the water to boiling. Tie a string to each leaf's petiole (little stems)

Table 10-3 Effects of environmental conditions on starch production

Plant number	Environmental condition	Prediction	Starch production (amount and location)
1.			
2.			
3.			
4.			

for easy handling. Immerse the leaves in the boiling water to remove any of the water-soluble deep-red pigments. Remove the leaves after a few minutes or when no red pigment remains in the cells.

Soak your leaves in ethyl alcohol poured into the smaller beaker. Warm the alcohol and leaves by placing this beaker in a container of hot water. This procedure will extract the alcohol-soluble pigments and should remove any remaining leaf color. After a few minutes, remove the leaves; then, briefly rinse and cool them with running water. Lay each leaf in a petri dish and test for starch by covering each leaf with I-KI.

Record your findings in the fourth column of Table 10-3. Include the location of the starch (if any) in each leaf, and the approximate amount of starch present with a 5 plus (+++++) to 0 system.

CONCEPTS/TERMS

Starch production and photosynthesis

Compare the results to your predictions of starch production. In which leaves were your predictions supported and in which leaves were they refuted?

19 ____

In what ways did other laboratory groups verify or refute your predictions?

20 ____

Explain any unexpected results.

21 ____

APPLYING

A variegated another leaf

Obtain a leaf with different-colored sections, such as a green-and-white variegated *Coleus* leaf. Predict where the starch storage areas are.

22 ____

In Figure 10-4, trace around the leaf and outline the different
color on your diagram. For future reference, you might outline the s
the different colors on the leaf with India ink. Treat the leaf as you
other leaves to remove pigments and test for starch.

Next to your drawing of the leaf with color sections identified (Figu
4), trace the leaf again and mark the starch storage areas. Explain your r
and indicate how it relates to your prediction.

23 ______________________________

____

____

____

Explain why you think this plant is or is not a good source of food for an **herbivore** (a plant-eating animal).

24 ______________________________

____

Environmental problems and photosynthesis

In magazines and newspapers, on television, and in schools, there are many discussions of environmental problems. For each of the following, express your thoughts about the effects of the environmental conditions on (1) the process of photosynthesis and (2) the long-term effects of these conditions on animals:

Figure 10-4
Starch stores in a variegated leaf.

Depletion of the ozone layer, which partially shields the Earth from the sun's radiation

25 ______

Accumulation of dust particles in the air from pollution, volcanic eruptions, or meteor collisions

26 ______

Increase in carbon dioxide from burning of fossil fuels such as gas, oils, and coal

27 ______

Cutting of the tropical rain forests

28 ______

Increase in hydrogen ion in the air from automobile emissions and fossil fuel burning

29 ______

LABORATORY 11

Tissues

Interacting specialized cells

Organization of organisms

What is the structure and the organization of cells that result in the functioning organism? Perhaps specialized cells help explain the different activities. Or perhaps there are special ways in which similar, simple cells are arranged that explain specific functions of the organism. Or perhaps there is a combination of both distinctive cell types and special ways in which these cells are arranged that are the structural bases for the functions of an organism.

In Laboratory 4, you explored some simple cells such as those of algal-type organisms *(Spirogrya),* those of single-cell organisms *(Paramecium* and *Hematococcus),* and your own cheek lining cells. You observed organelles, differences in organelles between animal and plant cells, and differences between prokaryotic and eukaryotic cells. In this laboratory, you will explore cells that interact with other cells, forming parts of complex organisms that have special functions. How are the structures and functions of these cell related? This is the primary question for the explorations in this laboratory.

EXPLORING

The cross section of a plant stem

Obtain your microscope, lens paper, and a box that contains the following commercially prepared slides:

Plant stem, x.s.
Cuboidal epithelium (amphibian kidney), sec.
Columnar epithelium (amphibian intestine), x.s.
Liver (mitochondria), sec.
Human blood, smear
Elastic cartilage (ear or epiglottis), sec.
Compact bone, x.s.
Skeletal muscle, l.s.
Smooth muscle, l.s.
Cardiac muscle (intercalated disks), sec.
Spinal cord (mammal), x.s.
Teased ox neurons

Locate a slide with a stained cross section (a thin slice taken across the structure) of a plant stem. Look at it briefly using the scanning objective of the microscope.

Before you observe with high power, think about a plant's activities and the role of the stem in these activities. The primary activity of the plant is photosynthesis, which requires water from the soil, carbon dioxide from the air, sunlight, and the photosynthesis machinery (chloroplasts) in the leaves. Considering the requirements for photosynthesis and what it takes to exist in windy and dry land environments, what are the functions of the plant stem?

1

Try imagining the kind of structure that would be necessary for the stem to carry out its functions. What would be the general characteristics of these structures? What kind of cell organelles and cell arrangements would give these characteristics to the stem structure? Let's take an example. You may think that one function of the stem is to store starch and this function requires a big space or large volume. So you might use *round* or *ball-shaped* as a descriptive word for the characteristic of the cells in the storage area and not the word *flat*. Then you may remember organelles such as vacuoles and plastids that enclose large spaces. So you may predict that the stem has round cells with large vacuoles or plastids.

In the first and second columns of Table 11-1, list the functions of the plant stem and the descriptive characteristics that are important to these functions. Then using your knowledge of organelles and cells, design a stem structure that carries out the particular function (third column).

Consult with your neighbors. Add other functions, stem characteristics, and predictions that arise from these discussions.

After thinking about the possible structures of the plant stem, return to the stem cross section and try to match the stem's structure with your predicted structures. (You may want to obtain a fresh-cut section of stem from a living plant to observe as you proceed with microscopic sections.) Use the high power objective of the microscope and begin with the cells at the edge of the stem that make up the stem's surface.

Which of the stem's functions listed in Table 11-1 can you assign to the cells at the surface of the stem?

2

Focus carefully on the cells at the edge of the stem cross section. What is the structure of each type of cell?

3

How are they spatially arranged? Do they have spaces between them or are they tightly packed? Are they in a simple (single) layer or are they stratified (multilayered)?

4

Table 11-1 Relating structural characteristics to stem function

Stem function	Required characteristics	Predicted structures

Now return to the question about the possible relationship between structure and function. If there is a relationship between structure and function, give several examples of how these cells are structured and arranged that allows them to carry out the functions you assigned to the cells at the stem surface.

5 ______________________________

____

____

____

Now look for cell structures that allow the stem to hold the leaves toward the sun against the gravitational force and withstand strong wind and pelting rain. Describe their locations and their structures.

6 ______________________________

____

____

Next consider the water-conducting function. Describe the locations and structures of stem cells that could carry out this function.

7 ______________________________

____

____

You may want to look at a longitudinal section of this same stem on the microscope at the demonstration table. The vertical spatial relationships may be important to some of the stem functions.

CONCEPTS/TERMS

Developing conceptual relationships

Relationship between structure and function

So far, you have explored only a few cells, but stop briefly to consider a fairly obvious generalization. Describe the ways in which you agree or disagree with the statement, "The function of a cell depends on its structure." In your explanation, give several examples that support your statements.

8 ______________________________

____

____

____

Also explain in what ways you agree or disagree with the statement, "Cells, the building blocks of life, have a variety of shapes and some differences in organelles." Or consider another form of this statement, "Many cells are specialized (differentiated)." Again use examples to illustrate your response.

9 ______________________________

____

____

____

One of the basic themes of life states that the functions and structures of organisms are intricately interrelated. In fact, this theme is not limited to life. Most people realize that the structure of an object is directly related to its function. (Can you imagine using a pencil with a layer of "lead" surrounding a central cylinder of wood with an eraser in the middle of its length!) If a structural component of an organism is altered, the function of that part and the function of the entire organism may change.

We rarely think about the relation of function to structure. Living organisms build structural parts and maintain them only if the proper metabolic reactions occur in the cells. Function is required for structure, and the two are intricately interwoven. Structure and function are each dependent on the other.

Interacting cells: tissues and organs

One cell alone does not cover the entire surface of a stem. Therefore it cannot act as a barrier to water loss from the underlying cells by itself. However, a number of cells with cell walls touching and without spaces between can cover the entire surface and prevent exposure of inner cells to dry air. In addition, if these surface cells have a water-impermeable material on their surface, they can provide better protection from drying. Similarly, one cell can not provide sufficient storage space for starch or photosynthetic pigments, but many cells can. In all multicellular organisms, cells interact. If interacting cells are of the same type, they are collectively called a **tissue.**

Biologists group similar tissues together to form several categories of tissues. Generally, they group the tissues of plants into vascular tissue (water and food conduction), ground tissue (cells surrounding the vascular tissue), and dermal tissue (outer protective covering). Biologists group tissues of animals with backbones into four major kinds of tissues: epithelial, connective, muscular, and nervous.

In the stem cross section, you can predict that edge cells protect underlying cells, thick-walled cells support and provide strength to the entire stem, and "empty" (thick-walled) cells form water-conducting tubes when placed one on top of another. These and other interacting tissues form the functioning stem, an **organ** of the plant.

Animals also have organs such as the skin, kidney, brain, and intestine, which are made of several interacting tissues.

APPLYING

Interacting cells of animal tissues

The general functions of some animal tissues are described in the following paragraphs. As you investigate sections of animal tissues, think about the structure and arrangements of cells that can provide each of the given tissue functions. Then observe each type of tissue thoroughly. The technique of differential staining has been used in preparing these tissues. Recall that biological stains are chosen to increase contrasts of some organelles and not others (see Laboratory 4). In these slides some expected organelles may be present but have insufficient contrast with their surroundings to be seen, or the organelle may be too small to see with your microscope.

Read the functions of the major types of tissues listed in the first column of the left side of Tables 11-2 through 11-4. In the second column, write your predictions regarding the structure (the shape and organelles) and arrangement of cells in the tissue based on its functions. In some cases, these functions may be performed by only some of the specific tissues within that tissue type. (For example, cells lining the surface of an internal organ might not have the structure to protect underlining tissues from drying as some of the other epithelial tissues do.) Therefore you might want to make a prediction about the appropriate structure for each of these functions separately. In other cases, all activities might be performed by just one type of cell. In these instances, you can lump all structural characteristics into one cell.

Thoroughly explore each slide listed in the right side of the tables while considering a few suggestions and questions in the following paragraphs. In the last column, draw a representative of these cells in detail and label its or-

Table 11-2 Predicted and observed structure in epithelial tissue

Function	Expected structure and cell arrangement	Slides to explore	Structure and spatial relationships
Protection		Cuboidal (amphibian kidney tubule)	
Absorption*			
Secretion†		Columnar (amphibian intestine)	
Excretion‡			
Sensation		Squamous (cheek cells)	
Surface transport			
		Mitochondria (amphibian liver)	

*In this context, *absorption* refers to moving substances from the fluid on one side of a cell through the cell to the fluid on the other side. It does *not* refer to the transformation of light energy into chemical energy that occurs in photosynthesis.
†Secretion is the release of products manufactured within the cells.
‡Excretion is the release of waste products from the organism.

ganelles. Then show the arrangement of two or three of these cells to one another by drawing outlines of several adjacent cells.

Epithelial tissues

Epithelial tissues have a number of functions in complex vertebrates (Table 11-2). Generally, epithelial tissues consist of cells with very little material between them (intercellular matrix). The cells exhibit three shapes: squamous (flat), cube like (cuboidal), and column like (columnar). Sometimes, they have just a single layer of cells, and at other times they may be stratified.

Cuboidal (amphibian kidney tubule). In this slide of kidney tissue, look for some circular or oval rings. Closely examine the cells that form these rings. Urine is modified inside the kidney tubes when specific substances move or are transported across the tubule cells by diffusion or active transport. Some substances move from the tubule cavity to the small surrounding blood vessels, and other substances move in the opposite direction. With these functions, do you expect the epithelial cells to be stratified or simple? Do you expect the cells to be packed tightly together or have spaces between them? Do you expect these cells to have a lot of mitochondria and large nuclei?

Columnar (amphibian intestine). Look at this cross section of the small intestine before you place it on the microscope; then view it with magnification. For this laboratory, explore the inner lining of the intestine, the intestinal epithelium. These cells actively secrete digestive enzymes into the intestinal cavity. Do you expect this intestinal lining to be stratified or simple?

10 ____

Explain your answer.

11 ____

How do digestive enzymes exit the cells that secrete them?

12 ____

How do the cells prevent the digestive enzymes from digesting the producing cells and the other epithelial cells?

13 ____

These cells produce a lot of proteins and require high amounts of energy. What organelles do you expect to be present in large numbers to carry out these functions?

14 ____

Considering the relatively small diameter of the intestinal tube, what is a possible advantage of the folds that extend into the cavity?

15 ____

Squamous (cheek cells). Look at your own cheek cell diagrams in Table 4-1. Considering the activities that occur in your mouth, which of the epithelial functions are probably associated with this lining?

16 ____

Do you expect the cheek lining to be simple or stratified?

17 ____

Mitochondria (amphibian liver cells). Liver cells actively remove food and toxic substances from the blood. After chemically changing these substances, the cells dump the products back into the blood or into ducts for elimination from the body. As blood vessels enter the liver, the blood flows into spaces. Can you distinguish these spaces from the ducts? What organelles are associated with a very high level of metabolic activity?

18 ____

Connective tissue

In general, connective tissues have cells that are separated from one another by a considerable amount of nonliving material that the cells manufacture (Table 11-3). The characteristics of these tissues depend on the nonliving intercellular material, the **matrix.** The matrix can consist of fluids, gels, or solids. The ma-

Table 11-3 Predicted and observed structure in connective tissue

Function	Expected structure and cell arrangement	Slides to explore	Structure and spatial relationships
Defensive		Human blood	
Structural (bind and shape)		Elastic cartilage (ear)	
Storage (isolate materials)		Compact bone	

trix of some connective tissues have fibers. Collagen fibers are strong and resist stretching. Elastic fibers can stretch and return to their original length. In the last column of Table 11-3, include the description of the matrix.

Human blood. This tissue has several types of cells, and each type has specific functions. The most numerous is the red blood cell (RBC). Of the functions listed for connective tissue, do you know which is associated with the RBCs? What is the three-dimensional shape of RBCs? Which of the listed functions is the responsibility of white blood cells (WBCs)? Which functions require the most energy? How do you describe the matrix of blood?

Elastic cartilage. Is your ear flexible? What does a rubber band have in common with this tissue? What are the characteristics of this matrix?

Compact bone. Are your leg bones flexible? How heavy are your bones? What are your expectations for the structure of this matrix?

Muscular tissue

Muscle cells are grouped into working units or organs called *muscles* (Table 11-4). Muscle cells contract, shortening the muscle. This shortening pulls the two body parts attached to the ends of the muscle closer together. Some muscles encircle or extend along soft structures such as the intestine and blood vessels. As muscles contract, the soft structure decreases in size. The membranes of some muscle cells are in very close contact with one another. Some muscles branch; others do not.

Skeletal muscle. First, thoroughly observe this slide on lowest power. Do you have a longitudinal section (along the length of the muscle) or a cross section? Is this a section of an entire muscle or only a portion of its length? (Hint: How long is the biceps muscle, which brings your hand to your nose?) Observe with the more powerful objectives. Considering all organelles of cells, which organelles are likely to be directly involved with the shortening process as muscles contract? How would these organelles be arranged within the cell?

Table 11-4 Predicted and observed structure in muscular and nervous tissues

Function	Expected structure and cell arrangement	Slides to explore	Structure and spatial relationships
MUSCULAR TISSUE Move bones		Skeletal muscle	
Decrease size of organ		Smooth muscle	
Pump blood		Cardiac muscle	
NERVOUS TISSUE Conduct electrical messages		Spinal cord (mammal)	
Insulate message cells		Teased ox neurons	

Smooth muscle. Observe briefly with low power and then high power magnification. How are these muscle cells sectioned?

19 ____________________

What is the general shape of the cells? Compare this tissue with skeletal muscle tissue. What are the similarities and differences?

20 ____________________

Cardiac muscle. Observe briefly with low power and then thoroughly with high power magnification. Compare this tissue with smooth and skeletal muscle. What are the similarities and differences between cardiac tissue and the other muscle types?

21 ____________________

Figure 11-1
Tissues of a mouse tail.

Nervous tissue

Some nerve cells conduct electrical messages between different parts of the organism. Some of these different parts are a meter apart. Other cells help the conducting cells.

Spinal cord (mammal). Look at this cross section of a spinal cord before placing it on your microscope. Can you distinguish different areas of the cord? Using the scanning and then higher power objectives, search in the more central H-shaped area. Are cells packed tightly within this area? Identify a large cell with a large nucleus. What is the general shape of this cell? If this cell has long extensions from it, what is a reason that you may not see them?

22 ______________________________

If you prick your finger, where is the message from the pricked finger relayed?

Observe the outer region of the spinal cord. Are the large parts of the cells located here? What are the small dark structures encircled by white areas?

23 ______________________________

Teased ox neuron. For the spinal cord section, you explored the cell bodies of the nerve cells, the neurons. On this slide, a short section of a nerve was pulled apart. What are you observing on this slide?

24 ______________________________

Observing interacting tissues in an organ

Obtain a prepared slide of the cross section of a mouse tail. Look for examples of its different tissues. In Figure 11-1, draw and label these tissues and indicate what they do for the functioning of this organ.

LABORATORY
12

Digestive system

Relation of structure and function

CAUTION
Many preservatives are used to prepare specimens for biological study. Follow your teacher's instructions for the safe handling of all preservatives. You will be using these pigs for a number of weeks. Therefore they require proper storage conditions to prevent the drying and decay of tissues. Your teacher will give you specific instructions for storing them.

Digestion in pigs and humans

Advertising of hamburger chains praise the virtues of their products. In all likelihood, whether you eat a "Whopper" or "Big Mac" makes no difference to your digestive system when it comes to the digestion of the beef, pickles, onions, ketchup, and bun. What does matter is whether you have the necessary equipment to carry out the digestive activity (the hydrolysis of polymers) on the proteins, fats, and carbohydrates found in the meat, bun, and condiments.

In this laboratory session, you will explore the digestive system of a fetal pig by examining its structure and observing the action of chemicals from mammalian digestive systems on certain foods.

Pigs are frequently used in medical studies because their structures and functions are quite similar to human's. Keep this in mind as you explore the organs of the pig's digestive system. Wild pigs feed on both plant and animal matter, just as we do. Pigs are called **omnivores** (OM-ni-vores), meaning "all eating." We are also omnivores, but our digestive systems differ in certain aspects from the pig's. You will examine the similarities and a few of the differences.

Fetal pigs come from meat-processing companies that find them in the uteri of adult females brought to market. Biological supply companies obtain these fetal pigs and prepare them for biological study. These supply companies inject preserving chemicals and colored liquid latex into the circulatory vessels through an opening in the neck of the pig.

EXPLORING

The external surfaces of a fetal pig

Working with a partner or as designated by your teacher, obtain the following materials:

- 1 pig
- 1 dissecting pan
- Dissecting tools
  - 2 blunt metal probes
  - 1 pair forceps
  - 2 dissecting needles
  - 1 scalpel or single-edge razor blade
  - 1 pair of sharp pointed scissors

Wash your pig under running water for a few minutes to remove excess preservative.

For a few minutes, explore the surface features of your pig. Some pigs may have cuts in the ventral (belly side of an organism) body wall that permit you to observe the inner organs. Some pigs may have a white secretion on their surfaces. All of them have a portion of the umbilical cord extending from the belly. In the uterus, this cord connects the pig to the nourishing placenta.

Using your knowledge of human anatomy, list the major parts of this pig such as chest (or thorax) and describe how they differ from the human parts.

1 ______________________________

____

____

Biologists classify, or group together, organisms that have hair and mammary glands, which produce milk for their young. This group of organisms is called **mammals.** Is your pig a mammal?

2 ______________________________

Cite evidence to support your answer.

3 ______________________________

____

Compare the appendages (legs and arms) of the pig to the human's. How many of your appendages do you use for walking?

4 ______________________________

How many does a pig use?

5 ______________________________

Locate the knee and ankle on the posterior appendage (hind leg) of the pig. What part of the appendage does the pig have in contact with the ground as it walks?

6 ______________________________

How does this compare to the parts you use?

7 ______________________________

CONCEPTS/TERMS

Anatomical directions

Biologists use the term **biped** to identify organisms that use two appendages for walking. They use the term **quadruped** to identify organisms that use four appendages for walking. The suffix *ped* means "footed," and *quad* and *bi* refer to "four" and "two," respectively.

Biologists use particular terms to describe the position of one body part to another part. These anatomical terms are illustrated on the pig in Figure 12-1. When you refer to the directions or positions of organs in organisms, use these terms. Note that most of the anatomical terms refer to directions or positions in reference to a part of the organism; they do not depend on the position of the organism. Other anatomical terms refer to directions on the organism as it pertains to the organism's "usual" position (for example, standing upright for humans). The meaning of more common terms such as *above* and *under* depend on the current position of the organism. Saying that "Your heart is cranial to your stomach" holds true whether you stand on your head or upright. However, saying "Your heart is above your stomach" is incorrect if you stand on your head or lie down.

Figure 12-1
A, Anatomical terms relating to a pig.
B, Anatomical terms relating to a human.

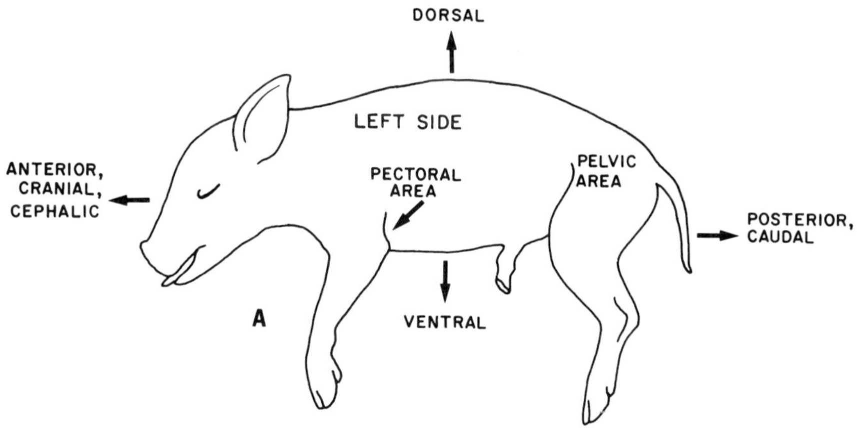

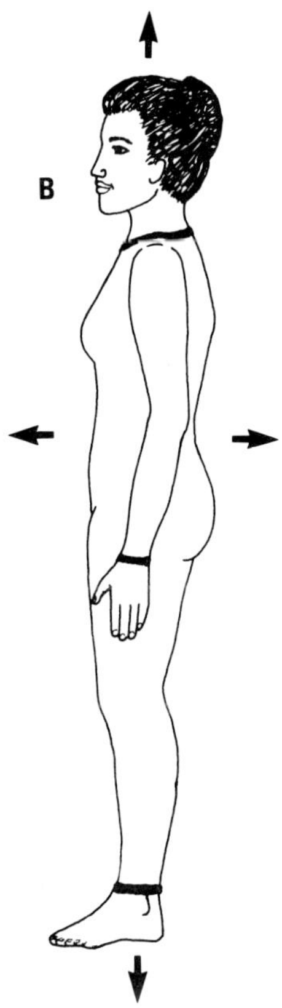

Label the figure of the human in Figure 12-1, *B* with the appropriate anatomical terms.

Anatomical vocabulary

Right or left—the organism's right or left
Anterior—toward the head (quadruped), toward the "front" (biped)
Cephalic—toward the head
Posterior—toward the tail (quadruped), toward the back (biped)
Caudal—toward the tail
Dorsal—toward the back
Ventral—toward the belly
Lateral—toward the side (right or left)
Medial—toward the midline or middle
Pectoral—referring to the chest region, particularly the breast
Pelvic—referring to the hip region
Superior—higher, toward the back (quadruped), toward the head (biped)
Inferior—lower, toward the belly (quadruped), toward the foot (biped)

EXPLORING

The fetal pig

Sex determination

After observing your pig's surface features, determine the sex of your specimen. Some but not all males have two "swollen areas" medial and caudal to the posterior legs. These are male reproductive organs covered by skin. However, in some male specimens, these swollen areas are inconspicuous. Therefore using the absence of these swellings as indication of a female may lead to a false sex determination.

Check with your neighbors until you find a specimen with a different arrangement of openings in the abdominal and pelvic regions. Compare these two pigs. In addition to the mouth, nostrils, and ears, fetal pigs have two other openings. One is at the posterior end of the digestive system (the anus) just ventral to the base of the tail. The other opening is either just posterior to the base of the umbilical cord or just ventral to the anus; it allows products from the reproductive system and urine from the excretory system to pass out of the pig. Using the positions of external openings as "determining factors," you can positively identify the sex of pigs. After exploring the placement of this opening, use your knowledge of human anatomy to predict the sex of each of these pigs. Which position of this urinary-reproductive opening identifies the male?

8 ______________________________

Does your partner agree?

9 ____

Ask your teacher to confirm your identification.

Opening the pig

If the ventral surface of your pig has not been cut open or the jaw has not been cut toward the ear, follow the instructions given in Figure 12-2 to open the pig. As soon as you have made the designated cuts, wash out the interior of the pig. You might find excess colored latex that escaped from the blood vessels when your pig was processed. Discard this material as directed by your teacher.

DISSECTION NOTES

Dissection and butchering are different techniques. The goal in dissection is separation of the structures of the organism from one another so that they can be carefully examined. The goal in butchering is separation of the organism into cooking portions or eatable chunks. Therefore the cutting tools of meat slicers, bone saws, scissors, and scalpel or single-edge razor blade are rarely used in dissections. You use the last three tools only for the initial opening of organisms or opening large internal structures. Only butchers use the two former tools.

For fetal pig dissection, use the blunt metal probe as your primary dissecting tool. For smaller organisms and small organs in the pig, use the dissecting needle. Many organs are anchored by a thin, shiny, connective tissue. You can separate the structures easily by simply moving the blunt probe around the structures.

Do not remove any structures unless you are instructed to do so.

Examining the digestive system

Locate a long, tube-like structure that is anchored securely at both the cephalic and caudal ends by solid tissues but is held loosely in place by the thin, connective tissue, called **mesentery,** between these anchored ends. Beginning at the anterior end with the mouth, the **oral cavity,** thoroughly explore this tube.

In the oral cavity, most "baby" teeth probably have not yet erupted through the gums. Cut into the gum at the most anterior portion of the mouth and in

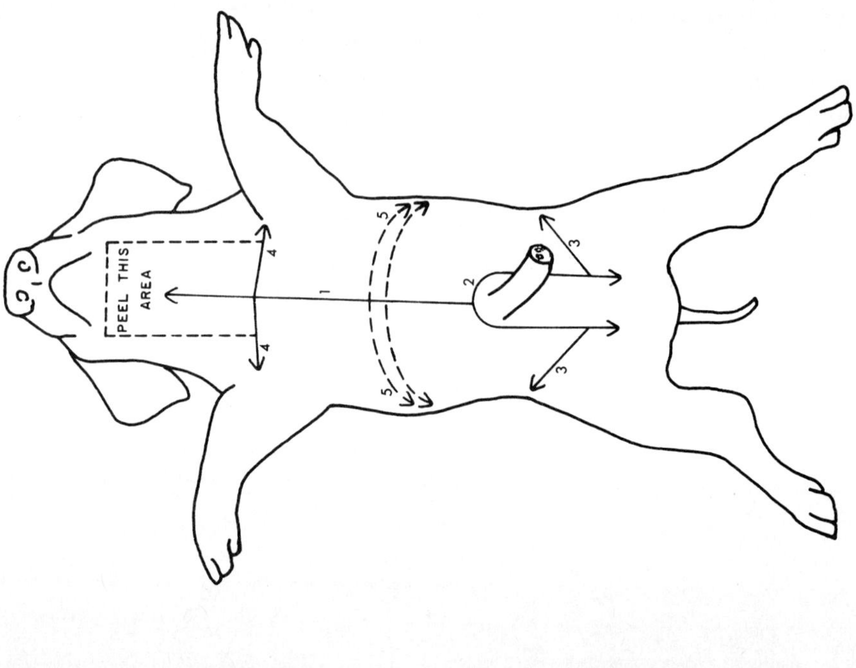

Figure 12-2
Opening a fetal pig. Keeping scissors parallel to the organism, cut the ventral walls in the designated order. Remove the ventral skin and underlying muscles in the neck. At the left joint between upper and lower jaws, cut through the bones with sturdy scissors.

the position comparable to your molar teeth. Locate a tooth at each gum location. Think about the differences in the structure of your teeth and relate them to their functions.

10 ____________________

Compare the structure and function of the two pig's teeth to your teeth that are in the same mouth position.

11 ____________________

____

If you were comparing the teeth of a plant-eating animal, a **herbivore,** and a meat-eating animal, a **carnivore,** what differences would you expect to find in the teeth?

12 ____________________

Explain your answer.

13 ____________________

____

Compare the teeth of several vertebrate skulls that are available in the laboratory. In what ways do their teeth differ from your expectations?

14 ____________________

____

For each skull, decide whether it belonged to a carnivore, omnivore, or herbivore and record this in the second column of Table 12-1. In the third column, explain the reasons for your decision.

Examine the pig's tongue. Look in a mirror to examine your own tongue. Compare the structure of the pig's tongue to your own. What are their structural differences and the similarities?

15 ____________________

____

Based on their structures, in what ways do you think the functions of the pig's and human tongues are similar and different?

16 ____________________

____

Remove the skin covering the jaw from the ventral part of the ear to the midline and the skin ventral to the tongue. Try locating a flocculent (wooly like) type of tissue and (if you are an excellent dissector) a small tube that empties fluid from this glandular tissue into the oral cavity. What could the function of this organ be?

17 ____________________

Table 12-1 Relating tooth structure to diet

Skull number	Type of eater	Explanation

Using a pencil, describe in detail each structure of the oral cavity in the second column of Table 12-2. (Three examples are provided to help you start.) In the first column, indicate the position of each structure to other digestive tract features by measuring the distance (in millimeters) of each structure from the anterior end of the tube.

After exploring all structures (including their openings and hard and soft surfaces) in the oral cavity of the pig, look in your mouth. What are some of the structural differences between the pig's oral cavity and yours?

18 ____

____

Explore the entire digestive tube from mouth to anus. Use as many of your senses as possible. Feel the tract. Note the thicknesses of its parts, expanded areas, constricted areas, and texture of the surfaces. Thoroughly explore openings into the tube visually, as well as with your finger and/or blunt probe. Compare parts of the tube, searching for structural differences. (Hint: Posterior to the throat [pharynx] are five major organs with a few additional structures that mark the boundaries of the organs.) Continue to fill in the first and second columns of Table 12-2. To measure curved surfaces, fit a string along the curved portions, then measure the length of the string. Do not bother with names for these parts. You can add names to these structures later as an optional activity.

Table 12-2 Digestive system of the fetal pig, *Sus scrofa*

Distance*	Structure and location	Possible function	Possible name
0	Soft, fleshy, muscular, flexible rim of opening	Aids in intake of food, keeps food in mouth	Lips
3	Chisel-like, hard surface in gum (3 × 1 × 9 mm)	Cuts	Tooth (incisor)
15	Hard "roots," with broad, somewhat flat surface in gum (4 × 8 × 6 mm)	Grinds against another surface	Molar

*Distance along tract is in milliliters.

Figure 12-3
Structure of organs in a mammalian digestive tract.
A, Digestive tract locations of **B** through **D**.
B through **D,** Sections of the digestive organs.

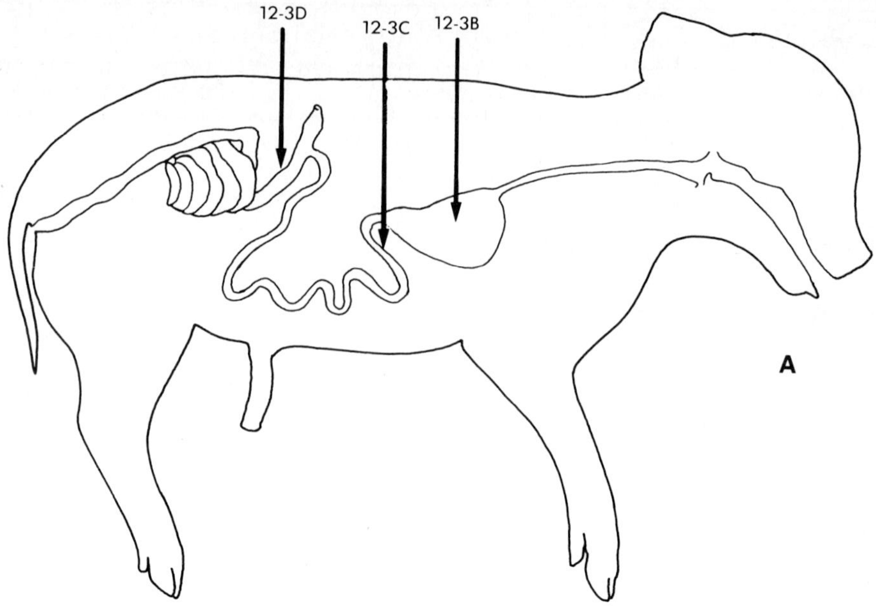

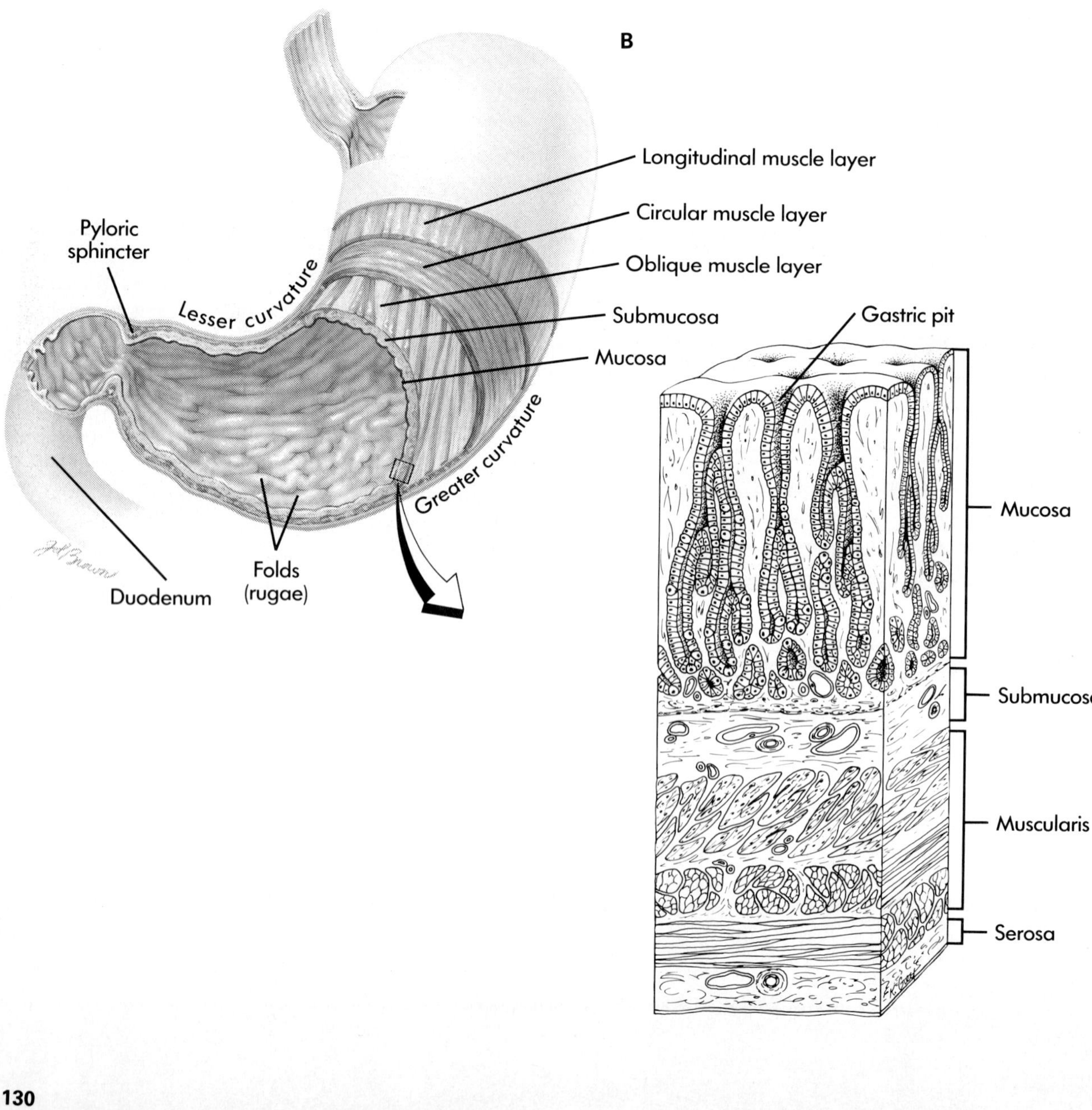

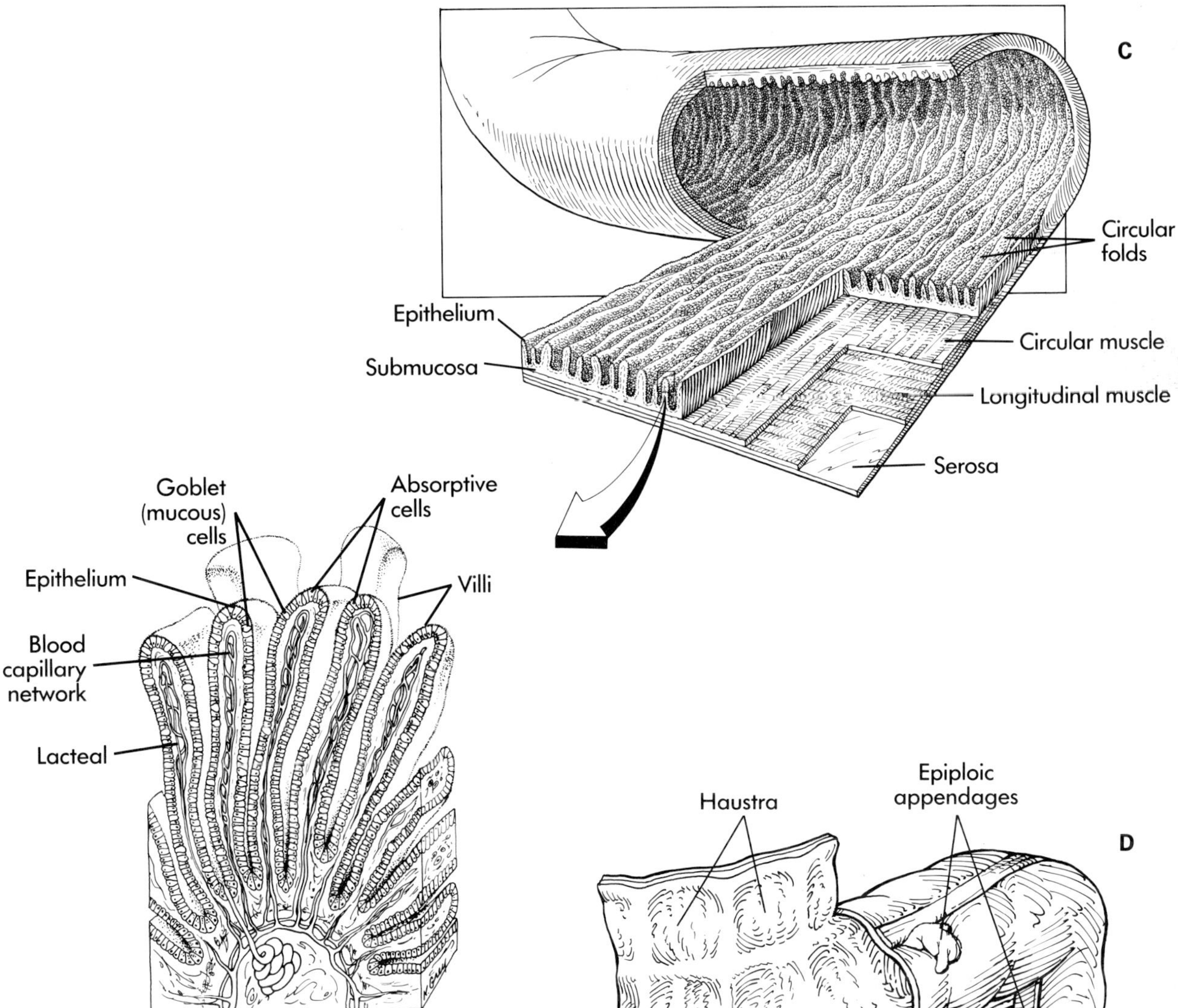
C
Circular folds
Epithelium
Submucosa
Circular muscle
Longitudinal muscle
Serosa
Goblet (mucous) cells
Absorptive cells
Epithelium
Villi
Blood capillary network
Lacteal

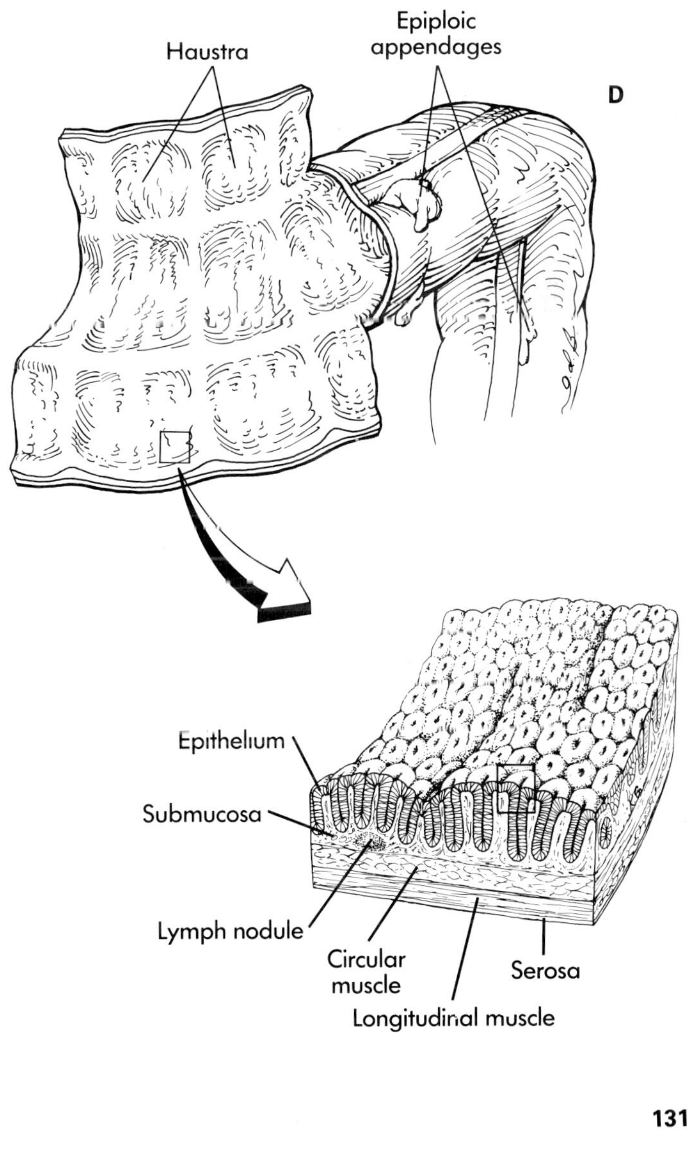
D
Haustra
Epiploic appendages
Epithelium
Submucosa
Lymph nodule
Circular muscle
Serosa
Longitudinal muscle

As you explore the entire tract, think of possible functions of each structure that you can later add to the third column of Table 12-2.

Using scissors, open a 2-cm length of the digestive tube in the neck (cervical) and chest **(thoracic cavity).** Begin a 14- to 15-cm cut just posteriorly from the point where the tube passes into the abdomen, or **peritoneal cavity.** Then open three portions of the tube posterior to this second cut: (1) along a section of the soft, irregularly folded portion; (2) along the tightly coiled portion; and (3) along the tube as it passes through the pelvic (hip) bones at the most posterior portions of the tube. Your teacher will help you expose the tube for this last cut.

Thoroughly inspect these inner surfaces and all openings into the tube. Remove small pieces of the digestive tube and examine their inner surfaces with a dissecting microscope. Are all estimates of thickness of the unopened tube that you made still acceptable?

19 ____

If not, modify your previous notations in Table 12-2. Note the types of tissues that form the tube's wall.

Check for other structures and accessory organs that are connected to the digestive tract by tiny tubes. Include these organs in Table 12-2.

Now, take a few moments to review the differences in structure along the digestive tube and think about the digestive processes. In the third column of Table 12-2, try to assign digestive activities to different parts of the tube, using the following list of digestive functions and your present knowledge of digestive processes. Some functions may be difficult to assign; match what you can now and complete this after you finish other laboratory activities.

1. Holds food for subsequent digestion.
2. Chemically breaks (hydrolyzes) proteins into their building blocks, amino acids.
3. Hydrolyzes carbohydrates into their building blocks, sugars.
4. Separates large units of lipids into smaller units.
5. Hydrolyzes lipids into their building blocks, glycerol and fatty acids.
6. Absorbs molecules of broken-down food into the blood transport system.
7. Stores undigested food for later elimination.
8. Physically chops or grinds food into smaller pieces.
9. Churns food with digestive juices.
10. Holds microorganisms that break down tough cellulose (a complex carbohydrate) found in plant material.
11. Is an accessory organ that produces starch, protein, and fat-hydrolyzing enzymes.
12. Is an accessory organ that produces bile salts for lipid emulsification.
13. Is an accessory organ that concentrates and stores bile.
14. Is an accessory organ that produces starch-hydrolyzing enzymes.
15. Moves food from one structure to another.
16. Moves food within a structure.
17. Absorbs water into the transport system.
18. Secretes mucus that coats the inner surfaces of the tube's wall.

Examining diagrams of the digestive system wall

So far, your explorations of digestive tracts have been at the macroscopic level. To complete the exploration thoroughly, you need to study this tube at the microscopic and submicroscopic (chemical) levels. Figure 12-3 and Plates 6 and 7 show three-dimensional diagrams drawn from cross sections of the digestive tube at positions indicated in Figure 12-3, *A*. Based on your knowledge of the functions of each type of tissue (see Laboratory 11), think of the possible activities of each tube section. List these activities below each diagram.

Observing in vitro digestive activity

Now observe some of the results of in vitro (in glass) digestive activities. Carbohydrates such as starch, proteins, and lipids are the three major kinds of food molecules. At the demonstration table, observe the effects of juices taken from different portions of the tract on the following foods:

1. Carbohydrate: solution of starch
2. Protein: cubes of hard-boiled egg white
3. Lipid: cream

The three juices are from the oral cavity, stomach, and pancreatic duct that connects the pancreas to the small intestine.

Starch and digestive juices. Each of these four bottles contains starch solution and one of the three juices or water. Conduct the I-KI test for starch on a sample from each of the four solutions of starch and on water alone (see Laboratory 7). Fill in Table 12-3 as you conduct each test. In the fourth column, identify digestive activities that occurred. Why is it important for you to test both the starch solution and water?

Table 12-3 Digestive activity

Fluid	Results	Possible digestive activity
STARCH Water		
Oral		
Stomach		
Pancreatic juice		
PROTEIN Water		
Oral juice		
Stomach juice		
Pancreatic juice		
LIPID Water		
Oral juice		
Pancreatic juice		

20

Protein and digestive juices. Observe each tube that initially contained several cubes of boiled egg white and one of the fluids. Continue to fill in Table 12-2.

Lipid and digestive juices: Observe the tubes that initially had cream, a blue-pink pH indicator, and one of three digestive fluids. (The very acid stomach juice is not tested.) The pH indicator, litmus, is blue in a nonacid environment and becomes pink as the solution turns acidic. If the fats in cream are digested to fatty acids and glycerol, what color do you expect the solution to be?

21

Why do you think this test for lipid digestion is not used for solutions containing stomach juice?

22

You cannot test the digestion of the fats by stomach juice today; however, biologists do agree that fats are not digested in the stomach.

Using the digestive activity in the third column of Table 12-3 and tissue information from the diagrams of the tube's walls, complete Table 12-2. Fill in any missing information and update your previous thoughts about digestive activities occurring in each listed structure.

Naming digestive structures (optional activity)

You can find names for these specialized sections of the digestive tube and the accessory organs by consulting references and labeled specimens available in the laboratory. You might consider some of the following names:

Appendix	Palate, hard and soft
Anus	Pancreas
Cecum	Pharynx
Epiglottis	Pyloric valve
Esophagus	Rectum
Gall bladder	Salivary glands
Glottis	Smooth muscle
Ileocecal valve	Stomach
Intestines (large and small)	Teeth
Liver	Tongue
Oral cavity	Uvula

CONCEPTS/TERMS

Synthesizing concepts

Induction of concepts

Think about your experiences with digestive systems. Take a moment to think about each of your individual observations of the pig's digestive tube. Try thinking of a general statement or an idea that relates all of these observations.

Review the second and third columns of Table 12-2. Have you developed any relationships between the structures of the different sections of the digestive tube and the digestive functions such as those recorded in Table 12-3?

23

Support your answer by giving several specific examples from your explorations of the pig's digestive tube.

24

From the specific to the general

Now, take an important step by examining your thinking process. Explain your thinking process as you matched digestive functions to digestive structures.

25

In one common thinking pattern, we form general statements, or concepts, by noticing the presence of a common idea in several specific observations. This mental process is called **inductive reasoning.** From your exploration of specific structures along the tube, examination of tissue diagrams, and observation of the results of digestion in the tubes, you matched many specific structures and functions together. Each time you did this, you identified a specific example of a structure-function relationship. From these specific examples you induced the general concept of an intricate relationship between structure and function when your mind realized this common theme among all examples.

Digestion and absorption

Using your knowledge of the pig's digestive system, including both structure and function, think about the most likely areas for the digestive products, the monomers, to be moved from the digestive tube cavity to the blood vessels of the circulatory system. Where is the absorption of these nutrients most likely to take place?

26

You have just used a form of thinking frequently used by scientists and non-scientists, called **inference.** In inference, an unobserved phenomenon (absorption) is predicted from items known or assumed. Scientists go further; they will test this inference. Examine your thinking process and explain how you came up with your answer.

27

APPLYING

Structure and function in the human digestive system

Concept of the structure and function relationship

You have already compared the structure of your mouth to that of the fetal pig's. Using your knowledge of the remainder of the digestive system of the fetal pig, infer the structure of the human tract. List several similarities and several differences you would expect to find.

28 Similarities

Differences

Test your predictions of similarities and differences by exploring the human models and/or charts available in the laboratory. What are five major similarities and three major differences?

29 Similarities

Differences

Special conditions of the digestive system

Using all of the information you have gathered, respond to the following conditions:

Sometimes the coating of mucus that lines the stomach is damaged by the body's response to stress and drugs such as aspirin and alcohol. If this damage results in a break in this mucous coat, what may happen as hydrochloric acid (HCl) and protein-hydrolyzing enzyme are produced by the stomach epithelium (parietal cells and chief cells, respectively) and released into the stomach cavity?

30

Diarrhea is characterized by watery feces and usually by intestinal gas and cramps. Thinking about the function of the lower parts of the digestive tube, give several possible causes of this condition.

31

Discuss the effects on digestion of failing to chew food thoroughly before swallowing.

32

Discuss the effects of a blocked pancreatic duct (the tiny tube from the pancreas to small intestine) on the digestion of starch, proteins, and lipids.

33

LABORATORY

13

Breathing system

Relationship of function and structure

PART ONE The function of the breathing system

Role of breathing

As cells of organisms respire, transferring glucose energy to ATP molecules, they use oxygen and produce carbon dioxide. Therefore animals as well as other organisms need a way to supply the body with oxygen and to eliminate carbon dioxide. In animals, this task is accomplished by oxygen and carbon dioxide diffusing in opposite directions across an "external" surface. Your external gas-exchange surface is deep within your lungs.

You may wonder about the advantage of having this exchange surface so far from the source of oxygen. In fact, not all animals have a "buried" exchange surface. For example, frog tadpoles and fish have their exchange surfaces, the gills, extending directly into their oxygen source, the water. However, land animals have a major problem if they expose their external exchange surfaces directly to the dry air of the land environment. These surfaces lose a great deal of water, thus dehydrating the organisms. Eventually, the exchange surfaces dry, which prevents gases from diffusing across them. Populations of land animals that evolved an infolding of this exchange surface, burying it deeply in the body, survived in the dry air better than animals with an exposed surface.

Burying the exchange surface deep within the body presents a new problem—the air with its oxygen must be moved to the buried external surface and air with the waste carbon dioxide must be removed. The process that solves this problem is **breathing,** which is frequently called **ventilation.** Since ancient times, breathing has been recognized as a critical sign of life.

EXPLORING

Breathing movements

Working with a partner, obtain a safety pin and a tape measure (or a length of string and a meter stick).

Take turns as an observer and a subject to make a series of ventilation measurements. Each observer will take a set of three before and after measurements each time the subject breathes in a particular way. As illustrated in Figure 13-1, a set consists of the following before and after measurements: (1) the circumference, or girth, of the chest immediately below the breast or the pectoral muscles, (2) the distance from the level of the first measurement to the top edge of the hip bone, and (3) the girth at the middle of the abdomen. To measure girth, pin one end of your measuring string or tape to the subject's clothing. Then loop the other end around the subject and through the pin. Pull the tape or string snugly but not tightly. Note the "before" measurement. Then as the subject breathes, let the tape slide back through the safety pin. Note the "after" measurement. The observer takes all measurements while the subject sits upright but relaxed in a chair.

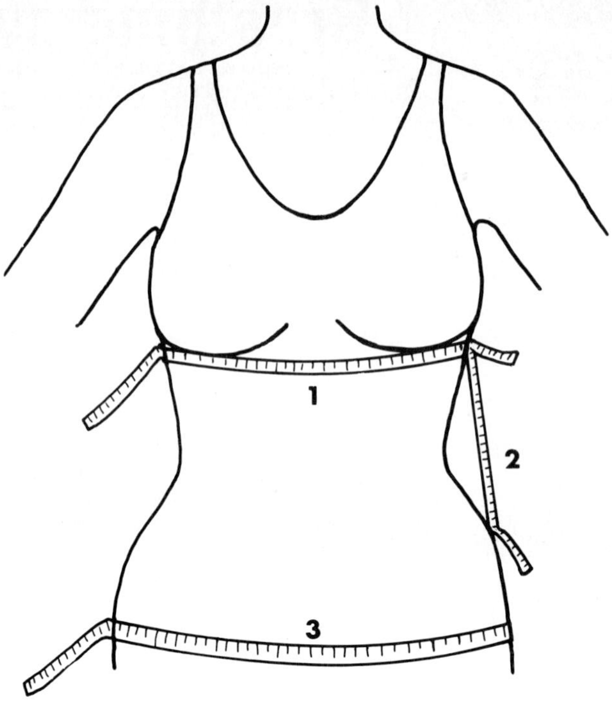

Figure 13-1
Measuring breathing movements.

The three before and after measurements are made each time the subject breathes in one of three ways. The observer takes the first set of measurements as the subject breathes in a normal, relaxed way. Take the *before* measurement just after the subject **exhales** (breathes out) and the *after* measurement just after the subject **inhales** (breathes in). Record the measurements in Table 13-1 in the reverse order of taking—the before measurement in the third column and the after measurement in the second column. Then subtract the before measurement from the after measurement to find the change in girth for the fourth column. Calculate the percent increase or decrease from the exhaled measurement, and put this result in the fifth column.

Take the second set of measurements for an extra-effort inhalation. Measure just after the subject has taken a normal inhalation (before measurement) and again just after the subject completes a *maximal* inhalation (after measurement). Record these measurements and the difference in the maximal and normal inhalation measurements in Table 13-1. Calculate percent increase or decrease for the extra-effort inhalation based on the normal inhalation measurement.

Take the third set of measurements for an extra-effort exhalation. Measure just after the subject makes a normal exhalation (before) and again just after a *maximal*, extra-effort exhalation (after). Calculate percent increase or decrease based on the normal exhalation measurement and record in Table 13-1.

After both of you have completed these measurements, describe how the percent changes in your measurements compare with those of your partner.

1 ________________________________

____

Now sit quietly in your chair and breathe normally while being completely aware of the breathing events occurring in your body. Then concentrate on your body's activities as you inhale and exhale maximally.

Table 13-1 Thoracic measurements on inhalation and exhalation

Measurement area	Inhale (cm)	Exhale (cm)	Change (cm)	Percent change
RELAXED INHALATION Rib girth				
Lower rib length				
Abdominal girth				
Measurement area	**Maximal inhale**	**Normal inhale**	**Change (cm)**	**Percent change**
EXTRA INHALATION Rib girth				
Lower rib length				
Abdominal girth				
Measurement area	**Maximal exhale**	**Normal exhale**	**Change (cm)**	**Percent change**
EXTRA EXHALATION Rib girth				
Lower rib length				
Abdominal girth				

CONCEPTS/TERMS

The process of breathing

Thinking about the activity of breathing

Think about what you have just experienced and the increase or decrease in the measurements you listed in Table 13-1. Based on these explorations, describe what happens in your body as you breath in the ways listed below. In your descriptions, include changes in such structures as your abdomen, chest, and air passageways and their possible effects on air flow.

2 Normal inhalation

3 Extra inhalation

4 Normal exhalation

5 Extra exhalation

6 Maximal exhalation after a maximal inhalation

Concepts of air movement

With your present knowledge, what do you think is the *direct cause* of air moving into the body?

7

How do your thoughts on the direct causes of air movement compare with your neighbors'?

8

As you think about the movement of air into the body, consider the following and describe the results: What effect does raising the ribs and lowering the abdominal organs have on the volume within the chest cavity and the lungs?

9

What effect will this volume change have on the air pressure within the chest cavity and the lungs?

10

On comparing the pressure within the chest cavity and lungs with the "outside" air, which has the lower pressure?

11

How does the pressure difference (a gradient) affect the movement of air?

12

This inhalation breathing pattern is called **negative-pressure breathing.** Explain how this term describes your breathing pattern.

13 ____

What are the sequence of events and conditions for the *exhalation* of air?

14 ____

Biological terms for lung (pulmonary) volumes

Biologists and medical doctors have specific terms to refer to the air in the lungs. The amount of air that you normally breath in and out is the **tidal volume.** You observed that the lungs can become filled with and expel more air than the normal tidal volume. The amount of air that you can inhale in addition to the normal tidal volume is called **inspiratory reserve volume.** The volume of air that you exhale in addition to the normal exhalation is called the **expiratory reserve volume.** The total volume that you exhale after a maximal inhalation is called **vital capacity.** The vital capacity is the sum of the tidal, inspiratory reserve, and expiratory reserve volumes (Figure 13-2). After birth and the first intake of air, some air remains in the lungs and can not be exhaled. This is the **residual volume.**

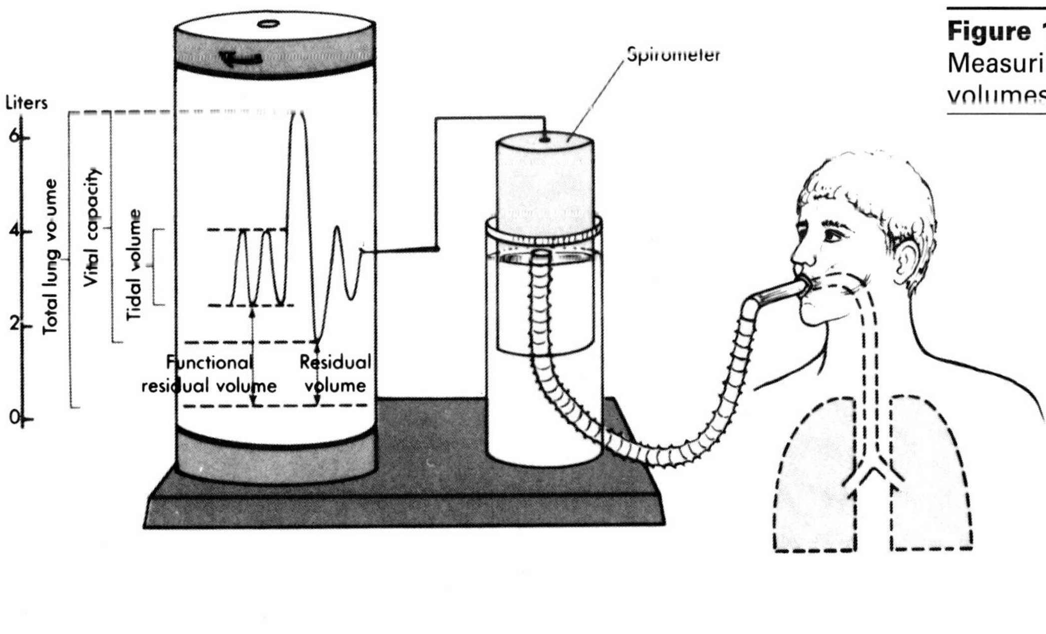

Figure 13-2
Measuring lung volumes

APPLYING

Spirometric measurements of lung volumes

Spirometers

Spirometers are instruments that measure lung volumes. Some have one-way valves that let you inhale "fresh" air and exhale "used" air through the same mouth tube. As the air is exhaled, it passes through a volume-measuring device. Others types of spirometers lack the one-way valves. With these spirometers, you inhale before you place the exhaling tube in your mouth. Some spirometers are connected to computers with programs that calculate measurements associated with lung volumes, sort subject data, and print out the data.

Your teacher will demonstrate the equipment available in your laboratory. With any spirometer you need to exhale all of the air directly into the collecting tube. Use a nose clamp to prevent air escaping through your nose and seal your lips around the tube.

CAUTION

Any equipment that is exposed to body fluids must be sterilized before you use it. After use, you must treat it as a potential biohazard until the equipment is resterilized. Disposable parts must be stored in labeled biohazard containers for future disposal. Your instructor will give you specific directions for sterilizing and disposing of materials.

Lung volume predictions

Based on your tape measure explorations, which lung volume do you expect to be the largest?

15 ______________________________

Which do you expect to be the smallest?

16 ______________________________

Measuring lung volumes

Measure the vital capacity, tidal (exhalation), and expiratory reserve volumes using available spirometry equipment. If the spirometer is not connected to a computer, add your data to Table 13-2 along with data from other class members.

Because spirometers measure only exhaled air, you need to calculate your inspiratory reserve volume for the fifth column. Calculate the inspiratory reserve volume by subtracting the tidal and expiratory reserve volumes from the vital capacity:

$$IRVc = VC - (TV + ERV)$$

$$\text{Inspiratory reserve volume (calculated)} = \text{Vital capacity} - (\text{Tidal volume} + \text{Expiratory reserve volume})$$

Calculate the class means (averages) for each of these lung volumes. Do this by adding all the data readings for each column and dividing by the number of subjects listed in the column. How do your own lung volumes compare to the class means?

17 ______________________________

____

Experiment: what factors affect lung volumes?

There are many factors that affect lung volumes. These factors might include gender, body size, age, smoking habits, level of physical fitness, and the presence or absence of disease. What other factors can you add to this list?

18 ______________________________

Working either in groups or individually as designated by your teacher, choose an independent variable from the factors that may affect one or more of the types of lung volumes. Propose a hypothesis concerning the effects of this condition on lung volume(s) and predict the effects of it on your classmates.

Table 13-2 Lung volumes of class members

Subject's initials	Vital capacity (ml)	Tidal volume (ml)	Expiratory reserve (ml)	Inspiratory reserve (ml)	Subject characteristics
Mean					

What is your hypothesis and your prediction?

19 ______________________________

Think about how you will define terms. For example, if smoking is the independent variable, you need to decide the minimum number of cigarettes, cigars, or pipefuls that a person smokes per week before that person is defined as a smoker. How are you defining your terms?

20 ______________________________

____

You already have most of the data necessary for testing your hypothesis in Table 13-2. However, you need to collect information about the state of each individual in regard to your independent variable. For example, if your independent variable is smoking, identify those who smoke and those who do not. Add this information to the sixth column of Table 13-2. To test your hypothesis, what information from Table 13-2 will you compare?

21 ______________________________

After double-checking your plans for the comparison with your teacher, carry out your analysis. What are the results?

22 ______________________________

____

What is your conclusion for this experiment?

23 ______________________________

CONCEPTS/TERMS

Limiting the generalization of conclusions in science

In this experiment you collected data only from your classmates. If you could run a second experiment that included all people in North America, would the results and conclusions be the same?

24 ______________________________

Explain your answer.

25 ______________________________

____

Your answer to the above question depends on how well you think your class members *represent* all North Americans in regard to lung volumes and your independent variable. For example, you have to ask the questions, "Are the smokers in this class comparable to all smokers in North America?" and "Are lung volumes in this class comparable to all lung volumes in North America?" If you answered "yes," your conclusion on smoking versus nonsmoking can be applied to all people in North America. If you answered "no" to either question, you cannot generalize from your specific data to all North Americans because your classmates are not representative of all North Americans.

Scientists apply conclusions of experiments only to larger groups that are similar to the group that provided the experimental data. Scientists call a subgroup of a large group a **representative sample,** if and only if the smaller group is similar to the larger group. If the experimental data do not come from a representative sample, the conclusion of the experiment cannot be applied to the larger group. For example, if a large group has 22% of its people smoking at least 40 cigarettes a day, 12% smoking between 39 and 20 cigarettes a day, 2% smoking 19 to 1 cigarette a day, and 64% smoking no cigarettes each day, a representative sample should have the same proportions of cigarette consumption as the large group. If not, data about smoking collected from the small group are not applicable to the larger group.

What would you do before you scientifically generalized the conclusions from your independent variable and lung volumes experiment to the larger group of North Americans?

26 ______________________________

____

SCIENCE NOTE

Scientists generalize conclusions of experiments only to those groups of organisms or objects accurately represented in the group that was measured in the collection of experimental data.

If the small group that is formed from a larger group accurately represents the larger group, then the small group is called a representative sample.

PART TWO Structures of the breathing system

EXPLORING

The structure of a breathing system

Predicting structure from a known function

You previously explored the structure of the digestive system and then predicted the functions of those organs and their tissues based on their structures. For the breathing system, reverse the order; predict its structures based on its functions. Using the functions of the breathing system, think about the organs and tissues needed to perform these functions.

List the structural characteristics that you expect to find in a gas-exchange surface.

27 ______________________________

____

If this exchange surface is buried in the animal's body, would the animal need some sort of plumbing pipes or structures through which the air is channeled to that surface?

28 ______________________________

If so, what characteristics do you expect the structures to have?

29 ______________________________

____

Table 13-3 Relating functions of the breathing system to structures

Structure's description	Function	Structure's name

If not, predict how the air reaches the exchange surface.

30 ____

____

What are the characteristics of structures involved in changing chest volumes?

31 ____

____

Searching for breathing structures in the fetal pig

Obtain your fetal pig, dissecting pan, and dissecting tools.

To explore the structures through which air moves to and from the gas-exchange surfaces, begin at the most anterior end of the pig. On only one side, remove about one half of the pig's snout with sturdy dissecting scissors using the following guides: Make a saggital cut through one nostril and extend it posteriorly to about the midlength point of the snout. Then, make one half of a cross-sectional cut to free this side of the snout from the animal. Explore the structure in this region. As you explore, link the structures to functions of the breathing system and fill in the first and second columns of Table 13-3. Another function of the breathing system in land animals is the moisturizing of the air before it reaches the gas-exchange surface.

Make a midline cut through the soft, dorsal roof of the oral cavity (the soft palate), exposing the structures between the nasal and pharynx (throat) areas. After exploring these structures, examine the ventral floor of the pharynx.

Examine passageways that lead from the pharynx to the lungs, which are located in the thoracic cavity. In this exploration, do not break or remove the latex-filled tubes that you will find in the thoracic cavity around the heart. (You will need these tubes in a future laboratory.) Using the blunt metal probe, gently separate the latex-filled tubes from the air passageways. Use your probe inside passageways to detect the extensions and structures of the passageways. Continue to describe the structures leading to the gas-exchange surface in Table 13-3.

Explore the lungs. After examining the lungs in the pig, remove one section of the pig's right lung for magnified examination. Observe this section of the organ with a magnifying lens or a dissecting microscope. At the demonstration table, observe the prepared microscope slides of the lung and the section of the air passageway in the neck. Add this structural information to the appropriate organ in Table 13-3.

Observe some muscles. Peel back a section of skin from the chest wall and examine the muscles between the ribs. Look at the structure that completely separates the thoracic and abdominal cavities. Continue adding information to Table 13-3.

Adding names to structures of the breathing system (optional activity)

After thoroughly exploring the passageways and gas exchange-surfaces of the fetal pigs, obtain names for these structures from the references and charts available in the laboratory. Some of the names you may want to use follow:

Alveolus	Hard palate	Pleura (lining of thoracic cavity and covering of lungs)
Bronchiole	Larynx	Soft palate
Bronchus	Lungs	Trachea
Diaphragm	Nasal cavity	Vocal cords
Epiglottis	Nostril	
Eustachian tube	Oral cavity	
Glottis	Pharynx	

CONCEPTS/TERMS

Linking structures to form a working relationship

Questions to consider

After these explorations of the fetal pig and slides, try to answer these questions:

1. How does the structure of the snout air passage prevent drying of the gas-exchange surfaces?
2. What structure of the nasal-pharynx area connects with the ears?
3. What structure is a passageway for both food and air?
4. What structure prevents food from entering air passageways?
5. What structure prevents the collapse of the air passageway in the neck?
6. What is the structure of the air passageway in the neck that aids the activity of the esophagus?
7. In general terms, what is the size of the gas-exchange surface?
8. What is the order of the structures that a volume of air travels through as the organism inhales?

Ordering the use of structures in exhalation

List in order the breathing structures through which a milliliter of air will travel from the lung to the external air at the nostril of the pig.

32 ______________________________

____

____

Representative sample

Do you expect to find major structural differences among fetal pigs?

33 ______________________________

Do you expect small structural differences among fetal pigs?

34 ______________________________

In what ways do you agree and disagree with the statement, "This fetal pig is a representative sample of all fetal pigs"?

35 ______________________________

____

Explain how you can or cannot apply the description of structures and functions of the breathing system of your fetal pig to *all* fetal pigs?

36 ______________________________

____

APPLYING

Comparing the breathing systems of the fetal pig and human

Using the available models and charts, compare the breathing system of humans to that of the fetal pig. What are some differences and similarities between these two organisms?

37

Comparing the breathing systems of humans and frogs

With water-dampened hands, obtain a live frog from the demonstration table. Place it on a damp, flat surface and cover it with a large, clear container to prevent jumping. As the frog sits quietly, observe its breathing movements.

What are several differences in the breathing movements between humans and adult frogs?

38

Two structural differences between a frog and a human are the absence of ribs and diaphragm in the frog. In addition, a frog can close its nostrils. A frog cannot decrease the pressure in its chest cavity on contracting nonexisting rib muscles and contracting a nonexisting diaphragm. Instead of establishing a lower pressure inside the chest cavity by increasing its volume as you do, the frog simply swallows air. It uses the muscles of its mouth region to force air into the lungs. This is called **positive-pressure breathing.** What do you think the sequence of events in the frog's positive-pressure breathing are?

39

Abnormal conditions of the breathing system

Asthma results from an allergic swelling of the lining of small air passages of the lungs. What is the effect of this condition on the functioning of the breathing system?

40

Emphysema results from a loss of elasticity of the lung exchange surfaces. What is the effect of this condition on the functioning of the breathing system?

41 ____

____

Describe the effects of inhalation and exhalation on a wound that opens an air passageway through the chest wall.

42 ____

____

____

____

LABORATORY 14

Circulatory system

The mammalian heart

The role of the circulatory system

Complex animals have elaborate transportation systems that move many types of materials. These materials include digested foods, oxygen, carbon dioxide, and other wastes. Transporting these substances is the function of the circulatory system. In our mammalian system, a pump provides the force that moves the materials through passageways that are similar in function to railways, highways, and water pipeline systems.

Today you will explore that remarkable organ, the heart. When we are relaxed, this dependable pump pushes on the circulating fluid (blood) about 70 to 75 times a minute. When we exercise, the heart can contract over 200 times a minute. This activity continues every minute of our lives.

EXPLORING

The mature mammalian heart

A fresh heart

With your classmates, explore the fresh beef heart. Using the simple diagrams in Figures 14-1 and 14-2, determine the ventral, dorsal, superior, inferior, right, and left sides of this heart. Use the following questions to provide a focus for your observations:

1. Does the external structure of the heart have differences? If so, what and where are they?
2. Do parts of the heart feel differently from other parts?
3. Are there indications that the heart is separated internally into compartments?

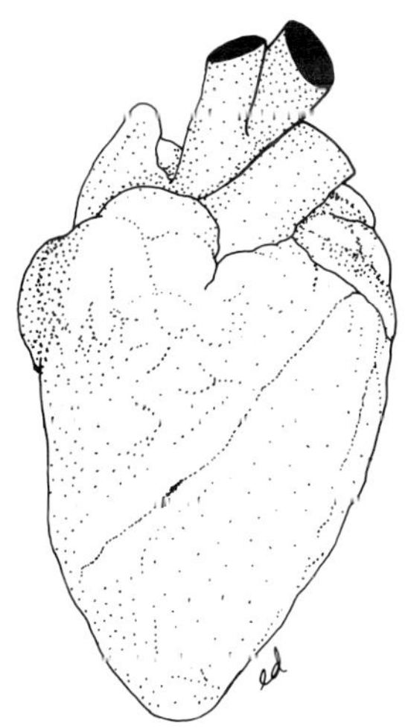

Figure 14-1
Ventral view of mammalian heart.

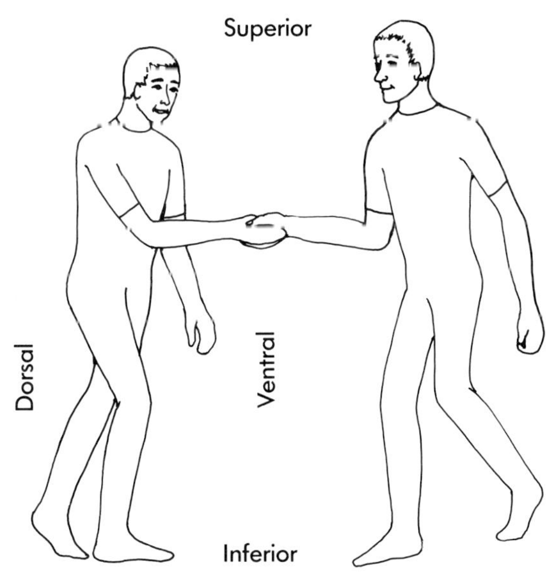

Figure 14-2
Anatomical directions.

4. What types of tissues are visible?
5. How many tubes or openings are in the heart?
6. Are all tubes constructed similarly or are they different from each other in diameter, thickness, and other ways?

A hint: If you are only familiar with the ventral view of the mammalian heart contained in most textbooks, you need to expand your mental picture. You should explore the actual material from all angles. After these explorations, you can integrate what you knew before the exploration with what you have just learned. This integrated information will enable you to create a three-dimensional image of the heart, which will greatly help you to understand the heart's structure and function.

Run water into each tube and opening of the heart. Observe the water flow. Some additional questions are the following:

1. How many tubes carry fluid into the heart?
2. How many tubes carry fluid out of the heart?
3. How many tubes carry fluid both in and out of the heart?

What other questions do you have that might be answered by exploring the fresh mammalian heart?

1

External examination of a pig heart

Working with a partner, obtain the following materials:

1 whole, mature pig heart
2 blunt metal probes
6 pieces of wooden dowel rod

Figure 14-3
External structure of the mammalian heart from the dorsal view.

Using your fingers and other tools, explore this mammalian heart. In Figure 14-3, diagram the whole heart from the dorsal view. Make your drawing about 10 cm (3 to 4 inches) in height.

Using information from your observations, predict which tubes are connected and which lack connections. Also predict which tubes are connected to particular compartments of the heart. Indicate your predictions by dotted lines in Figure 14-3. What observations led you to these predictions?

2 ____

____

Internal examination of a pig heart

Obtain a frontally sectioned, mature pig heart. Explore this heart, internally and externally. Observe the large internal spaces and major tubes. Examine the wall thicknesses and construction, strands, flaps, and other details of the heart structure. Push on the tip of the heart muscle from the exterior to roughly simulate the results of heart muscle contraction on the heart's internal structures. Lay the ventral side of the heart on the table and view the ventral half from the inside *by raising the dorsal half of the heart.* In Figure 14-4, draw the internal

Figure 14-4
Internal structure of the pig heart from the dorsal view.

Table 14-1 Structures and functions of the pig heart

Number of the heart structure	Description of structure and location in heart	Possible function

structures of the heart. At this point, do not try to determine names of these structures. Just identify each structure with a number.

Have you found the structures that explain the water flow patterns you observed in the fresh mammalian heart?

3 ______________________________

If needed, run additional water in the tubes of the demonstration heart or in your preserved whole heart. Place arrows on Figure 14-4 to indicate directions of blood flow through the heart. Label the right and left sides of the heart as it is positioned in your body. When completed, ask your teacher to check your diagram.

Describe your observations of heart structures in the second column of Table 14-1. Give a probable function for each structure in the third column. In other words, infer the function of each structure based on its anatomy. Be sure your descriptions and inferences are logical. Identify structures using numbers that correspond to the numbers on Figure 14-4.

Source and destination of blood in vessels

Examine the human model, wall charts, and diagrams available in the laboratory. Follow the major vessels (tubes), tracing them until they disappear into a leg, a head, or an organ. Apply what you learned from the fresh heart concerning the direction of blood flow in each of these vessels. Return to Figure 14-4, and with arrows, indicate the flow through these major vessels to and from the general body areas and to and from the lungs. Identify the body areas using general terms such as *head, abdomen, legs,* and *lungs.*

Using reference sources available in the laboratory, name the numbered structures in Figure 14-4 and Table 14-1. Add the names to the figure and table. Some names you may want to add are the following:

Aorta
Apex
Atrium (right and left)
Abdomen
Base
Chordae tendinae
Papillary muscle
Pulmonary artery and vein
Valve (bicuspid, semilunar, and tricuspid)
Vena cava (posterior or inferior as well as superior or anterior)
Ventricle (right and left)
Red blood cells

Sequence of structures in blood flow

Think of the structures that a drop of blood passes through as it travels from the right atrium to the aorta. List them in order of use.

4 ______________________________

____

CONCEPTS/TERMS

Blood flow in the mammalian pump

Based on your study of the preserved pig heart structure and observation of the fresh heart, consider the following questions:

How many different blood pathways flow through this heart?

5 ______________________________

Do these blood pathways merge at any point?

6 ______________________________

Is the flow through these pathways one way or two way?

7 ______________________________

What is the general structural pattern of the blood vessels as they channel blood away from the heart?

8 ______________________________

What is the general structural pattern of blood vessels as they return blood to the heart?

9 ______________________________

Which side of the heart pumps blood to the lungs?

10 ______________________________

Which side of the heart receives blood from the lungs?

11 ______________________________

Which side of the heart pumps blood to the head?

12 ______________________________

Which side of the heart pumps blood to the hind legs?

13 ______________________________

Which side of the heart receives blood from the head and legs?

14 ______________________________

Explain the ways in which you agree and disagree with the statement: "Mammalian circulation has the configuration of the number 8 with a double pump in the middle."

15 Agree ______________________________

____

Disagree ______________________________

____

Which diagram in Figure 14-5 most closely resembles the human circulatory system?

16 ______________________________

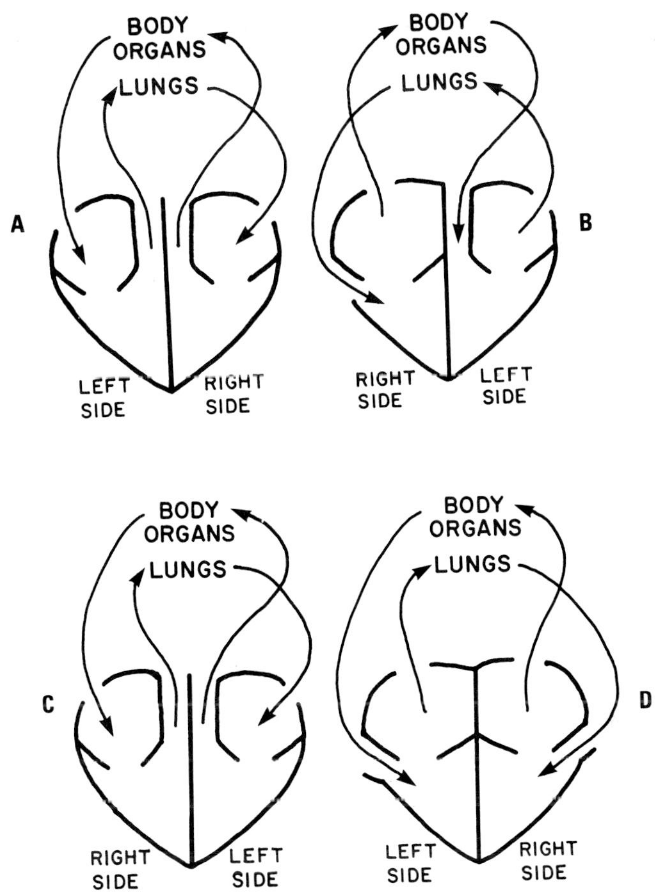

Figure 14-5
Circulatory schemes.

Do you agree with your neighbor? If the general concepts of the mammalian circulation pattern are not clear to you or your partner, ask the teacher to guide you to additional study materials.

The loop of circulatory vessels that conducts blood from the right side of the heart to the lungs and then back to the left side of the heart is called the **pulmonary system.** The loop that conducts blood from the left side of the heart to all other body cells and then back to the right side of the heart is called the **systemic system.**

APPLYING

The fetal heart

An in situ pig heart

Explore the external surfaces and vessels of the in situ ("in original place") fetal pig heart. Using a blunt probe, gently remove the membranous sac, soft thymus glands, and fat surrounding the heart. (The membranous sac is called the **pericardium,** meaning "around the heart." The disease pericarditis is an inflammation of this membrane.)

Carefully uncover the vessels that extend from the heart. Separate other tissues from the vessels as you follow them into body parts. Follow the major tubes from the heart until you can determine the general area of the pig served by each major vessel.

Using the names of the vessels as given in Figure 14-4 and Table 14-1, describe the areas of the fetal pig served by each of these major vessels.

17

Did you locate an area near the heart where there is a connection between two thick-walled vessels that conduct blood away from the heart?

18

If not, return to the fetal pig and search for it. This connection is the **ductus arteriosus** (DUCK-tus ar-TER-e-O-sus).

Using your fine-pointed scissors, cut the lateral walls of the right and left sides of the fetal pig heart. These cuts should open the atria and ventricles on both sides of the heart, exposing the inner chambers. With forceps, pull out any latex filling the heart chambers. As you do this, note the spaces from which you are removing this material. As you tug on the latex, some of the latex in adjacent vessels will likely be removed as well. Wash out any loose material.

Explore the inner compartments and any openings with your probe. Be sure you can identify the openings into and from the major vessels. Can you locate any vessels that bring blood to the heart tissue itself?

19

(You may want to return to the frontally sectioned mature pig heart to answer this last question.)

Search for an opening, the **foramen ovale,** between the two atria of the fetal pig. You may also be able to locate an opening between the two ventricles. For the former opening, gently probe in the atrial walls dorsal to the beginning of the aorta. For the latter opening, gently probe in the interventricular wall near the atrioventricular (AV) valves.

Effects of the ductus arteriosis and atrial and ventricular passages

What are the effects of these special fetal structures on blood flow through the heart?

From birth through adult life, blood normally flows into the right atrium, through the tricuspid (right AV) valve into the right ventricle, and then through the pulmonary semilunar valve to the pulmonary artery, which conducts blood to the lungs. In the lungs, the blood takes up oxygen and releases carbon dioxide. From the lungs, blood returns to the left atrium through the pulmonary veins and passes through the left AV valve into the left ventricle. From the ventricle, blood passes through the aortic semilunar valve into the aorta, which distributes oxygenated blood to the body.

From birth through a mammal's lifetime, blood is oxygenated in the lungs. Where does the fetal blood pick up oxygen?

20

Take a piece of the fetal pig's lung and drop it in a container of water. If the fresh mammalian heart has any attached lung tissue, remove a small piece and also drop it into the container of water. Does the fetal lung have any air in it?

21 ____________________

On what evidence did you base your answer?

22 ____________________

Fetal blood picks up oxygen in the placenta, not from its own lungs. (The fetus does not inhale air.) Consequently, all blood flowing through the lungs is "wasted effort" for the fetus. The loop of the figure 8 blood circuit that takes blood from the right heart to the lungs and back to the left heart (the pulmonary system) is used very little. The ductus arteriosus, the foramen ovale, and the opening between the right and left ventricles of the fetal heart bypass all or part of the pulmonary system.

Compare the blood flow in the fetal pig to an adult mammal such as a human. In Table 14-2, list in order the structures through which a drop of blood would travel from the right atrium to the aorta in the fetus using (1) the ductus arteriosus and (2) the foramen ovale bypasses.

At birth, these bypasses must close. When they do not, the baby has a very blue (cyanotic) skin from the lack of oxygen in the tissues. These defects frequently need to be surgically closed at an early age. About 1 in 5500 babies has patent ductus arteriosus, the nonclosure of the ductus arteriosus.

Heart problems

Our dependable pump sometimes fails to function properly. For each abnormal condition described, give the effect of each condition on blood flow as the heart contracts and relaxes. Then use your knowledge to infer the effect of the defect on the individual.

Aortic stenosis

This is a condition in which the leaflets—or cusps—of the aortic semilunar valve become scarred and somewhat fused together. (This condition sometimes results from the effects of rheumatic fever, deposition of calcium in the cusps, or a birth defect.) The opening of the valve is narrowed, and the cusps do not meet in the midline on valve closure.

23 ____________________

____

Table 14-2 Fetal use of the pulmonary system

Bypass	Order of structures used
Ductus arteriosus	Right atrium→ →aorta
Foramen ovale	Right atrium→ →aorta

Mitral prolapse

This is a condition in which the cusps of the left AV valve are enlarged, thickened, and scalloped. The chordae tendinae may be elongated also. The cusps billow upward into the left atrium on valve closure.

24 ______________________________

____

Myocardial infarction

This is a condition in which the blood supply to the heart muscle is interrupted and the muscle cells die. The interruption can be caused by a blood clot lodged in a vessel taking blood to the heart muscle or by fatty deposits in the vessel wall that close the vessel. What may be the effect if cells die in a small area of an atrium muscle?

25 ______________________________

____

What may be the effect if cells die in a large area of the left ventricle muscle?

26 ______________________________

____

LABORATORY
15

Circulatory system

The heart beat and its regulation

The pump

Your heart is a pump of living tissues. When you are resting, your heart pumps about 5 liters of blood per minute. When you are exercising vigorously, your heart pumps over 24 liters a minute. How does the heart work as a pump? What causes the changes in its rate of pumping? These are some of the questions you will explore today.

EXPLORING

Heart tissues

Obtain your microscope and a prepared slide of a longitudinal section through the atrium and ventricle of a rat heart.

Carefully observe tissues of the two chambers. In Figure 15-1, draw the two chambers and identify the location of at least three tissue types: epithelial, muscular, and connective. What other tissues, if any, does the heart have?

1 ______________________________

Figure 15-1
Tissue of the mammalian heart.

Table 15-1 Comparison of the tissues of the atrium and ventricle

Chamber	Tissue type (small [+] to large [+++++])			
	Connective	Muscular	Epithelial	Other
Atrium				
Ventricle				

Compare the atrium and ventricle with respect to the amount of the three tissue types. Using a 1 (small) to 5 (large) plus system, rank the relative amount of each tissue in each chamber's wall. Fill in Table 15-1.

Based on tissue types and the relative amounts of each tissue in the wall of the chambers, what do you think is the function of each of these chambers?

2 ______________________________

What differences do you predict in the activity of the two chambers?

3 ______________________________

A functioning frog heart

Working together as a class and with the help of your teacher, carefully observe the functioning heart of a frog with the aid of physiological equipment called a **physiograph.** (Before proceeding in this section, read Appendix H.)

A question for our society: what is death?

Your teacher will supply the class with a "double-pithed" frog. This means that the brain and the spinal cord were destroyed. Is this frog dead or alive?

4 ______________________________

What is your definition of death?

5 ______________________________

What are some of the different definitions used by your classmates?

6 ______________________________

____

What is the current definition of death used by the medical profession?

7 ______________________________

In what ways has the medical definition of death changed in the last 50 years?

8 ______________________________

____

The question "What is death?" is a current topic in our society. Technological advances in medicine have produced equipment that can provide some of the necessary functions of an organism. These machines keep tissues functioning even though some vital organs have ceased functioning. When to use life-support equipment is a question asked by many families, the medical profession, and many state lawmakers. The term *living* as applied to a mammal can no longer be defined by such easily determined criteria as the presence of breathing and a beating heart. Currently, most members of the medical profession use the absence of brain activity as the defining criterion of **death.**

Is there any rhythmic movement of the ventral wall of the frog? Using the most current medical definition of death, is this frog dead or alive?

9 ______________________________

Electrical activity, the electrocardiogram

Using physiograph equipment, the class will record the electrical activity occurring in the frog heart and explore the movement of its chambers.

With your assistance, the teacher will attach pickup electrodes to the frog's anterior legs (negative to the right, positive to the left) and a grounding wire to its left posterior leg. Note that the cables connect these electrodes through an amplifying system to a display device (chart recorder or oscilloscope). If any electrical activity occurs between the two pickup electrodes, it is amplified and converted to the movement of a pen (for a chart recorder) or movement of a point of light (for an oscilloscope). At what speed is the paper or the light beam moving?

10 ______________________________

Record any electrical activity occurring in the frog. If a hard copy is made, divide the paper tracing into several segments and distribute them to class members. This record, or tracing, is called an **electrocardiogram (ECG).**

Examine the ECG. Is there a regularity (repeating sequence) to the tracing?

11 ______________________________

If so, mark off and label one of the repeating sequences. Tape or staple the record of one or two repeating sequences in Figure 15-2.

When the ECG is made at a specific paper speed, you can determine the length of time for each sequence with proportional reasoning. You set the paper speed (mm/sec) equal to the distance (mm) of one repeating unit/unknown time interval (*x* sec). Then you can convert the answer to the number of sequences per minute, which is the usual time unit for presenting the rate of this heart activity. How many of these repeating sequences (events) occur in 1 minute?

12 ______________________________

If you have problems setting up proportional equations, use the equations given below as a guide. Show your calculations.

Proportional reasoning to determine the length of time for one event:

$$\frac{\text{mm of paper moved under the pen}}{\text{1 sec}} = \frac{\text{mm for one event}}{\text{x sec (time for one event)}}$$

Solving this proportional relationship for x gives you the time (in seconds) for one event. Convert one event occurring in seconds to number of events per minute by dividing the length of time (sec) for one event into 60 (number of seconds per minute). This is another proportional relationship.

Figure 15-2
Electrocardiogram of a frog.

Contraction of the frog heart

Your teacher will expose the heart by removing a portion of the ventral wall of the frog. Observe the beating heart. Describe how the heart contracts.

13 ____________________

____

In the intact frog the heart is bathed with body fluid. To keep the heart from drying, bathe it with a fluid that contains the normal concentration of salts found in the frog's body. This isotonic solution is called *frog ringer's*. Your teacher will help you or a classmate attach the apex (caudal tip) of the heart to a lever of the displacement transducer. With your finger, gently push the heart lever downward. What happens to the recording trace line each time the displacement lever is pushed downward?

14 ____________________

What happens to the recording pen each time the frog heart contracts?

15 ____________________

The physiograph equipment records movement by converting (transducing) the mechanical energy of the lever movement into electrical energy. It amplifies this electrical message and converts it to the movement of the recording pen against the moving paper; this last part of the equipment set up is the same as in the ECG. Record the movement of the contracting heart for 1 or 2 minutes. As you record this activity, observe the heart of the frog as it contracts. Does the entire heart contract simultaneously or do portions contract separately? Describe the type of contraction motion you see.

16 ____________________

____

Figure 15-3
Heart contraction.

With some students watching and describing the key events of the heart contraction while others are watching the movement of the recording pen, match the heart's contraction events to the "hills and valleys" of the tracing.

Cut and distribute sections of the tracing to classmates. In Figure 15-3, attach a section of the tracing with two or three heart contractions or accurately draw the shape and size of the contraction record. Label the hills and the valleys of your tracing with the appropriate contraction events of the heart.

What advantage may this wave of heart contraction give to its pumping function?

17 ______________________________

____

If you have any difficulty with this question, obtain a sausage-shaped balloon. Place your hands next to each other around the balloon. Try "pumping" the balloon with both hands at the same time. Compare the resulting air movement inside the balloon to the air movement when you pump by alternating your hands.

The relationship between ECG and heart contraction

If a two-channel recorder is available, record both the electrical activity of the heart and the heart contractions simultaneously. This ECG will be somewhat distorted from the other ECG tracing because the heart is now pulled up toward the heart lever. Not as much of the electrical current reaches the forelimbs as before. Distribute tracing segments to classmates.

On your tracing, draw vertical lines from the end of each "hump" (wave) on the ECG to the contraction record. These vertical lines emphasize the relationship between electrical and contraction activity. Place this tracing in Figure 15-4. Using this record, match the electrical events occurring in the heart with the contraction events. Describe the relationships among specific waves of the electrical sequence of the heart and the contraction events.

18 ______________________________

____

Figure 15-4
Relation of electrical and contraction activity.

In general, how many sequences of repeating electrical activity occur for each complete heart contraction?

19 ______________________________

CONCEPTS/TERMS

Cardiac cycle

The sequence of events that occurs in the heart from the initiation of one beat, or contraction, to the initiation of the next beat is called the **cardiac cycle.**

As you observed, a specific set of electrical activities is associated with each heart contraction. A wave of electrical change (a message) occurs in all types of muscle cells; it triggers the shortening of these cells, which contracts the muscle. Review the structure of cardiac and skeletal muscle tissue (Laboratory 11). What do you think is the structural difference between cardiac and skeletal muscle that permits the cardiac muscle to pass electrical messages from one cell to another when skeletal muscle cannot?

20 ______________________________

Figure 15-5 shows the electrical changes occurring in the heart of a human. The changes designated as *P* and *QRS* waves are immediately followed by the shortening of heart muscle cells in the atria and ventricle, respectively. Ventricle relaxation follows the T wave. The wave of electrical change begins spontaneously at a specific area of the right atrium called the **pacemaker** (sinoatrial [SA] node) of the heart (see Figure 15-6). (You will not be able to see the pacemaker and other special electrical conducting fibers in your dissection.) As this electrical message spreads from one atrial muscle cell to others, it is re-

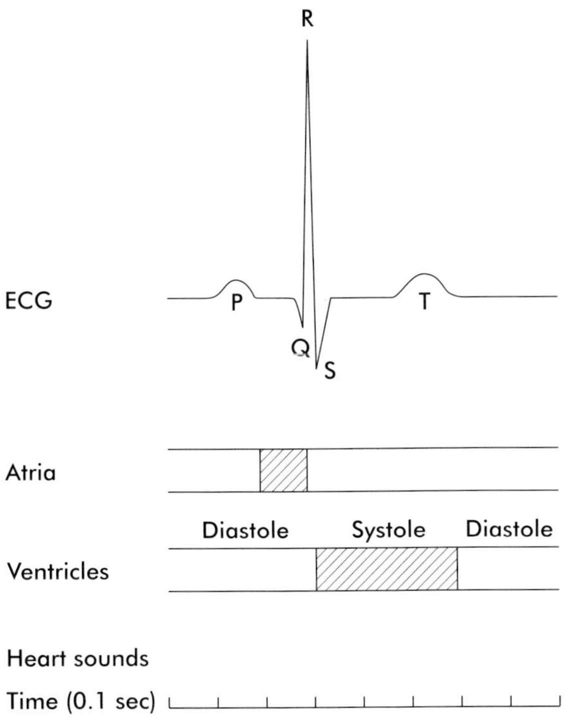

Figure 15-5
The human cardiac cycle.

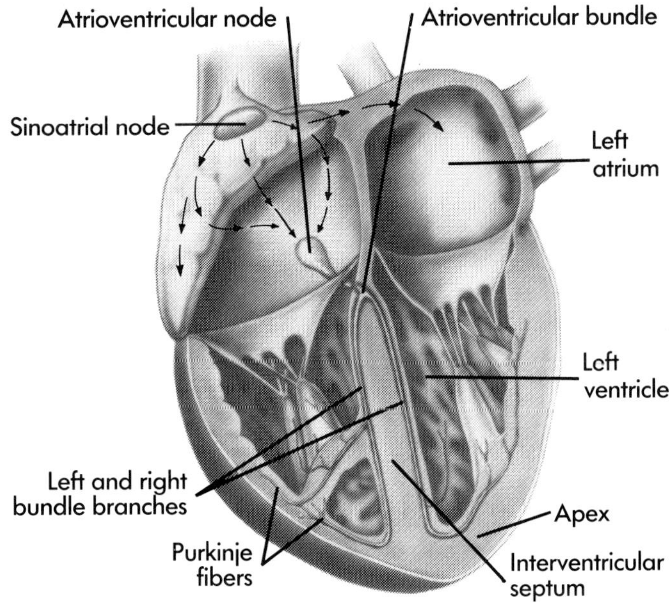

Figure 15-6
Electrical message pathway in the heart.

corded as the P wave. After passing through the atria, the electrical message pauses slightly at the atrioventricular (AV) node between the right atrium and the wall between the ventricles. Then this wave passes down a bundle of specialized fibers that divides into two bundles in the wall between the ventricles and then into additional branches at the heart apex. These branches distribute the wave, and it spreads out over the ventricular muscle. The wave of electrical changes in the ventricle is the QRS component of the ECG sequence. After contraction of the ventricle muscle, the muscle electrically reorganizes. This is the T component that marks the beginning of the ventricular relaxation in the cardiac cycle. Shortly thereafter, the cardiac cycle begins again.

In the heart the contracted state is called **systole** (SIS-toe-lee) and the relaxed is called **diastole** (di-AS-toe-lee). Mark the systolic and diastolic phases for the atria and ventricles on Figures 15-3 and 15-4.

When the atrium contracts before the ventricle, it pumps the blood remaining in the atrium into the ventricle. Then, the ventricles contract, forcing the blood into the vessels that conduct blood away from the heart.

EXPLORING

Effects of two body chemicals on heart contraction rate

Now return to the contracting frog heart. Continue to bathe the frog heart with the ringer's solution. In the first line of the second column of Table 15-2, record the time for one cardiac cycle based on the average of three or four contractions. Convert this average time interval to contractions per minute, and record this number in the third column.

With an eyedropper, remove the normal saline solution and replace it with fresh salt solution that contains the drug epinephrine (adrenalin). Mark the point on the tracing at which the epinephrine was added. This chemical is naturally formed by the organism at the ends of some types of nerve fibers and by the adrenal medulla, one of the endocrine glands. After a few minutes, record the effects of this chemical on the contraction rate of the heart. Calculate heart rate in the presence of epinephrine and add it to Table 15-2.

Remove the epinephrine saline and replace it with a saline containing **acetylcholine.** This chemical is produced at the ends of certain types of nerves.

Table 15-2 Drug effects on the frog heart rate

Drug in frog ringer's	Average time for a cardiac cycle	Contractions/min
Ringer's only		
Epinephrine (2 parts/10,000)		
Acetylcholine (2 parts/10,000)		

After a short period, record the contractions on Table 15-2, and calculate the heart rate. Compare the effect of these two chemicals on the contraction rate.

21 ______________________________

____

Explain the ways in which you agree and/or disagree with the statement, "Certain chemicals can change the rate of the heartbeat."

22 ______________________________

____

____

CONCEPTS/TERMS

Regulation and homeostasis

Living organisms regulate their cell and organ activity producing an internal steady state called **homeostasis.** This regulation can occur by glands releasing antagonistic (having the opposite effect) chemicals into the blood or by the ends of nerve cells releasing the chemicals onto the responding organ such as a muscle. When epinephrine is released from the ends of some nerve fibers near the heart pacemaker, what happens to the heart rate?

23 ______________________________

When acetylcholine is released at the ends of other nerve cells close to the pacemaker, what happens to the heart rate?

24 ______________________________

The action of these antagonistic chemicals are a major control mechanism of the pumping rate of the heart.

APPLYING

Valve closure

Using your knowledge of heart structure and the one-way blood flow in the heart, think about the opening and closing of the heart valves during the cardiac cycle. In general, what regulates the opening and closing of the heart valves?

25 ______________________________

Use Table 15-3 to list when these valves are opened and closed, controlling the one-way pumping of blood.

Table 15-3 Functioning of values during the cardiac cycle

Chamber state	State of valves: AV valves	State of valves: Semilunar valves
Atrium systole		
Atrium disatole		
Ventricle systole		
Ventricle diastole		

Heart sounds

Obtain a stethoscope, a sound-amplifying instrument used by doctors and nurses. Before using, clean the ear pieces with alcohol.

Place the bell of the stethoscope just to the left of your breastbone (sternum) at midchest level. Search for and then listen to the sounds of your contracting heart. Describe what you hear.

26 ______________________________

Think about the possible sources of the sounds. Considering all of the events occurring during the cardiac cycle, what is the most likely cause of each of these sounds?

27 ______________________________

____

Check with your neighbor. If you do not agree on the cause of heart sounds, talk with your teacher. Return to the cardiac cycle of humans in Figure 15-5. Indicate the time of each heart sound on the lower portion of the diagram.

Experiment: effect of exercise on heart contraction rate

Low levels of carbon dioxide in the blood activate a part of the brain that sends messages to the acetylcholine-releasing nerves near the pacemaker of the heart. After considering this information, what do you think are the effects of increased blood carbon dioxide levels on the heart contraction rate?

28 ______________________________

Give two alternative hypotheses that propose the effect of exercise on the ECG of humans.

29 ______________________________

____

What is the effect of hyperventilating (excessive breathing) on the ECG?

30 ______________________________

Give two alternative hypotheses that propose the effect of hyperventilation on the ECG of humans.

31 ____

____

Working in small groups or as designated by your teacher, design an experiment to test predictions from one set of alternative hypotheses (for exercise or hyperventilation) on willing classmates.

CAUTION
People who volunteer to be subjects for this experimental test should be in good physical condition and have no signs of illness.

Your group will need to decide on the activity level of the independent variable, how often and when you should record the subject's ECG, and what you plan to compare and analyze in the ECGs to obtain information about the alternative hypotheses.

Human ECGs can be made with physiographs or computer interfacing devices. Your teacher will demonstrate how to use the equipment in your laboratory. With either type of recording equipment, the quality of the contact between the subject's skin and the surface pickup electrodes is important to the successful ECG. Clean an area on both wrists and the left ankle with alcohol. Apply a small amount of electrode gel to each of three electrodes. Place the appropriate electrode on each of the cleaned skin surfaces, moving each electrode slightly until the gel makes good contact with the skin and electrode. Hold each electrode in place with a perforated rubber strap or electrode tape.

Following the directions from your teacher for your particular equipment, connect cables from the electrodes to a lead selector box (use only one setting, lead II, for this experiment), then to the electrical activity amplifier, and finally to the recording device (chart recorder or computer).

What is your independent variable?

32 ____

Describe your planned procedures including the controlled conditions.

33 ____

____

____

____

____

____

____

When your group has designed the experiment, present it to your teacher for any suggested changes. After receiving your teacher's approval to proceed, conduct the experiment.

Describe your data. Attach tracings to support your statements.

34 ____

____

____

Explain in what ways the data support or refute each of your alternative hypotheses.

35

What are your suggestions for verifying the results of this experiment?

36

What do you conclude from this test of the alternative hypotheses?

37

Was homeostasis demonstrated in the subject's response to the independent variable?

38

Explain your answer.

39

Figure 15-7
ECG with a split QRS wave.

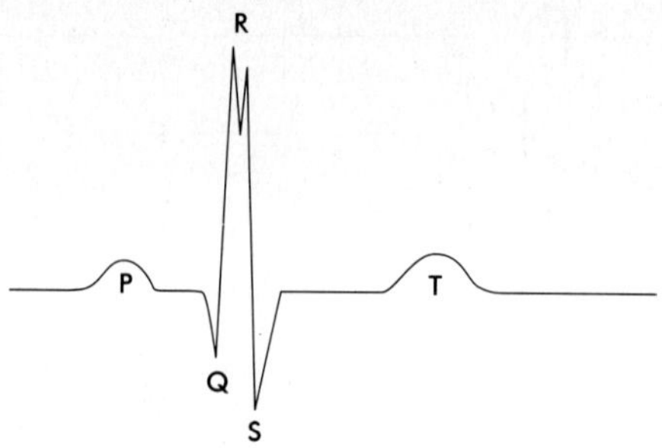

Abnormal heart conditions

What happens to blood flow if the conduction of the electrical activity is greatly delayed at the AV node?

40 ______________________________

____

What difference(s) would you expect in the ECG of such a subject compared to a normal person?

41 ______________________________

____

Figure 15-7 shows an ECG with a split QRS wave. What abnormal activity causes this kind of QRS wave?

42 ______________________________

____

If a person's heart suddenly stops (cardiac arrest), what drug would you recommend giving to the patient?

43 ______________________________

What other procedure might be used?

44 ______________________________

LABORATORY

16

Circulatory system

Structure and function of vessels

PART ONE **Vessel structure**

Transport

Bloods flows from the pumping heart to the body, and then it returns and is pumped again and again. As it circulates, the blood transports nutrients and wastes. In this laboratory, you will explore the structure and function of the organs associated with blood flow from and its return to the heart.

EXPLORING

The vessels of the fetal pig

Vessels of the forelimb

Obtain your fetal pig and dissecting tools.

Remove the skin from the right foreleg of the pig. Make a shallow cut circling the limb as close to the trunk as possible; grasp the edge of the cut skin and tug the skin off the limb as you might remove a tight glove.

Locate the blood vessels at the cranial portion of the heart. Identify each of these vessels and place the names in the first column of Table 16-1 (see Laboratory 14). Also include the chamber of the heart to which each vessel is attached. In the second column, indicate whether the vessel carries blood to or away from the heart.

From the heart, follow the vessels and their branches that extend into the right foreleg. At some locations such as in the chest wall, you may find vessels extending into or through muscles. At other locations, you may find vessels extending through the connective tissue between muscles. Most of the blood vessels are filled with either pink or blue latex. In addition, you will see tough, creamy-white "cords." These cords are nerves.

Using the blunt metal probe, carefully separate structures. Leave all structures intact except for the chest wall muscle. With scissors, make one lateral cut in the middle of the chest wall from the breastbone to the vessels that extend into the foreleg to expose the vessels as they pass through the chest wall.

Follow all the visible vessels into the foreleg. In Figure 16-1, draw the general outline of the right forelimb (2 to 3 times actual size). On your diagram, draw the vessels of the forelimb. You may want to use two colors corresponding to the colors of the latex in these vessels. (In some pigs the latex may not have been evenly pumped throughout the system of vessels. Consequently, pink latex may be in vessels that should have blue latex, or vice versa.) Indicate which vessels are deeper and which are more superficial. Also identify any part of the leg that lacks blood vessels.

Describe the general branching pattern of vessels in the foreleg.

1 ______________________________

____

Table 16-1 Organs served by major blood vessels

Vessel's name/attachment	Direction of flow	Body areas/organs served

Major vessels of the trunk

Return to the trunk of the pig. From the heart, follow all the large vessels in the chest and abdominal cavities until they enter organs. As you explore, consider the following suggestions and record your observations in Table 16-1.

Begin by following the aorta (large white vessel) that extends from the left ventricle. In the third column of Table 16-1, list in order all areas and organs that have branches from this vessel.

Explore the other major vessels directly connected to the heart. As before, list in order the areas or organs served by each of these vessels.

Consider the following questions:

What areas and organs of the pig lack blood vessels?

2 ______________________________

Figure 16-1
Circulatory vessels of the foreleg of the pig.

What happens when you prick the skin of your ankle with a pin?

3 ____________________

What happens when you prick the skin of your forehead, finger, or abdomen.

4 ____________________

Are nails and hair living material?

5 ____________________

Does all living material need a blood supply?

6 ____________________

If so, explain.

7 ____________________

____

From your fetal pig vessel explorations and your common knowledge, explain the ways in which you agree and disagree with the statement, "All parts of the body receive blood from the heart and return blood to the heart through a continuous system of vessels."

8 ____________________

____

Preliminary examination of the vessel structure

Locate one vessel that takes blood from the heart and one of equal diameter that brings blood to the heart. Cut each half way through with the tips of small scissors. You may want to remove a small section of the latex to expose more of the inner surfaces of the vessels. Explore the walls with a hand lens or dissecting microscope. Describe the similarities and differences in the structure of the walls of the two vessels.

9 ____

____

At this point, how do you think the structural differences in these vessels affect their function in carrying blood either to or away from the heart?

10 ____

Comparing the vessel system and highways

On heavily traveled routes, interstate highways are built instead of single-lane roads, which carry two-way traffic. Interstates consist of two roads, side by side, with vehicles moving only in one direction on one and in the opposite direction on the other. Which road design carries traffic most efficiently?

11 ____

To which of these types of road systems is the circulatory system most similar?

12 ____

Describe these similarities.

13 ____

____

CONCEPTS/TERMS

System of vessels

Closed circulatory system

The heart continually circulates blood throughout the body in a continuous system of vessels. Because blood is confined within vessels, the fetal pig's circulatory system is called a **closed circulatory system.** In a closed circulatory system, thick-walled arteries branch into smaller arteries as they carry blood away from the heart to body organs. The smaller and thinner-walled of these vessels are called **arterioles,** meaning "little arteries." However, arterioles are not the end of the branching pattern. They continue to branch into the smallest and thinnest-walled of all vessels, the **capillaries.** Larger-diameter vessels called **venules** collect blood from the capillaries. Venules fuse, forming larger vessels called **veins.** The veins return blood to the heart.

A dual pathway

The arteries to and the veins from each organ improve blood flow as interstate highways improve vehicular movement. With two vessels, blood moves in only one direction in each vessel. Therefore blood flow is not impeded by "mixing" opposite forces, and smoother, more-rapid flow results. Blood flowing to the organ with nutrients also does not mix with the blood carrying wastes from the organ. This helps maintain higher concentration gradients of nutrients between the blood and cells of the organ, promoting diffusion of nutrients out of the blood to the tissues. This flow pattern also maintains a greater concentration gradient for the cellular wastes to diffuse into the vessels, which promotes the removal of wastes from the organ.

APPLYING

Vessel structure

Recalling your exploration of the mammalian heart in Laboratory 14 and your preliminary examination of the walls of an artery and a vein, predict which types of tissues make up the walls of these vessels.

14 ______________________________

Examine the slides at the demonstration table without magnification. Decide whether each preparation is a small piece of a vessel, a part of an organ, or a cross section of one or more types of vessels.

Observe the slides with magnification. Using the high power objective, observe the slides of a small piece of a vessel. Try to identify the tissues forming the walls of the vessel.

As you observe the material from various organs, look for circular or oval cross sections of arteries, capillaries, and veins. (These may be collapsed.) Are these cut vessels scattered individually throughout the organ, found in pairs, or arranged in some other way in the tissues?

15 ______________________________

In Table 16-2, give the relative amount of each type of tissue that you can identify in the walls of the vessels. Use the one to five "plus" system to indicate the relative amount of each tissue.

Based on your microscopic examination of vessel structure, in which type of vessel do gases, nutrients, and wastes move between blood and the cells of the body?

16 ______________________________

Circulatory vessels in organs

Examine latex-injected preserved organs, models, and wall charts of organs that are available in the laboratory. In what ways are their circulatory vessel systems similar to and different from the vessels of the pig foreleg?

17 ______________________________

____

____

____

Table 16-2 Tissues of the circulatory vessels (+ to +++++)

	Relative amount of the tissues			
Structure	**Connective**	**Muscular**	**Epithelial**	**Others**
Artery				
Capillary				
Vein				

Constructing a model of a vessel system

Obtain a box of T- and Y-shaped glass or plastic tubes, short tubes, and several lengths and diameters of flexible tubing. Construct a simple vessel pathway representing the vessel system that carries blood away from one side of the heart and returns blood to the other side of the heart.

Describe the important features of your model.

18 ______________________________

____

Unless you are told otherwise by your teacher, keep your constructed vessel system intact for later use.

PART TWO Vessel function

EXPLORING

Pulse

The heart alternates between contraction and relaxation. What are the effects of this heart action on the volume and pressure of blood in the arteries near the heart?

19 ______________________________

Predict the effect of arterial structure on these volume and pressure changes.

20 ______________________________

Sit quietly with your arm resting in your lap and the palm of your hand facing upward. Place the tips of two fingers (not your thumb) on the wrist artery (the radial artery) that is close to the base of your thumb and just to the thumb side of the tendons (cords) that run lengthwise in the wrist. This artery crosses the end of the lateral bone of your forearm. If you have trouble locating this artery, seek help from your teacher. Describe what you feel.

21 ______________________________

Using a watch or clock, count the number of pulses and base it on a 1-minute time interval. You can count for 1 minute, or for 15 seconds and multiply by four, or you can count the number in 30 seconds and multiply by two. How does the length of the timing interval affect the accuracy of the pulse measurement?

22 ______________________________

What is your pulse rate based on a 1-minute time interval?

23 ______________________________ /min

Also measure your pulse rate in the carotid artery in the neck. Using your two fingertips, search for the artery just lateral to your larynx. Measure your pulse rate in this vessel. In what ways is the carotid artery pulse rate and strength similar to and different from the one in your wrist artery?

24 ______________________________

____

Predict the similarities and differences for the pulse in an artery on the anterior surface of your ankle compared to the pulse in the wrist and carotid arteries.

25 ____

Explain your prediction.

26 ____

Blood pressure

Measuring blood pressure

Working with a partner, obtain a sphygmomanometer (SFIG-mo-man-OM-e-ter) and, if needed, a stethoscope. Each sphygmomanometer has a cuff that you wrap around the upper arm and pump full of air. It also has a pressure gauge from which you read the pressure in the inflated cuff. Some sphygmomanometers have internal pressure sensors that connect to a digital display of pressures. With some sphygmomanometers, you use the stethoscope to listen for sounds of the blood pulsing through the brachial artery (Figure 16-2). (The brachial artery runs close to the anterior surface of the elbow.) Other sphygmomanometers have an internal microphone that replaces the stethoscope. Your teacher will give you information about the particular equipment available in your laboratory.

CAUTION

When the cuff is fully inflated, it stops the blood flow to and from the arm. Do not leave the cuff inflated for longer than 1 or 2 minutes.

Measure your partner's blood pressure. Have the subject sit quietly with the arm supported at heart level. If your sphygmomanometer requires the use of a stethoscope, do the following:

1. Clean the ear pieces of the stethoscope with alcohol.
2. Place the inflatable portion of the cuff on the anterior surface of the subject's upper arm. Wrap the sphygmomanometer cuff around the subject's upper arm just above the elbow.
3. Place the ear pieces in your ears and the stethoscope's bell just below the "hollow" of the elbow and slightly to the inside of its midline.

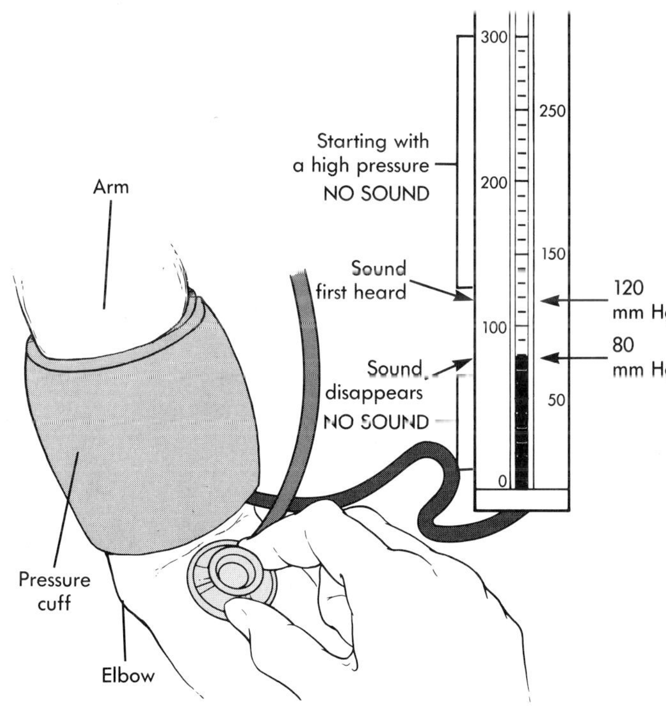

Figure 16-2
Measuring blood pressure.

4. Close the valve at the sphygmomanometer's air-pumping bulb.
5. Pump air into the cuff until the pressure is about 100 mm of mercury as indicated by the scale on the manometer. Listen for the thumping sounds of blood flowing in the artery.
6. When you hear the sounds, inflate the cuff to about 170 mm of mercury.
7. Slightly open the valve of the air pump, slowly releasing air from the cuff (a decrease of 2-4 mm/sec).
8. As the pressure drops, listen for thumping sounds in the artery and note the pressure at which the sounds begin.
9. Continue to listen to the sounds until they disappear, noting the pressure at which you can no longer hear the sounds.

If your sphygmomanometer has the digital readout, just wrap the cuff around the subject's arm with the sensor above the artery and inflate the cuff as described. Wait while the air is released slowly and the cuff takes and displays the readings. Its readings are the pressure at which the sounds are first heard and the pressure at which the sounds first disappear.

What causes an artery to pulse?

27 ______________________________

Why are sounds absent when the cuff pressure is very high?

28 ______________________________

Why are sounds absent when the cuff pressure is very low?

29 ______________________________

Blood pressure is written similarly to a fraction, with the pressure at which you first hear the sounds over the pressure at which you can no longer hear the sounds. What is your blood pressure?

30 ______________________________

What is your partner's pressure?

31 ______________________________

CONCEPTS/TERMS

Pulse and blood pressure

When you inflate the cuff, its pressure pushes against the artery. When the cuff pressure is greater than the pressure pushing blood through the artery, the artery collapses and the blood flow is stopped. At this point, you hear no sounds. When the pressure in the cuff is lowered below the pressure of the blocked blood, the blood pushes through, partially opening the artery. At this time, you hear the thumping sounds of the turbulent blood as it flows through the narrowed artery. Finally, the cuff pressure drops below the pressure in the artery even when the blood is not pulsing. At this point, you hear no sounds.

The pressure at which the blood first pushes through the completely blocked artery is called the **systolic pressure.** The pressure at which the blood first flows freely through the artery, making no sounds, is called the **diastolic pressure.**

The pulsing of blood in arteries reflects the contraction and relaxation of the heart. When the heart forces the blood into the arteries during contraction, the arterial pressure increases; this is the arterial systolic pressure. When the heart relaxes (diastole), it pumps no blood into the arteries and the arterial pressure drops; this lower pressure is the diastolic pressure. The alternating systole and diastole of the heart causes the pulsing pressure changes in the arterial blood.

The pressure in the arteries does not drop to zero when the heart is in diastole. The elastic tissue in the arterial walls pushes on the blood when it recoils from being stretched by the blood pumped into the arteries. This recoil force is the source of the pressure in the arteries when the heart is in diastole.

Now consider the differences in the strength of the pulse you may have noticed as you measured pulse rate in the neck and wrist. As the heart pushes blood through the vessels, the walls of the vessels resist the push. This resistance decreases the force pushing the blood. As the blood moves farther from the pump, it continues to encounter resistance and the blood pressure continues to decrease. Therefore the pressure in the neck, which is closer to the heart, is higher than the pressure in the wrist, which is more distant from a person's heart.

APPLYING

Pressure in an in vitro system

Based on the concept "There is a inverse relationship between distance from the pump and blood pressure," answer the following questions: Which type of blood vessel has the lowest pressure?

32 ______________________________

Which type of vessel has the highest pressure?

33 ______________________________

Test your answers by using the model vessel system with some modifications. In one vessel of each type (the artery, the vein, and one capillary), replace a single length of straight tube with an upside down T-shaped tube. Connect your vessel system to a faucet by a long, flexible hose from the arterial end. The water pressure at the faucet will provide the force to push the water through your vessel system. Put your vessel system in an area with a drain to remove excess water, and then run water through the vessels. Observe the "fountains" of water squirting from the open ends of the three upright T tubes. Rank the three types of vessels according to the heights of their fountains.

34 ______________________________

Assuming that the height of each fountain is a rough measure of the pressure in that part of the vessel system, did this test support the distance-pressure concept?

35 ______________________________

Explain your answer.

36 ______________________________

____

Leave the vessel system attached to the faucet.

Abnormal blood pressure conditions

Nicotine and epinephrine (adrenalin) cause the smooth muscles of the arteries to contract, decreasing the size of the vessels. What effect does smoking have on blood pressure?

37 ______________________________

What effect does stress, which causes the release of epinephrine, have on blood pressure?

38 ______________________________

Explain your answers.

39 ______________________________

____

Arteriosclerosis is a disease characterized by abnormally thick and stiff arterial walls. What effect does this condition have on blood pressure?

40 ______________________________

Explain your answer.

41 ______________________________

What do you think the effects of low blood pressure are in humans when they suddenly stand up after lying on a bed?

42 ______________________________

EXPLORING

Rate of flow

What do you think happens to the flow rate of blood as it moves through the different kinds of circulatory vessels? (Hint: What happens to the rate of water flow in a narrow section of a river or stream?)

43 ______________________________

On what information did you base your answer?

44 ______________________________

____

Return to your model vessel system. Run water through the vessels. For a specific length of time (10 sec, for example), catch all the water from each of the three fountains in similar jars. Compare the amount of water collected from each type of vessel. Although you lost some water from the system as you collected it from each fountain, what general statement can you make about your results?

45 ______________________________

How do you explain the results?

46 ______________________________

____

Calculate the total area of the cross sections for each type of your system's vessels: arteries, capillaries, and veins. Measure the inside diameter (d) of the tubes that represent each type of vessel. Then calculate the cross-sectional area of the tube by the following formula: area = $3.14(d/2)^2$. After calculating the cross-sectional area for each type of vessel, multiply the area by the number of vessels of that type in your system. This is the total cross-sectional area for each vessel type. What is the total cross-sectional area for the system's arteries?

47 ____________________

Its capillaries?

48 ____________________

Its veins?

49 ____________________

What is the relationship (if any) between the total cross-sectional area of a vessel and the rate of flow within that vessel?

50 ____________________

CONCEPTS/TERMS

Cross-sectional area and flow rate

At the heart there are two large arteries receiving blood from the ventricles. There are four veins returning blood to the atria. In the branching pattern of the arteries, the arteries become smaller but increase in number. By the time the vessels are capillaries, the vessels are about 8 μm in diameter and the number of capillaries is very large. (Each cell of the body has to be within diffusion distance of a capillary to receive nutrients and eliminate waste products.) In the return vessels, the diameters of the veins become larger, but the number becomes much smaller as the vessels fuse. The total cross-sectional area of all arteries is small, and the total cross section of all veins is small, but the total cross section of all capillaries is very large. Under these conditions, the flow of blood in capillaries is very slow. Blood flow in the arteries and veins is faster than in the capillaries.

What do you think is the effect of flow rate on the exchange of nutrients and waste products between the blood and cells?

51 ____________________

APPLYING

The frog web or the goldfish fin

In general terms, what flow rates do you expect to find in the vessel system of the frog or fish?

52 ____________________

At the demonstration table, observe the web between the toes of a frog or the fin of a fish with the dissecting microscope. Focusing at different depths, identify the different types of vessels based on their size and thickness of walls. In which type of vessel do you see pulsating flow?

53 ____________________

In which type of vessel do you observe the fastest flow?

54 ____________________

In which type of vessel do you observe the slowest flow?

55 ____

Do you and your partner agree on the identification of vessel type and the flow rates in each? If not, seek clarification from your teacher.

LABORATORY

17

Urinary system

Kidney and homeostasis

Role of the urinary system

The kidney, the major organ of the urinary system, is intricately involved with maintaining the steady state of the body. As your body takes in materials, it accumulates excessive amounts of some substances. As the body cells metabolize, they produce products that may accumulate in excessive amounts. The urinary system removes excess substances such as glucose, various ions, and toxic nitrogen-containing compounds from the blood and retains other substances that are in low supply. It constantly regulates the water volume of the body; water regulation is necessary to maintain osmotic balance in the cells and their surrounding fluids. The function of this system is vital to the survival of all vertebrates.

EXPLORING

Dialysis

Working in small groups or as designated by your teacher, obtain the following materials:

2 small beakers
1 middle-sized beaker (200 to 250 ml)
1 funnel
String
1 piece of cellophane dialysis tubing, 15 cm long (6 in)
1 spot plate
1 dropper bottle of I-KI
Protein reagent test strips
Glucose reagent test strips

Handling the dialysis tubing carefully, soak it in distilled water until it is soft and flexible. At one end, fold the sides together several times, and tightly tie this end closed with string. Open the opposite end of the tube by rubbing it between your fingers. Fill the tubing bag with distilled water to test for pinhole tears and water leakage at the tied end. If you find leaks at the tied end, retie it. If you find tears, obtain and soak a new piece.

At the materials supply table, add 15 ml of each of the following solutions to a small beaker: glucose, starch, salt (NaCl), and albumin (protein). Mix these solutions together. Leaving enough room to tie the untied end of the tubing bag, pour the mixture of solutions through the funnel and into the bag. Squeeze the bag slightly to remove excess air. Fold and tie the open end securely. Rinse the outside of your dialysis bag with distilled water. Put the bag into the middle-sized beaker and add fresh distilled water (about 150 ml) to just cover the bag.

Within 2 minutes, remove about 10 ml of the water from the beaker fluid and place it into a clean, small beaker. Test for protein and glucose using the protein and glucose reagent test strips (follow directions on their containers).

Table 17-1 Dialyzing substances

Time (min)	Substances in diffusate			
	Albumin	Glucose	Starch	Chloride
2				
20				
40				
60				
80				

Test for starch using I-KI solution (put a few drops of beaker fluid in a spot plate and add 1 drop of I-KI as in Laboratory 12). Set the small beaker on a dark background. Test for chloride ions by adding one to two drops of silver nitrate ($AgNO_3$) to the remaining fluid in the small beaker. When chloride ions are present, they react with the silver nitrate to form a white precipitate (silver chloride). Every 20 minutes, remove additional fluid (diffusate) from the original dialysis beaker and repeat these tests. If the results show a positive test in two successive testing intervals, do not test for this substance again. Record your results in Table 17-1, using a + for the presence of the substance and a − for its absence.

As you collect the data, determine to which of the substances the dialysis tubing is permeable. What substances pass through the tubing?

1 ______________________________

Do these substances move with their concentration gradient (from high to low) or against it?

2 ______________________________

Which substances do not pass through?

3 ______________________________

Which of the molecules inside the bag are very large?

4 ______________________________

Proceed with other laboratory activities between testing intervals.

The urinary system

Obtain your fetal pig and dissecting tools. If you have not already identified the kidney from previous explorations, do so now. Look for two large structures anchored to the dorsal wall of the abdominal cavity by thin connective tissue. Separate one from its "bed" but do not remove it from the pig or cut any attached blood vessels or tubes. Describe its external appearance, noting such characteristics as color, shape, size, texture of its covering, and blood vessels.

5 ______________________________

____

____

What other organs and tissues, if any, are loosely attached or located close to the kidneys?

6 ______________________________

Find a tube connected to each kidney that is not a blood vessel. In what anatomical direction do the two tubes extend?

7 ______________________________

What structure is at the other end of each tube (the one not attached to the kidney)?

8 ______________________________

Cut open the elongated bag that lies in the posterior, ventral midline of the abdominal cavity. Examine its wall. Inside the bag, locate the openings of the tubes from the kidneys. Predict the function of this organ based on its structure.

9 ______________________________

What tissues would you expect to find in the wall of this organ if you were to observe it microscopically?

10 ______________________________

Follow the tube that extends posteriorly from the base of this bag. Where does this tube lead?

11 ______________________________

Can you locate any areas where the cavity of this tube is narrowed or constricted?

12 ______________________________

If so, where are these constricted areas located?

13 ______________________________

What similar constricted areas did you find in your exploration of the digestive system?

14 ______________________________

What is the probable function of the constricted areas?

15 ______________________________

Explore a neighbor's pig of the opposite sex. Compare this most posterior tube of the urinary system of your pig and the corresponding tube of the opposite sex. What differences do you find?

16 ______________________________

Return to the kidney. With a sharp single-edge razor blade or scalpel, make a frontal section of the left kidney from the lateral edge, separating the dorsal and ventral halves. Examine the internal structure of the kidney. In Figure 17-1, draw the frontal section of the kidney (two to three times normal size). If the colored latex injected into the circulatory system reached your specimen's kidney, note the location of the pink and blue latex-filled vessels. Indicate these on your diagram. Are there any tiny concentrations of the colored latex within the kidney tissues? Locate the portion of the kidney that is closest to the tube extending from the hollow of the bean-shaped kidney. What is the probable function of this inner space?

17 ______________________________

Examine the solid tissues of the kidney with a hand lens or dissecting microscope. Can you identify any latex-filled "nodules" or clump-like structures?

18 ______________________________

If so, add them to your drawing in Figure 17-1.

If a large, latex-injected mammalian kidney is available in your laboratory, examine its tissues with a hand lens or dissecting microscope. Can you identify colored nodules in this kidney?

Figure 17-1
The kidney.

19 ____________________

If so, in what parts of the kidney are they located?

20 ____________________

In what areas are most small arteries located?

21 ____________________

Add any additional information you obtained from this kidney to Figure 17-1.

Obtain reference materials available in the laboratory and identify parts of your drawing of the kidney. You may want to use the following terms:

Aorta	Renal capsule	Renal vein
Bowman's capsule	Renal cortex	Ureter
Hilum	Renal medulla	Urethra
Nephron	Renal pelvis	Urinary bladder
Renal artery		

Sections of a kidney

At the demonstration table, observe two letter-identified microscope slides of the same type of kidney section. The section extends from the lateral to medial edges of the kidney. Identify the medial side by the presence of a portion of the ureter or the hilum. The microscopes are focused on different parts of the section. One is focused on the outer portion—the **cortex,** and the other on the middle portion—the **medulla.** Between the two fields of view, find structures that look like a nest of capillaries—the **glomerulus.** Also try to identify sections of microscopic tubules.

Figure 17-2 (Plate 8A) shows the microscopic functional unit of the kidney, a **nephron.** Each nephron consists of a long, looped tubule with a cup-shaped end and is associated with arterioles, capillaries, and a venule. On Figure 17-2 near the glomerulus of the nephron, place the identifying letter of the microscope slide that shows the glomerulus. In what area of the kidney are the glomeruli located?

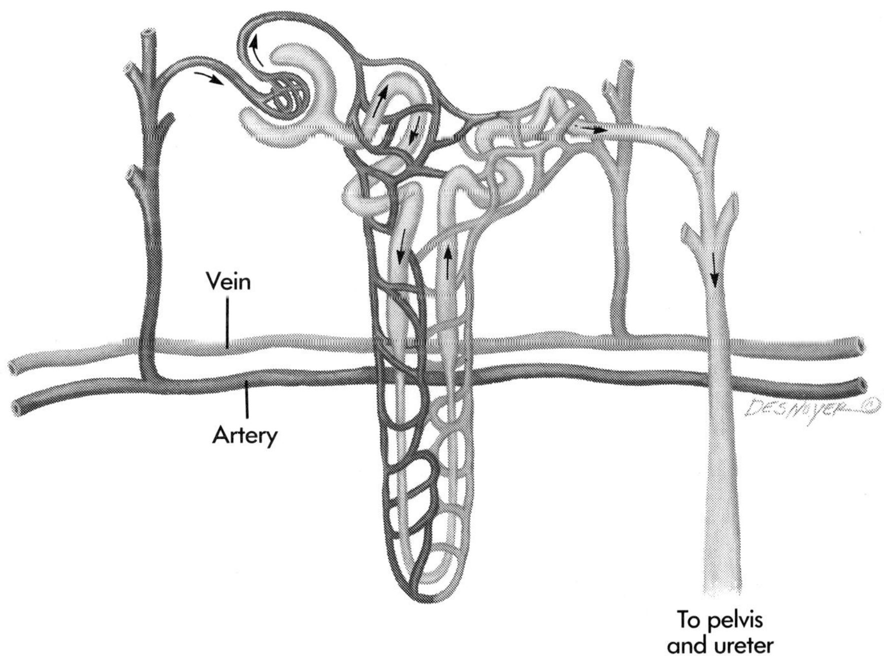

Figure 17-2
The nephron.

22

Use the letter of a microscope slide that shows kidney tubules to identify at least one corresponding area of the nephron. In what part of the kidney is the long tubule loop located?

23

How many different sets of capillaries are in the vessel system of a nephron?

24

Where in the human body is another vessel system that has this type of structural arrangement?

25

The artery-capillaries-artery or vein-capillaries-vein arrangement is called a **portal system.** In general terms, what is one possible functional advantage of two capillary beds within one loop of vessels (from and to the heart) over the single capillary bed found in most vessel loops?

26

CONCEPTS/TERMS

Relating structure and function

Each of the 1 million nephrons in a human kidney uses three processes to regulate substances in the blood as urine is produced. These three processes are filtration, reabsorption, and secretion.

Filtration is the movement of materials through a sieve-like membrane from an area of higher pressure to an area of lower pressure. The source of high pressure in the kidney is the blood pressure within the glomerular nest of capillaries. The low pressure is in the cavity of a cup-shaped part of a tube that surrounds the glomerulus, **Bowman's capsule.** The walls of the capillaries and the inner wall of the Bowman's capsule are very porous. Together, they act as a filtering membrane. The plasma of the blood except for its large molecules is pushed from the glomerular capillaries into the lumen of Bowman's capsule. Materials that are bigger than the pores in the filtration membrane, such as large proteins and blood cells, do not pass into the tubule. The materials that do pass through the membrane are collectively called the **filtrate.**

Reabsorption is the movement of filtrate substances from the tubule to a second set of capillaries that surrounds it. This reabsorption is a selective process. Like the dialysis tubing, cells that make up the wall of the tubules are selectively permeable. Some substances are reabsorbed passively by the processes of diffusion and osmosis (water). Other substances are reabsorbed by active transport against their concentration gradients. Still other substances remain in the tubule. Movement is affected by differences in the permeability along the tubule's wall. Some but not all areas of the tubule are permeable to water, and some but not all areas are permeable to sodium and chloride ions. In your textbook, you can find additional information about the permeability of different parts of the tubule.

Secretion is the active transport of specific substances into the cavity of the tubule from the blood in the second set of capillaries.

At this point in your understanding of the kidney function, what is the general functional advantage of a portal system of blood vessels?

27

Urine is the product of the processes of filtration, reabsorption, and secretion in the nephron tubule and the collecting duct leading from the nephron. The collecting duct drains urine from several nephrons into the kidney's pelvis. From the pelvis the urine flows through the ureter to the urinary bladder where it is stored. When the bladder is full, valves (sphincters) in the urethra open and urine flows through the urethra to the outside of the body.

APPLYING

Inferring nephron processes from concentration changes

Table 17-2 lists the concentrations of various substances in the blood, filtrate in Bowman's capsule, and urine. From the changes in the concentrations or volumes of each substance as it moves through the kidney, determine the process or processes acting on the substance in the nephron. For example, water has the same volume (180 L) in the blood and the filtrate. From this observation, you can reason that water is *filtered* through the filtration membrane. However, of the 180 L of water in the filtrate, an average of only 1.4 L appears in urine. Therefore most of the water must be *reabsorbed* by the cells in the walls of the nephron tubule. List these processes in the fifth column.

Because most of the water that filters into Bowman's capsule from the glomerulus is reabsorbed, many filtrate substances are concentrated simply by removal of water from the tubule and not because of secretion. In such question-

Table 17-2 Changing concentration of major solutes in the nephron

Substance	Blood	Filtrate	Urine	Nephron processes
Water (L)	180	180	1.4	
Organic Molecules (mg/100 ml)				
Protein	3900-5000	6-11	0	
Glucose	100	100	0	
Urea*	26	26	1820	
Ions (mEq/L)				
Sodium	142	142	128	
Potassium	5	5	60	
Chloride	103	103	134	

Adapted from Seeley R, Stephens T, Tate P: *Anatomy and physiology,* St Louis, 1992, Mosby.
*The urea molecule contains waste nitrogen.

able cases, list both processes that you think may be involved. Add a question mark beside them to indicate that insufficient data are available to determine whether one or both processes are involved.

Movement of materials through the kidney

Trace the movement of each of the following materials from a small artery near a glomerulus to the structure in which each material leaves the kidney: a red blood cell (RBC), a molecule of glucose, and a molecule of a waste (inulin) that is completely filtered and excreted. List the structures in the order encountered by each of the materials in Table 17-3.

Constructing a nephron model

Working with a partner or as designated by your teacher, obtain a box of equipment that contains the following materials:

- Dialysis tubing (or plastic wrap)
- Filter paper
- Funnel
- Ringstand with ring
- Rubber tubing pieces
- String or twisties
- Syringes (2)
- Thread, red and blue

Use these materials to construct a model of a nephron. You may want to use the ringstand for holding the nephron, the string or twisties for holding things together, and syringes with their plungers in or out to represent secretion and reabsorption, respectively. What does the funnel represent?

28 ____________________

The filter paper?

29 ____________________

The dialysis tubing?

30 ____________________

The rubber tubing?

31 ____________________

The thread?

32 ____________________

Table 17-3 Movement of materials through the kidney

Material	Structures encountered
RBC	
Glucose	
Inulin	

When you have completed the construction of the nephron, explain to your neighbors and teacher why you chose particular pieces of equipment to represent specific parts of the nephron and how these things are coordinated to function as a nephron.

Abnormal conditions of the urinary system

What part of the nephron is probably damaged if protein appears in the urine?

33 ______________________________

Explain your answer.

34 ______________________________

____

Cystitis (bladder infection) is an infection commonly caused by bacteria from the digestive tract. Why do more women than men suffer from cystitis?

35 ______________________________

What nephron process is not working normally if large quantities of very dilute urine are produced (as in diabetes insipidus, a nonsugar-related diabetes)?

36 ______________________________

Sometimes substances crystalize within the urinary system, forming kidney stones. If a round stone of 1-mm diameter formed in the pelvis of your fetal pig, would it pass through the urinary system or would it be lodged in a part of the urinary system?

37 ______________________________

Explain your answer.

38 ______________________________

____

Where would you expect to find a 3-mm stone formed in the pelvis of the fetal pig's kidney that had begun to move through the urinary system?

39 ______________________________

Explain your answer.

40 ______________________________

____

What effect do you expect this stone to have on urine production?

41 ______________________________

____

On urine flow?

42 ______________________________

Obtain a small bead and a needle and thread. Implant the bead in the pig's urinary system where you would expect to find a stone of this size obstructing urine flow. Anchor the stone in place with thread. Consult with your neighbors. Do they agree with your placement?

43 ______________________________

LABORATORY

18

Nervous system

Neurons and reflexes

Role of the nervous system

You can sense the cold of a winter day and the heat of a summer day and respond to temperature changes. When you are cold, your surface blood vessels constrict (decrease in size), which shunts blood to inner body parts, conserving your body heat. When you are hot, surface blood vessels dilate (expand), which increases blood to the body surfaces, releasing heat through your skin. In addition to these automatic responses, you respond consciously. When cold, you add layers of clothing. When hot, you wear less clothing. Living organisms can monitor environmental changes and respond to them. In animals, the nervous system senses the environmental changes, integrates sensory information, and initiates responses. In this way, the nervous system is deeply involved in the maintenance of the steady state of your internal environment.

EXPLORING

A simulation of a neuron

Working with a partner or as designated by your teacher, obtain a set of dominoes. Stand the dominoes on end in a row with each close enough so that when one falls, the rest fall. The question to explore is the following, "What factors affect whether the last domino in the row falls?" Test different factors such as path shape, whether the first domino falls, and how hard the first domino falls. What are at least two other factors that could affect the falling of the last domino?

1 ______________________________

What changes in the path's shape can you test?

2 ______________________________

Determine and test each type of factor that may affect the falling of the last domino. Describe the changes in the factor (such as different path shapes) in the second column of Table 18-1. In the third column, indicate how the factor affects the last domino.

CONCEPTS/TERMS

Functions of neurons

You have just observed that factors either cause dominoes to fall or they do not. A similar phenomenon happens in the individual cells of the nervous system, the **neurons.** Neurons are paths for traveling messages called **action potentials** or **nerve impulses.** Just as the dominoes either fall or do not fall, the action potential either travels along the neuron or it does not, there is no in between. This is the **all-or-none response** of the neuron.

Table 18-1 Will the last domino fall?

Type of factor	Changes in the factor 1st	2nd	3rd	Effect on last domino
Path shape				
How hard the first domino falls				
Whether the first domino falls				
Other factor				
Other factor				

If the first domino only wobbled but did not fall, nothing happened to the rest of the dominoes. You had to give the first domino a push of sufficient strength before the message traveled down the line of dominoes. In the neuron, the action potential occurs, or "fires," only if a specific type of change in the environment is of sufficient strength. In this simulation, the push of your finger is the change in the environment, or the **stimulus,** that triggers the domino action. The minimal stimulus required to make the first domino fall is called the **threshold stimulus.** Once the threshold is reached, does it matter in the falling of the last domino if the stimulus is stronger?

3 ______________________________

Explain your answer.

4 ______________________________

____

What is the effect on the movement of the action potential if several of the dominoes were removed or if the space between dominoes was larger than one domino's length?

5 ______________________________

EXPLORING

Neurons

At the demonstration table, examine a prepared slide of a piece of nerve that has been pulled apart. Observe it under both low and high power magnification. Also observe the cross section of the spinal cord that you may have explored in Laboratory 11. The nerves and spinal cord consist of at least two distinct types of cells. The large cells are the message-transmitting cells. The others are "helper" cells. What organelles are in the large, transmitting cells?

6 ______________________________

What are the distinctive features of the transmitting cells?

7 ______________________________

____

If these neurons (1) transmit messages along their length, (2) transmit messages only if they are intact, and (3) link parts of the body, do you think you can see all parts of these neurons?

8 ______________________________

CONCEPTS/TERMS

The structure of neurons

You should be able to detect the following features of a typical neuron:

1. **Cell body:** This is the large, expanded portion that contains a large, round nucleus with a nucleolus. It contains many mitochondria.
2. **Dendrites:** These are many highly branched, tree-like extensions from the neuron's cell body. The length of dendrites varies from very short to very long. In the intact nervous system, action potentials travel from the ends of the dendrites toward the cell body.
3. **Axon:** This is a single, thicker extension from the cell body. The length of the axons varies from short to very long. In the intact nervous system, action potentials travel away from the cell body to the tip of the axon.

EXPLORING

Simulating a reflex arc, the functional unit of the nervous system

To simulate a functioning unit of interacting neurons, a **reflex arc,** you need at least one **receptor,** two **neurons,** a **synapse** with its **transmitter,** and one **effector.** Table 18-2 lists these components, their counterparts in the domino model, and their roles in the reflex arc.

Based on the information in Table 18-2, set up the following materials to simulate a reflex arc:

Dominoes (14 for each neuron)
Rubber or golf ball
6 in ruler
Mousetrap

CAUTION
Be very careful cocking the mousetrap.

When your model is successful, the mousetrap should respond to a threshold stimulus of your finger toppling the first domino (receptor) of the sensory neuron. Describe the action of a reflex arc using your model.

Table 18-2 Components of a simulated reflex arc

Component	Model counterpart	Role in reflex arc
Receptor	First domino	Initiates the action potential when stimulated
Sensory neuron	Row of dominoes	Carries the action potential toward the spinal cord
Synapse	6-in ruler	The junction between neurons (in the model—a sensory neuron and motor neuron)
Transmitter	Ball	A chemical that carries the message from one neuron to the next
Motor neuron	Row of dominoes	Carries action potential away from the spinal cord
Effector	Mousetrap	Changes the action potential into action

9

Questions to consider

Nearest which neuron did the transmitter have to be placed for the model to work?

10

Explain your answer.

11

Did the reflex arc work in both directions (first domino to mousetrap and mousetrap to first domino) or only in one direction?

12

Explain your answer.

13

Which was faster, the falling of the dominoes or the movement of the transmitter?

14

Once you toppled the first domino, did it take any more effort on your part to spring the mousetrap?

15

How could you have stopped the action potential after it was initiated?

16

Using the components of your model, list at least three ways you could decrease the time between the application of the stimulus and the response of the effector.

17

CONCEPTS/TERMS

Reflex arcs

You observed the following features of the action potential as the reflex arc carried the message:

1. The impulse travels one way because of the structure of the synapse. This phenomenon is called **one-way transmission.** Dominoes can fall in either direction.
2. Once a stimulus is at or above threshold, no additional effort is needed to get a maximal response. This is the all-or-none response, just as in a single neuron.
3. Movement across the synapse was much slower than movement down a nerve cell. This is known as **synaptic delay.**

Structure of a simple reflex arc

Interneurons

You previously examined two microscope slides containing neurons that are necessary components of the reflex arc. Most reflexes involve additional neurons between the sensory and motor neurons. These additional neurons are called **interneurons.**

Junctions

You probably did not see synapses on the microscope slides because the neurons had been moved from their original locations in the organism and because they are too small. (The synaptic space is only about 0.02 μm across.) The electron micrograph and a drawing in Figure 18-1 (Plate 8B) show a synapse between two neurons.

A third slide at the demonstration table shows another type of junction, the **myoneural** or **neuromuscular junction.** The myoneural junction is between the end of a motor neuron's axon and the effector (responding organ). The branching tip of the axon forms junctions with several skeletal muscle cells. These axon tips also release transmitters. Explain how you think the axon-muscle junction may function.

18 ____

____

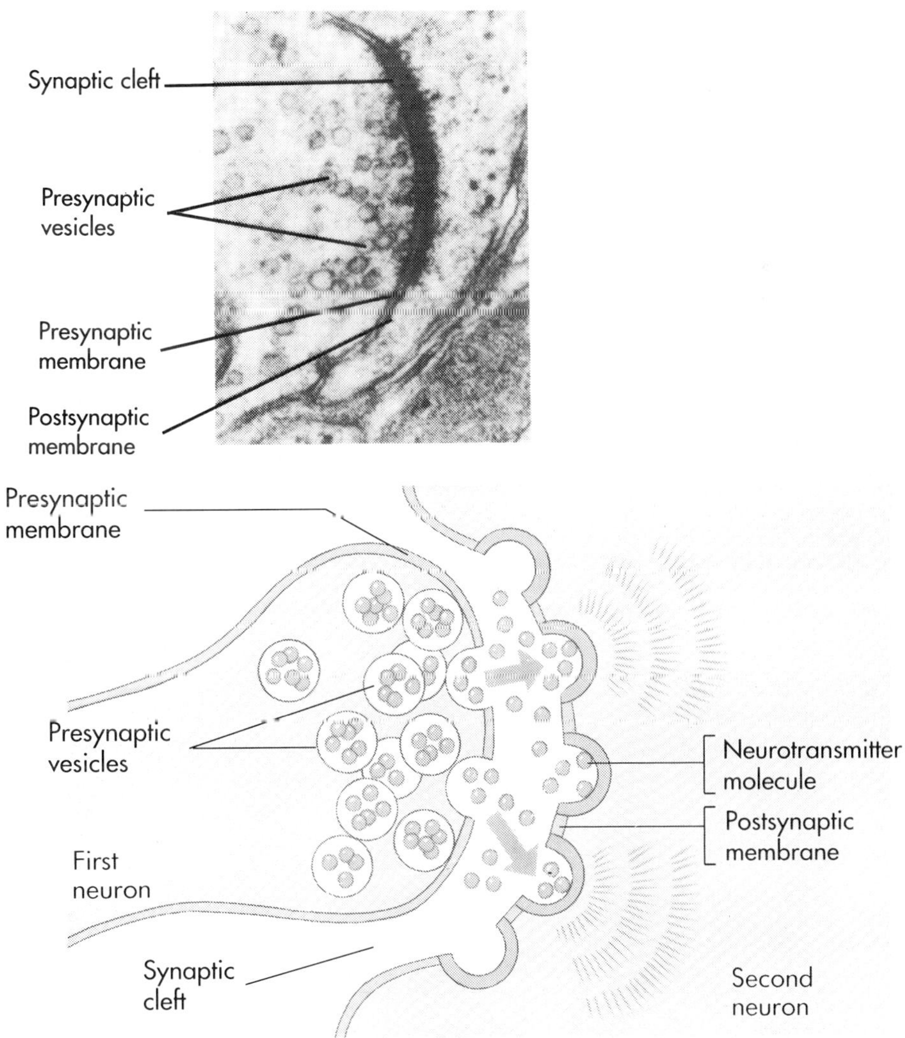

Figure 18-1
Synapse between two neurons.

Figure 18-2
Components of the reflex arc.

In the intact nervous system, a dendrite is the part of the neuron that receives the message at the synapse from the preceding neuron or from the receptor. The dendrite transmits the message toward the cell body. The axon transmits the message away from the cell body to the synaptic connection with the next neuron or to the myoneural junction with the effector.

Interaction of components, the functioning reflex arc

Now focus on the interaction of the reflex arc's components as they transmit the message from the receptor to the effector. Keeping in mind your domino simulation of a reflex arc and your observation of nerve tissue, draw your conception of the interacting components of the reflex arc (including an interneuron) in Figure 18-2. Identify components and their structure such as axons, cell bodies, dendrites, effector, motor neuron, myoneural junction, receptor, sensory neuron, synapses, and transmitters. (In future laboratories, you will explore receptors. For now, just draw the receptor as a box.)

Draw a large arrow above Figure 18-2 that indicates the direction in which the action potential travels. When you have completed Figure 18-2, explain it to your neighbor. If the two of you do not agree, seek assistance from your teacher. Complete the following statements: The end of the

19 ______________________________

contains the transmitter that moves across the

20 ______________________________

and initiates the impulse in the next neuron. The impulse cannot go the other way because the

21 ______________________________

lacks transmitter chemicals.

APPLYING

Questions to consider

Botulism (BOT-u-li-sum), a severe type of food poisoning, is caused by a bacterium that produces a toxin (poison). This toxin prevents the release of transmitter substances at myoneural junctions. It produces this toxin when growing under anaerobic conditions such as an improperly processed jar of food. Using your knowledge of the ways that the human body functions, explain how this toxin can cause death if it is consumed.

22 ______________________________

____

Some South American native hunters dip their arrows in curare (cu-RAR-ee). Curare prevents the transmitter chemical in myoneural junctions from attaching to the muscle membrane. What effect would this have on the animal?

23 ______________________________

Explain your answer.

24 ______________________________

What is a possible medical use of curare?

25 ______________________________

The black widow spider's venom causes an excessive release of transmitter chemical at the myoneural junction. What are the most dangerous results of a bite from this spider (see Plate 16B, *left*)?

26 ______________________________

A reflex arc in action: the knee-jerk

Working in pairs, obtain a reflex hammer. Find the depression on the anterior side of your leg just below the kneecap; inward from this depression is a cord-like tendon that is attached to a large muscle in your anterior thigh, to your kneecap, and to a bone in the lower leg. One partner, the subject, should sit on a tall chair or a table so that both feet are above the floor and dangling freely. While the subject sits in an upright but relaxed position, moderately tap the kneecap tendon at the depression with the hammer or the edge of your hand. What is the subject's response?

27 ______________________________

If the lower spinal cord within the backbone is damaged, this knee-jerk reflex does not occur.

Tap the tendon again while the subject interlocks the fingers and tries to pull the hands apart. What is the subject's response?

28 ______________________________

This is one of the simplest reflexes in humans. As the tapper hits the knee-cap tendon, it stretches an anterior thigh muscle. A specialized receptor within this muscle initiates action potentials when stretched. What is the effector in this reflex? In other words, what is the direct cause of the lower leg movement?

29 ______________________________

Messages from other parts of the nervous system can affect the activity of this reflex.

In Figure 18-3, draw your conception of the reflex arc responsible for the action of the lower leg when the anterior thigh muscle is stretched. Does the spinal cord play a role in this reflex?

30 ______________________________

As in Figure 18-2, identify axons, cell bodies, dendrites, effector, motor neuron, myoneural junction, receptor, sensory neuron, and synapses. (This is one reflex that does not involve interneurons.) Also include the general location of these components of the reflex arc. For example, indicate that the stretch receptor is within the anterior muscle of the thigh.

The spinal cord plays an important role in the knee-jerk reflex. It is the place where the sensory and the motor neurons synapse. In fact, all reflexes involve the spinal cord or brain. (The brain and spinal cord together are called the **central nervous system [CNS].)** The sensory neuron extends from the receptor to the CNS, even if the receptor (such as a cold temperature receptor in your toe) is located quite a distance from the CNS. When the receptor initiates the action potential in the sensory neuron, the message travels along the sensory neuron to the CNS, where it crosses the synapse to one or more other neurons. The second neuron can be a motor neuron (as in the knee-jerk reflex) or it can be an interneuron (such as an interneuron connecting parts of the CNS or an interneuron between the sensory and motor neurons). Eventually, an interneuron delivers a message to the motor neuron. The message travels along the motor neuron and crosses the myoneural junction to the effector (such as the muscle of the cold toe). Then the effector responds.

Figure 18-3
Knee-jerk reflex.

Pupillary reflexes

Continuing to work with your partner, observe reflexes of the pupil. The pupil is the dark center of the eye, which is an opening in the center of the iris (the colored part of the eye). In medical examinations, doctors frequently look through this opening to see internal parts of the eye and inspect the blood vessels at the back of the eye. The iris contains two sets of muscles: one set is arranged radially as the spokes of a wheel and the other set forms a ring-like arrangement within the iris. Observe the size of your partner's pupils under the following conditions and record your observations in Table 18-3:

Far and near vision. While your partner is looking over your shoulder at the wall across the room, measure the approximate diameter of the pupils by holding a millimeter ruler below the eyes. Ask your partner to focus on a nearby object, such as your shoulder, while you continue to observe the pupils.

Table 18-3 Pupillary reflexes

Condition	Diameter of pupil (mm) Left	Diameter of pupil (mm) Right
Focus on far object		
Focus on near object		
Before light in left eye		
With light in left eye		
Before light in right eye		
With light in right eye		
Before left eye is covered		
With left eye covered	covered	
Before right eye is covered		
With right eye covered		covered
Before conscious effort		
With conscious effort		

(The two objects on which the subject focuses should be of the same brightness.) If a change has been observed, how much time did it require?

31 ______________________________

Located just behind the pupil is the lens of the eye. The eye lens refracts, or bends, light (just as the microscope lens does), focusing the light rays on the light-sensitive retina at the back of the eye. The change in the pupil diameter helps you to see near objects clearly.

Bright and dim light. Measure pupil diameters before and after shining a small flashlight into one eye of the subject. Repeat with the other eye. As much as you can, explain the ability to see more clearly if the pupil constricts in bright light and dilates in dim light. (Think of what happens to the ability to see when you first walk from a dark room into a brightly lighted room.)

32 ______________________________

____

Covering one eye. Cover one eye. Observe the response of the pupil in the uncovered eye. Cover the other eye and observe the pupil in the uncovered eye. Propose an explanation for this response.

33 ______________________________

____

Conscious control. While looking at the far wall, try to consciously change the size of the pupils by controlling the muscles of the iris. Do not contract the muscles of the forehead or those surrounding the eye. Is there conscious control of the pupil diameter?

34 ______________________________

The size of pupils changes as the two sets of muscles in the iris of the eye contract and relax in opposition to each other. What happens to the pupil diameter when the radial muscles relax and the circular muscles contract?

35 ______________________________

What happens to the pupil diameter when the radial muscles contract and the circular muscles relax?

36 ______________________________

Each set of muscles is innervated by motor neurons in a different nerve. A third nerve—the optic nerve—also plays a role in the pupillary reflexes. It contains the sensory neurons that transmit action potentials to the brain. The action potentials are initiated when light shines on the light receptors in the retina of the eye. Draw and label your conception of these reflex arcs associated with pupillary light responses in Figure 18-4. Also indicate the environmental stimuli that cause each one of the two sets of pupillary muscles of the iris to contract. For example, connect "far vision" with the nerve that innervates the radial iris muscle. Place arrows on your drawing indicating the direction of action potential movement.

Figure 18-4
Reflex arc of pupillary reflexes.

The homeostatic function of the nervous system

Next consider the relationship of the pupillary reflexes to the maintenance of an internal steady state. Using examples from your observations to illustrate and support your statements, explain how the nervous system is involved in homeostasis.

37 ____

____

____

____

Low blood pressure in the aorta and high levels of carbon dioxide in a neck artery speed up the heart rate. High blood pressure in the aorta and low levels of carbon dioxide result in a decreased heart rate. Describe a possible mechanism by which these stimuli can affect heart rate.

38 ____

____

____

LABORATORY
19

Nervous system
Organization and receptors

PART ONE

Organization of the nervous system

Complex organization and development

The neurons of the nervous system are highly organized. A highly organized structure is necessary to integrate sensory input and initiate an appropriate response. Scientists are just beginning to unravel the complex structural and functional relationships of the nervous system.

In the early prenatal development of vertebrates, the nervous system develops from a layer of cells that forms the dorsal surface of the embryo. Folds at the edge of a flat plate of cells roll up until they meet in the midline, forming the neural tube. The primary parts of the vertebrate nervous system develop from this tube. At the future head end of the organism, the tube expands, forming the brain. The remainder of the tube forms the spinal cord. Together, these two structures are called the **central nervous system (CNS).** Long, fiber-like structures grow out from or into the central nervous system. Collectively, these structures, the **nerves,** are called the **peripheral nervous system (PNS).**

EXPLORING

The central nervous system

The brain

Working with your partner or as designated by your teacher, obtain the following materials:

Dissecting tools
Fetal pig
1 sheep brain, whole
1 sheep brain, saggitally sectioned
1 string, about 30 cm
1 meter stick

Figure 19-1 (Plate 9A) shows some of the larger brain areas of several animals. In humans, the **cerebrum** is the area of the brain involved with thinking, memory, sensory perception, and other higher functions. Using Figure 19-1 as a reference, locate the cerebrum on the whole sheep brain. Describe the structure of its surface and its number of sections **(cerebral hemispheres).**

1 ______________________________

____

Measure the length of the sheep's cerebrum and compare it to the total brain length. How does the proportion of the cerebrum's length to the brain's length in sheep compare to the organisms in Figure 19-1?

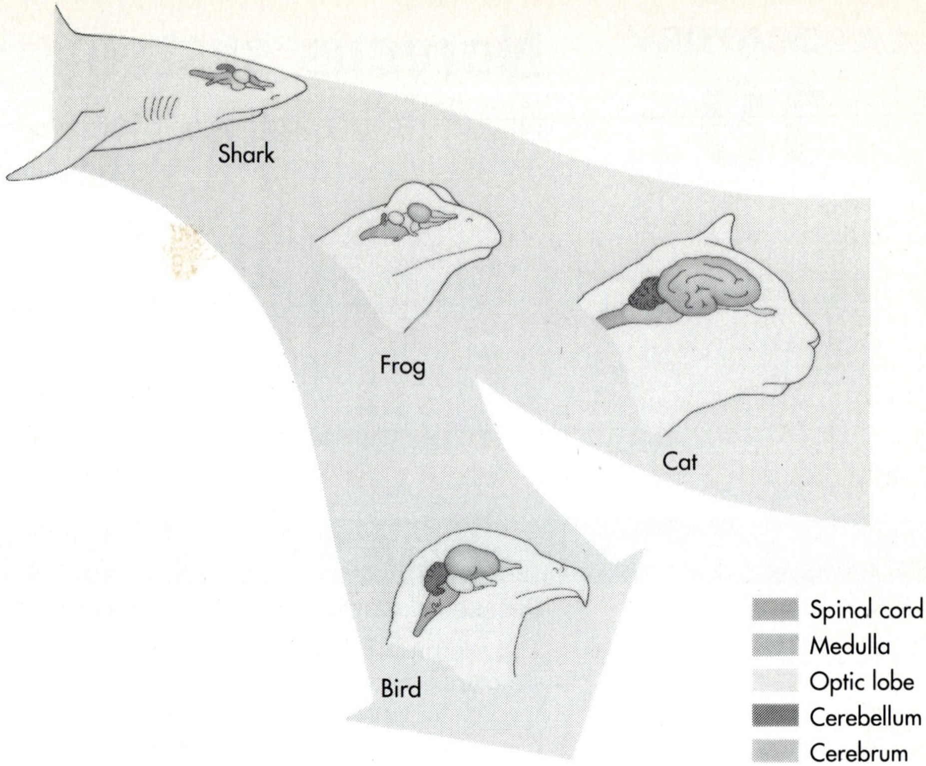

Figure 19-1
Relative sizes of different areas in the brain of vertebrates.

2 ______________________________

____

Where do you rank humans in this cerebrum/brain proportion compared to these other vertebrates?

3 ______________________________

What is a possible advantage of the ridges and valleys structure of the cerebral surface layer?

4 ______________________________

(Do you remember the primary advantage of the villi in the intestine of the digestive system?)

Examine a cross section of the sheep cerebrum at the demonstration table. Does the cerebrum vary in texture and color or is it relatively consistent?

5 ______________________________

Describe the cross section.

6 ______________________________

____

____

Many pathways connect parts within and between each cerebral hemisphere. (Although the cerebral hemispheres are connected, some functions are associated with only one hemisphere.) Gently separate the two cerebral hemispheres at the groove until you can see the base between them. Also find this area in the saggitally sectioned brain. This area is called the **corpus callosum.** What is a possible function of this structure?

7 ______________________________

The color of this area is due to fat containing helper cells that surround the dendrites and axons of the action potential–transmitting neurons.

Look for spaces ventral to the corpus callosum and in other brain locations. Describe the shapes and locations of the spaces.

8 ______________________________

____

Considering the early development of the central nervous system, what is the possible origin of these spaces, the **ventricles?**

9 ______________________________

Posterior to the cerebrum is the **cerebellum,** which coordinates the subconscious activity of skeletal muscles. Describe its surface.

10 ______________________________

Describe the color patterns in the saggitally sectioned cerebellum.

11 ______________________________

____

Examine other parts of the whole sheep brain and the saggitally sectioned brain. Is there a consistent arrangement of color within the brain?

12 ______________________________

Describe the colors and their arrangement.

13 ______________________________

____

Continue exploring the saggital section. The walls surrounding the space (third ventricle) below the "base" of the cerebrum is the **thalamus** (THAL-a-mus). Are the two parts of the thalamus connected?

14 ______________________________

Considering its color, central location, and closeness to the cerebrum, what is a possible function of the thalamus?

15 ______________________________

The most ventral part of this area is called the **hypothalamus.** It coordinates many nervous and endocrine system interactions. Does your specimen have a bulb-like structure hanging on a stalk from the hypothalamus?

16 ______________________________

This is the pituitary gland, which releases chemical messengers called *hormones.* Your specimen may lack the pituitary because it is frequently broken off when the brain is removed from the skull. Chemicals from the hypothalamus regulate the production and release of hormones from the pituitary.

The **medulla oblongata** is the most posterior part of the brain. The medulla contains tracts (bundles of nerve fibers) that connect the spinal cord with other brain parts. It also controls the rate of breathing and heart contractions. What is its color pattern?

17 ______________________________

Just ventral to the cerebellum and anterior to the medulla is a thick, "crosswise" (and longitudinal) bundle of fibers called the **pons.** Does the pons contain nerve tracts?

18 ______________________________

What evidence supports your answer?

19 ______________________________

____

If the pons has tracts, what parts of the brain does it connect?

20 ______________________________

The medulla, pons, and **midbrain** (a small area located between the pons and thalamus) together are called the **brainstem.**

Compare parts of the sheep brain with corresponding parts of the other vertebrate brains shown in Figure 19-1 (Plate 9A). What are some of the most prominent differences among these brains?

21 ______________________________

____

If you have time, carefully remove the crown of the fetal pig's skull with sturdy scissors. What tissues protect the brain?

22 ______________________________

Describe the layers of materials **(meninges)** between the brain and the skull.

23 ______________________________

Based on the structure of the meninges, what is one of its likely functions?

24 ______________________________

The spinal cord

Now explore the remainder of the neural tube, the **spinal cord.**

At the demonstration table, examine a bone from the vertebral column and a vertebrate skeleton. Where do you think the spinal cord is located in reference to the vertebral column?

25 ______________________________

CAUTION

You will be cutting through bone to expose the spinal cord. Take particular care to make these cuts safely.

With your finger, feel the spine of the pig to identify the vertebrae. Remove the skin over the vertebral column in an area extending cranially from the hip bones. There are two ways to expose the spinal cord from this point. Each has its own advantages.

The first method exposes the cord in situ ("in place"). Curve the pig over a beaker on its side so that the pig's midback is easily accessible. Using a sharp single-edged razor blade or scalpel and starting with the most dorsal layer, slice thin layers of muscle and other tissues from the midregion of the back. As you remove the first few slices, you will see a row of whitish ovals in the midline. What part of the soft fetal vertebrae are you removing?

26 ______________________________

Continue slicing thin layers until you see pairs of tan ovals and you can feel a soft area between the members of the pairs. Put your thumbnails into the soft groove and pull the tan ovals apart, moving them laterally. Pull several of these adjacent pairs apart and describe what you find in the exposed area.

27 ______________________________

____

The second method removes a section of the vertebral column. With sturdy scissors, cut through the attached ribs on each side of the midback about 1 cm from the vertebral column. Cut between vertebrae anteriorly and posteriorly to free a 1 to 2 cm section of the vertebral column. What is the cylinder of soft material running through the center of the vertebrae?

28 ______________________________

From the freed section, gently remove one vertebra at a time, taking care to preserve the spinal cord and any of its attachments.

Is the surface of the spinal cord attached to the vertebrae?

29 ______________________________

Draw the relationship of the spinal cord to the vertebrae in Figure 19-2. Identify all parts of your drawing.

Describe any color patterns that you observe in a freshly cut cross section of the spinal cord.

30 ______________________________

____

Compare the color pattern of the spinal cord with the brain's. Describe any general differences.

31 ______________________________

____

What structures do you think make up the different colored areas?

32 ______________________________

What is the evidence that the spinal cord developed from a tube?

33 ______________________________

Figure 19-2
Spatial relationship between vertebra and the spinal cord.

CONCEPTS/TERMS

Integrating information about the central nervous system

In the fresh nervous system, accumulation of neuron cell bodies appear gray. In areas consisting only of neuron processes (dendrites and axons), the areas appear creamy white because many of the processes have a fatty covering (myelin sheath) formed from helper cells. The common terms *gray matter* and *white matter* originated from these color differences. Gray matter nonscientifically refers to the intellectual function that results from the integrating activity of many synaptic connections within the gray clump of cell bodies. These color differences may be somewhat obscured in the preserved pig spinal cord and the sheep brain. The pinkish-gray, H-shaped internal area of the spinal cord is similar in color to the surface layers **(cortex)** of the cerebrum and cerebellum. You may also find other smaller accumulations of cell bodies in the inner portions of the cerebrum, thalamus, midbrain pons, and medulla oblongata. In the brain, these smaller gray areas are called **nuclei.** When cell bodies of neurons accumulate outside the central nervous system, the clumps are called **ganglia.** The white matter of the brain includes the major portions of the inner cerebrum, corpus callosum, pons, medulla, and thalamus. It connects all parts of the brain and the brain and spinal cord.

All voluntary actions are controlled by the cerebrum, whereas involuntary actions are usually controlled by the cerebellum, pons, medulla oblongata, and spinal cord.

APPLYING

The effects of alcohol on the CNS

Alcohol first affects the most cephalic part of the central nervous system; the effects then proceed caudally. Rank the following functions in the order that you expect alcohol to affect them.

_______ Breathing and heartbeat
_______ Thought, imagination, vision
_______ Knee-jerk reflex
_______ Balance and muscle coordination

At what point in this progression do you think alcohol causes death?

34 ______________________________

On what information did you base your answer?

35 ______________________________

EXPLORING

The peripheral nervous system

The cranial nerves

Return to the ventral side of the sheep brain. Locate string-like fibers attached to various parts of the brain, particularly the brainstem. Include anteriorly projecting flat ones and stubs of some that come together in an "X." Are these fibers found in pairs or singly?

36 ______________________________

As the brain is removed from the organism, many of these fibers are broken. How many remain on your specimen?

37 ______________________________

Examine the human brain model and charts that are available in the laboratory. Which brain area has the greatest number of fibers (nerves)?

38 ______________________________

The nerves attached to the brain are called **cranial nerves.**

The spinal nerves

Return to the foreleg of the pig on which you explored the branching circulatory vessels. Examine the tough, cord-like nerves of the foreleg. Follow them as far as you can into the chest of the pig. What areas of the body do these nerves link?

39 ______________________________

Return to the spinal cord of the pig and the **spinal nerves,** which extend laterally from the spinal cord. (If you did not indicate these spinal nerves on Figure 19-2, add them to your drawing.) Follow a few of these nerves as far as you can. If you can follow the spinal nerves of the thoracic region far enough, you will find that some of them innervate the forelimbs. Does each nerve have the same diameter from beginning to end? Describe one of these nerves as it extends 1 cm or more from the spinal cord.

40 ______________________________

____

CONCEPTS/TERMS

Integrating information about the peripheral nervous system

Peripheral nervous system

Together the cranial and spinal nerves form the peripheral nervous system (PNS). These nerves contain the neuronal processes that extend between the central nervous system and the rest of the organism. Humans have 12 pairs of nerves that carry messages to and from the brain and 31 pairs of spinal nerves that carry messages to and from the spinal cord. Each nerve consists of many neuronal processes. A nerve is similar to a telephone cable that contains many individual wires.

Both sensory and motor neuronal processes are packed into all of the spinal nerves and into some of the cranial nerves. One example of a cranial nerve that contains both sensory and motor neuronal processes is the vagus (cranial nerve X), which innervates organs of the thoracic and abdominal cavities, such as the heart and the intestines. Other cranial nerves carry primarily only one kind of neuronal process, either sensory or motor. Two cranial nerves that contain only sensory neuronal processes are the olfactory nerve (I) from the nose and the optic (II) from the eyes. The oculomotor nerve (III) is primarily a motor cranial nerve. This nerve innervates four of the six muscles that move each eyeball, the muscles that raise the upper eyelid, and the circular muscles of the iris.

Although the sensory and motor neurons make up the spinal nerves, they separate into two "roots" close to the spinal cord. Figure 19-3 and Plates 10A and 10B shows spinal nerve roots extending from the sides of the spinal cord. The sensory neurons enter the dorsal section of the spinal cord through the **dorsal root** of the spinal nerve. From the receptor, the sensory neuron's dendrites extend through the spinal nerve to the sensory neuron's cell body in a ganglion on the dorsal root. This ganglion is called the **dorsal root ganglion.** The short axon of the sensory neuron extends from the dorsal root ganglion into the **dorsal horn** of the cord's H-shaped gray matter. The cell body of the motor neuron is in the **ventral horn** of the spinal cord's gray matter. The short dendrites of the motor neuron synapse with the sensory axon or with interneurons that synapse with the sensory neuron. From the motor neuron's cell body, the motor axon passes through the **ventral root** of the spinal nerve. The two roots of the spinal nerve join. From this junction to the periphery, both motor axons and sensory dendrites form the spinal nerve.

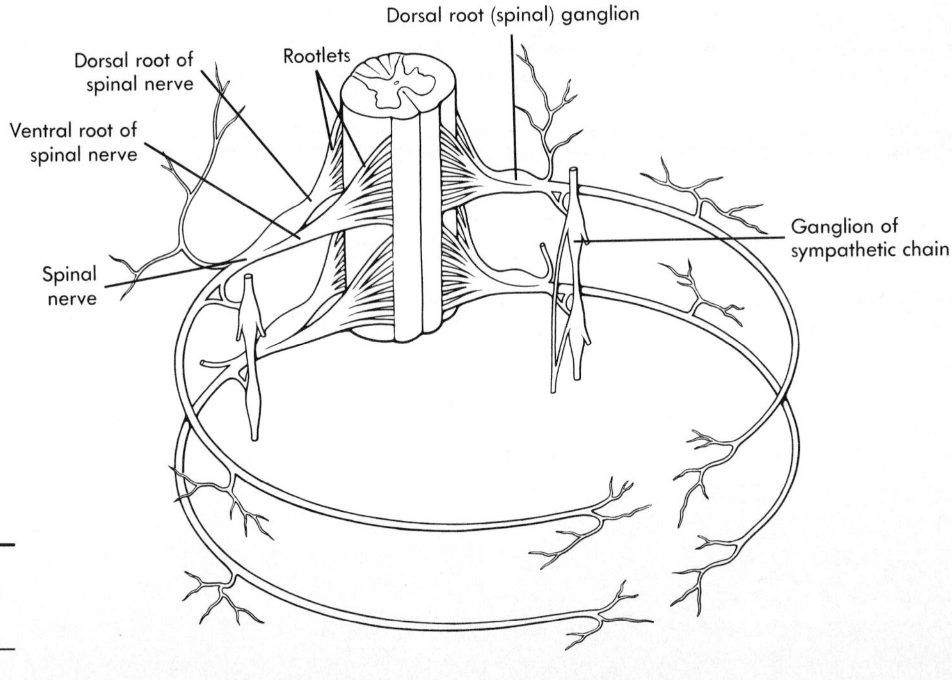

Figure 19-3
Typical spinal nerves of the thorax.

Frequently the sensory neuron of a reflex arc is called the **afferent** ("to bear inward") neuron, and the motor neuron is called the **efferent** ("to bear outward") neuron.

The somatic and autonomic nervous systems

The motor branch of the reflex arc with one motor neuron from the central nervous system to the effector is only one of two motor arrangements in the body. This one neuron efferent branch is called the **voluntary** or **somatic nervous system.** The axons of these single motor neurons innervate skeletal muscles, which can be voluntarily controlled.

In reflexes that involve other effectors such as glands, smooth muscles, and cardiac muscles, the motor branch is composed of two chains, each consisting of two neurons. These two sets of motor neurons are in different nerves. One is in nerves from either the brain or the lower spinal cord; it is called the **parasympathetic nervous system.** The other set runs in thoracic or lumbar (small of the back) spinal nerves; it is called the **sympathetic nervous system.** Together these efferent systems are called the **autonomic nervous system.**

When nerve impulses travel to the effector through these two parts of the autonomic nervous system, they initiate opposite effector responses by releasing antagonistic neurotransmitters.

APPLYING

Abnormal conditions of the central and peripheral nervous systems

Sometimes a bacterium or other microorganism infects the meninges (soft covering) of the central nervous system. As in an infected cut, white blood cells migrate to the area, the blood vessels dilate, and the tissues swell. Based on the anatomical relationship between the central nervous system and the vertebrae and the skull, give a likely explanation for the frequently quick and deadly results of meningitis if it is not promptly treated with the appropriate drugs.

41 ____

____

Keeping in mind the types of neurons in spinal nerves, describe your expectations of the effect of an injury that severs the dorsal roots of spinal nerves innervating the leg of an individual.

42 ____

____

PART TWO Receptors: sensors of change

Receptors and stimuli

Receptors are the first component of the reflex arc. Each responds to a specific kind of change (stimulus) in its environment by initiating action potentials in sensory neurons. Some respond to an external environmental stimulus, others to an internal body stimulus, and others to a change in the body's position. Receptors can vary from very simple structures such as the dendritic tips of a sensory neuron to the very complex, many-celled structure of the eye, ear, or equilibrium receptors.

EXPLORING

Taste receptors of the tongue

Working in pairs, obtain the following materials:

- 8 cotton-tipped applicator sticks (or 4 double-tipped applicator sticks)
- 4 small paper cups, each containing about 2 ml of one of the following solutions: sugar (5%), vinegar, quinine sulphate, salt.

Table 19-1 Taste receptor distribution

Solution	Tongue locations: Back	Middle	Sides	Tip
Salt (salty)				
Sugar (sweet)				
Vinegar (sour)				
Quinine (bitter)				

CAUTION

Treat all materials that have encountered body fluids as potential biohazards. Immediately after using, place all applicators and other contaminated materials in biohazard containers as directed by your teacher.

The subject should rinse the mouth with distilled water and dry the surface of the tongue with a paper towel. (Place the towel in the proper disposal container.) Dip the applicator into the salt solution and remove excess fluid by pressing the applicator against the side of the cup. Apply the salt solution to the tongue's surface at the back, middle, sides, and tip. In Table 19-1, identify with a + those areas where the subject detects salt.

Using the same procedure, determine the location of taste receptors for sugar (using only the 5% sugar concentration), vinegar, and quinine.

Are all areas of the tongue equally sensitive to each substance?

43 ______

Explain your answer using examples from your taste explorations.

44 ______

______

Obtain 2 ml of 1%, 2%, and 3% sugar solutions and three more cotton-tip applicators. Without letting the subject know which solution is being used, test the less-concentrated sugar solutions on the sugar-tasting area of the tongue. What are the results?

45 ______

Change places with your partner and repeat the taste receptor tests. Remember to properly dispose of used materials.

What is the least concentrated sugar solution you can detect?

46 ______

What is the least concentrated solution your classmates can detect?

47 ______

Try to explain the sugar concentration results.

48 ______

______

Apply dry sugar to the sugar-tasting parts of the subject's dry tongue. What are the results?

49 ______________________________

Let the sugar dissolve in the fluid of the mouth. Can the subject taste the sugar now? From these results, what would you say is a necessary requirement for a substance to stimulate taste receptors on the tongue?

50 ______________________________

CONCEPTS/TERMS

Chemoreceptors

Taste receptors detect four general types of chemicals: sweet, sour, bitter, and salty. Each receptor responds to one of these chemical types more readily than the others. As you observed, the taste receptors are not equally distributed on the surface of the tongue. The chemoreceptors are stimulated only when the chemical is in solution. These chemoreceptors are organized into structures called *taste buds* that are embedded in small projections of the tongue called *papillae*. As with other receptors, chemoreceptors require a stimulus of a certain strength, the threshold stimulus, before they initiate the action potential in the sensory neuron. The concentration that serves as the threshold stimulus varies with the type of substance and the individual.

What other areas of the body have chemoreceptors?

51 ______________________________

EXPLORING

Touch receptors

Continuing to work in pairs, obtain a compass (for making circles) and a millimeter ruler.

The subject should sit with one arm resting comfortably on a table, palm down, and *the with eyes closed.* Adjust the compass so that the points are about 2 cm apart. Gently press two points simultaneously against the skin of the forearm *or* only one point to check the subject's perception of the number of touching points. The question to determine is the following: "What is the shortest distance between the two points at which the subject still detects a two-point touch?" Bring the compass points closer together as you continue to ask the subject to discriminate between two-point and one-point touching. Record the shortest distance between the points at which the subject can still detect two points in Table 19-2 and in the class data table on the chalkboard. Repeat the testing on the tip of the middle finger and record your results.

Decide on two other body areas to test for two-point discrimination and add the locations to the fourth and fifth columns of Table 19-2. Predict whether these areas will be more or less sensitive than the arm and the fingertip.

52 Area 3 ______________________________

Area 4 ______________________________

Test these two areas and record the results.

What are some advantages and disadvantages of having differences in two-point touch discrimination on the body surface?

53 ______________________________

Table 19-2 Two-point touch sensing

Subject	Shortest distance for two-point sensing (mm)			
	Arm	Fingertip	Area 3 ____	Area 4 ____
Partner 1				
Partner 2				

How do you and your partner compare?

54 ______________________________

What is the longest and the shortest two-point discrimination distance for class members' arms? In other words, what is the range of class responses for this discrimination?

55 ______________________________

Try to explain the difference in two-point touch discrimination in the same body area of different classmates.

56 ______________________________

CONCEPTS/TERMS

Mechanoreceptors

Touch receptors are mechanoreceptors responding to pressure that changes their shape. Human touch receptors are modified endings of the dendrites of sensory neurons. One type (Meissner's corpuscles) is located more deeply in the skin than the other type (Merkel's disks) (Figure 19-4). As with the chemoreceptors, they are not equally distributed in the skin. They also require a stimulus of a specific minimum strength to fire the sensory neuron.

EXPLORING

Effects of balance receptors

Balance reflexes in the frog

Obtain the following items:
- 1 live frog
- 1 movable flat surface
- 1 clear container to cover the frog

Keeping the skin of the frog moist, place the frog on the flat surface, and cover it with the container to prevent it from jumping. Let the frog settle into a sitting position, and then observe the frog as you turn its resting surface. First turn it slowly and then rapidly in a clockwise direction. Then turn it in a counterclockwise direction. Tilt the front of the frog surface up and down. Tilt the surface to each side. Test other types of movement as well. Record the frog's responses to each of these movements in Table 19-3.

A model of balance receptors

Obtain a disposable cup that has threads through the side of the cup. Add water until the threads are covered. Add two or three small beads to the cup.

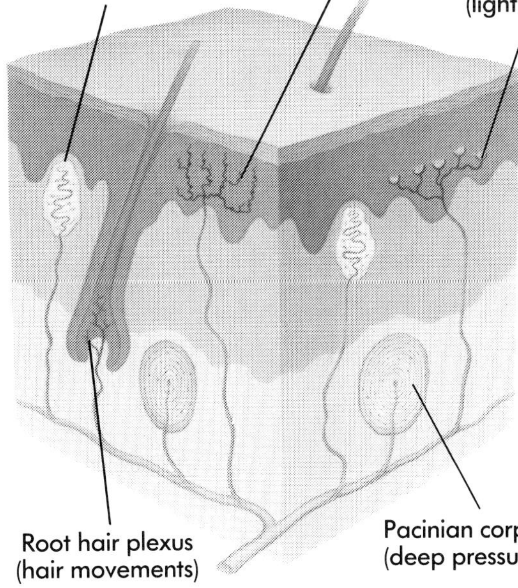

Figure 19-4
Touch receptors in the skin.

Table 19-3 Frog's responses to movement

Type of movement	Frog's response
CLOCKWISE Slow	
Fast	
COUNTERCLOCKWISE Slow	
Fast	
TILT TO FRONT Up	
Down	
TILT TO SIDE Right	
Left	
OTHER	

Anchor the cup with clay to the center of a spinnable disk such as a lazy-susan. Rapidly spin the cup. Describe the direction of water flow compared to the direction of cup rotation both on the acceleration of the cup and after the rotation is at full speed.

57 ______________________________

____

Describe the bending of the threads in regard to the direction of fluid flow and cup rotation.

58 ______________________________

____

Slow the rotation speed. Describe the effects of this change in rotation speed on the water direction and the bending of the threads.

59 ______________________________

What happens when you stop the rotating movement?

60 ______________________________

CONCEPTS/TERMS

Activity of balance receptors

The mechanoreceptors that detect motion are ciliated cells. When the cilia are bent in one direction, they stretch the cell membrane. This change initiates additional nerve impulses in the sensory neuron. When the cilia are bent in the opposite direction, the nerve impulses initiated in the sensory neuron decrease.

The sensory neurons transmit these nerve impulses to the brain, usually through the vestibulocochlear cranial (VIII) nerve to a nucleus within the medulla oblongata. From this nucleus, neuronal paths run to other areas of the central nervous system. Some of these areas are the spinal cord, cerebellum, cerebral cortex, and nuclei associated with motor neurons of the eye muscles. In addition, receptors in skeletal muscles and the eye send information to integrating brain centers.

APPLYING

Experiment: balance responses in humans

Predicting balance responses in humans

The balance receptors for humans are encased within the bones of the skull at the inner ear. Using Figure 19-5 and the models and charts available in the laboratory, explore the anatomy of the balance organs: the utricle, saccule, and semicircular canals. Within the balance organs are ciliated cells called **hair cells.**

The **saccule** and **utricle** provide information about the position of the head relative to gravity and change in speed. The cilia of their cells extend into a gelatin-like mass that contains small "stones" (otoliths). The stones shift position in response to gravitational force, triggering a response in the ciliated cells.

The three **semicircular canals** provide information about movements of the head. Within the base (ampulla) of each canal are ridges with ciliated cells embedded in a gelatin mass. When the fluid in the canal moves, the gelatin moves, the cilia bend, and nerve impulses are increased or decreased in the connecting sensory neurons.

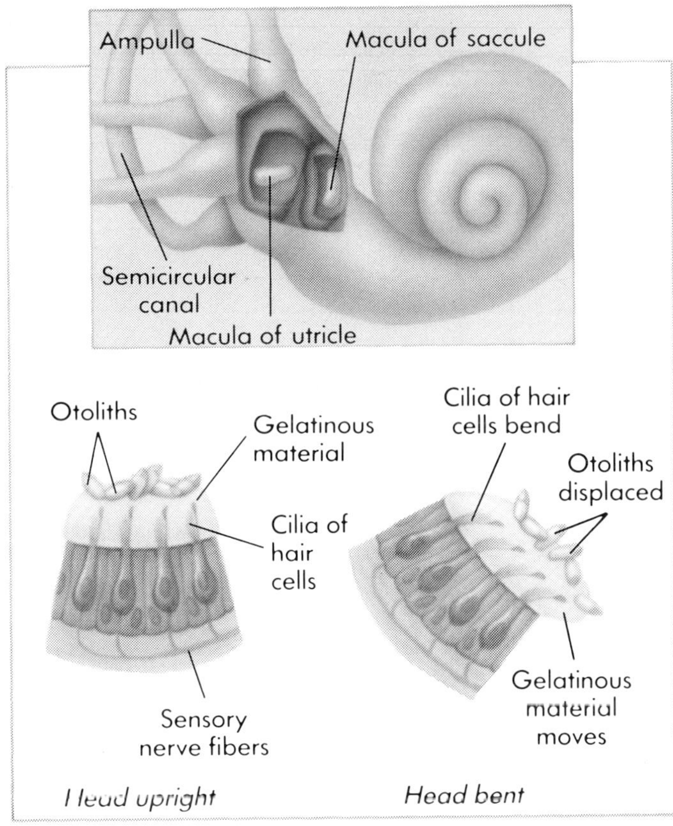

Figure 19-5
Balance organs in humans.

With the information you have about the complex balance process in animals, form alternative hypotheses about the effects on eye movement and body position when you tilt and/or rotate humans. What are your hypotheses?

61 ______________________________

____

Testing predictions

Working in small groups or as designated by your teacher, design a test of the hypotheses. What are your predictions?

62 ______________________________

____

CAUTION
Anyone who volunteers to be a subject should be in good physical health. The subject should have no illness or injury that could be aggravated by serving as a subject in this experiment.

A swiveling stool and a tiltable chair are available if you want to use them in your experiment. After checking the design of your experiment with your teacher, test your predictions.

Attach the written report of your experiment to this manual. Include the hypotheses, predictions, procedure, results, discussion of results, and the conclusion. In the discussion section, include information about the role of the receptors and the nervous system in the results and indicate how the results support or refute the predictions and hypotheses. In the conclusion, briefly state whether the experimental results support or refute the hypotheses.

LABORATORY 20

Musculoskeletal system

Roles of the skeletal and muscular systems

One very important role of skeletons is to maintain our basic form against the force of gravity. Bones also protect soft tissue, produce blood cells, and are important in movement. Skeletal muscles and bones interact, moving us from one chair to another or propelling atheletes up and over the high-jump bar.

Muscle cells contract whether they are smooth, cardiac, or skeletal muscle. If contracting muscles are sheets of smooth muscle surrounding the digestive tract, they mix and move the contents of the digestive cavity. If contracting muscles are the mass of cardiac muscle surrounding the blood-filled spaces of the heart, they propel the blood through and out of the heart. If contracting muscles are longitudinally arranged bundles of skeletal muscle between two bones, they pull the bones closer together.

Today you will focus on the skeletal system and the muscles that move its parts. These interacting systems are called the **musculoskeletal system.**

EXPLORING

The structure of long bones

At the demonstration table, explore the preserved or fresh long bones. When helpful, use a magnifying lens.

Figure 20 1 presents a long bone with a section removed showing some of the interior features. Describe the texture, color, surfaces, and other features of the lettered bone areas.

CAUTION
When material is at room temperature, harmful bacteria may grow on or in its cells. Therefore handle these materials with gloved hands.
Later you will explore bones that are fastened by bolts and other hardware. If these become loose, you could be injured and the specimen damaged. If you find a loose fastener, report it to your teacher.

1 a. ______________________________

____

____

b. ______________________________

____

____

c. ______________________________

____

____

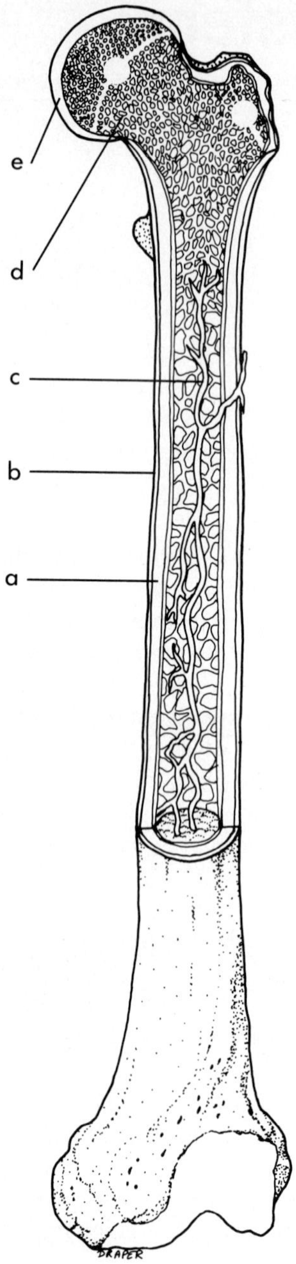

Figure 20-1
Long bone.

d. ______________________________

____

____

e. ______________________________

____

____

Thoroughly examine the bone surfaces. What evidence do you find that blood vessels and nerves have access to the bone-forming cells and the bone interior?

2 ______________________________

What evidence do you find of other tissues attaching to the bone?

3 ______________________________

Add some of these attachment sites to the drawing in Figure 20-1.

Which lettered areas of the bone have the structure that can support the body against the force of gravity?

4 ______________________________

What structural characteristics are important to the support function of bone?

5 ______________________________

Which lettered areas of the bone have the structure that can manufacture blood cells?

6 ______________________________

What structural characteristics do you think are important to blood cell production?

7 ______________________________

At the demonstration table, observe the microscope slides with the sections of bone taken from the areas identified by letters *a* and *d* in Figure 20-1. (You may want to review your explorations of bone tissue from Laboratory 11.)

Also examine three long bones of the chicken wing. The first bone (A) has been soaked in an acid solution for several days. Acid removes calcium and other minerals from the bone. The second bone (B) has been heated at high temperatures, which removes protein. Describe these two bones and compare each one to the untreated chicken bone (C).

8 ______________________________

____

____

____

CONCEPTS/TERMS

General structure of bones

The cylindrical shaft of the long bone is composed of the outer, solid-like **compact bone** and a cavity. The cavity is filled with stored fat called **yellow bone marrow** (Figure 20-1, *c*). The cylindrical construction with a central cavity gives this bone greater strength for the same length and mass than a solid, rod-like construction.

In compact bone, immature **osteocytes** ("bone cells") produce an intercellular protein matrix. The cells deposit calcium in this collagen matrix. You observed the flexibility of the acid-treated chicken bone, which lacked calcium, and the brittleness of high temperature–treated bone, which lacked protein. What do you infer from these observations about the role of calcium in bones?

9 ______________________________

What do you infer about the role of protein in bones?

10 ______________________________

The ends of the long bone are primarily composed of bone material interspersed with many small cavities that contain red bone marrow. The red marrow is the site of blood cell production. This **spongy bone** is covered with a thin layer of compact bone. A layer of cartilage covers surfaces that contact, or articulate, with other bones.

The **periosteum** (pear-e-OS-te-um), the tightly adhering outer layer (Figure 20-1, *b*), covers the bone except at the articulating surfaces. This layer is continuous with the **ligaments** that connect bones to bones and with the **tendons** that connect muscles to bone.

EXPLORING

The skeleton

Obtain access to skeletons of human and four-legged vertebrates and/or charts that are available in the laboratory. Explore these assembled bones that set the basic form of these animals. What are the common names for the two groups of interacting bones forming the central axis of the skeleton?

11 ______________________________

Using the terms *ventral* and *dorsal*, describe the direction of the curves of the column of stacked bones in the central axis in the cranial to caudal order.

12 ______________________________

What other groups of bones are associated with this central axis?

13 ______________________________

Turn your attention to the general shapes of individual bones throughout the skeleton. In Table 20-1, list at least three locations where you find bones of the following shapes: long, short (as wide as they are long), flat, and irregular (bones that do not fit into any of the above categories). Also indicate how much flexibility there is in the corresponding areas of the body that are supported by these four types of bones. An example of this flexibility and type of bone relationship is the highly flexible wrist that is composed of short bones.

CONCEPTS/TERMS

The skeleton

The bones that form the skull and the vertebral column with its attached thoracic rib cage are called the **axial skeleton.** The bones that form the arms and legs and the articulating bones of the shoulder and hip girdles are called the **appendicular skeleton.**

Table 20-1 Location of differently shaped bones and their flexibility relationship

Bone shape	Location	Flexibility of the body area
Long		
Short		
Flat		
Irregular		

EXPLORING

Joints

Focus your attention on the joints as you continue to examine the skeleton. **Joints** are the surface areas where two or more bones articulate. While you are examining the joints of the skeleton, move your own corresponding joints to determine the limits of their movements. By looking at the structures of the articulating bone surfaces and associating them with the movability of your joints, rank at least 10 joints based on their flexibility. Record this ranking (1 being immovable to 5 being freely movable) in the second column of Table 20-2.

Return to the demonstration table and examine the preserved, fresh, and model joints. For two joints, identify the location of each joint and describe its structural characteristics.

14 a. ____________________

____

b. ____________________

____

Figure 20-2 diagrams and classifies a number of different types of joints. Using this information, classify each of the joints that you have listed in Table 20-2.

CONCEPTS/TERMS

Joints

Generally, joints are classified by the degree of their flexibility. The **immovable** and **slightly movable** joints can also be classified by the material that binds the bones together. The immovable joints between bones of the skull are called **sutures.** Fibrous connective tissue tightly binds these flat skull bones. Cartilaginous connective tissue bind the ribs to the **breastbone** (sternum).

Freely movable joints are called **synovial joints.** In synovial joints, the cartilage-covered articulating surfaces of two bones are separated by a fluid-filled space. Tough, surrounding connective tissues hold the two articulating bones together.

EXPLORING

Interaction of skeletal muscles and bones

At the demonstration table, review your previous exploration of the microscopic structure of skeletal muscle. Examine longitudinal and cross sections of muscle tissue.

Table 20-2 Flexibility of joints

Joint location	Flexibility ranking	Type of joint

Figure 20-2
Joint classification.

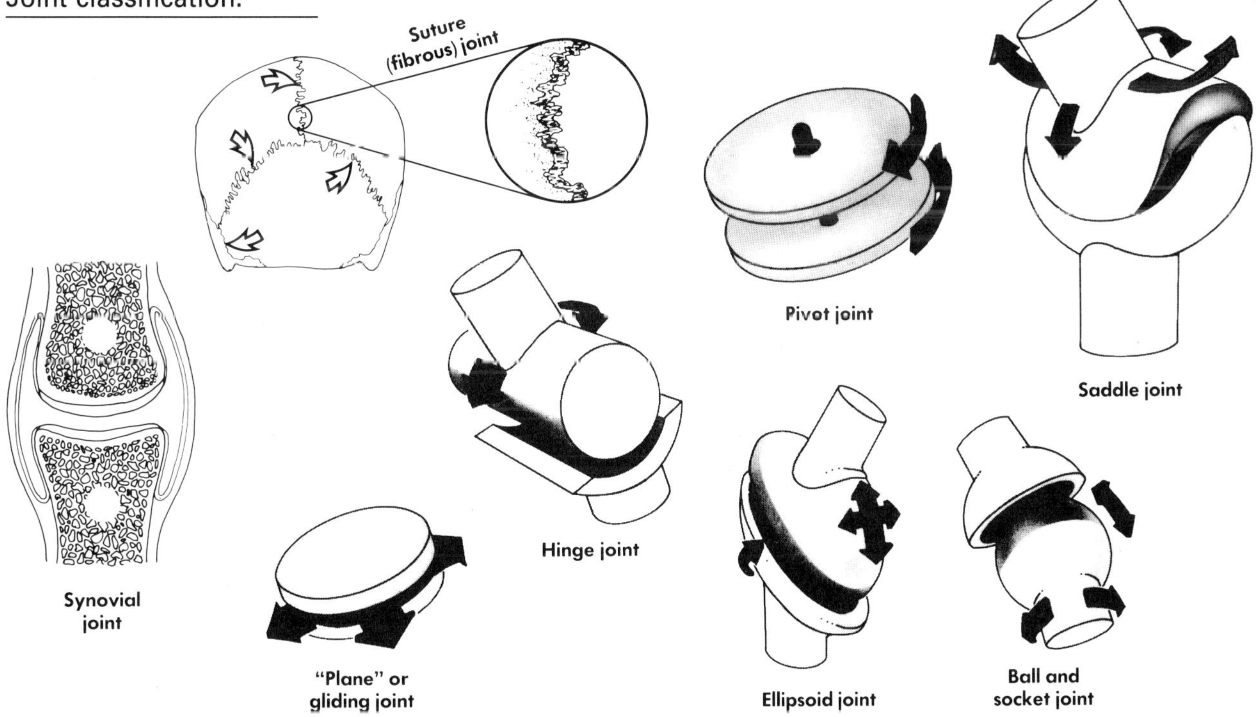

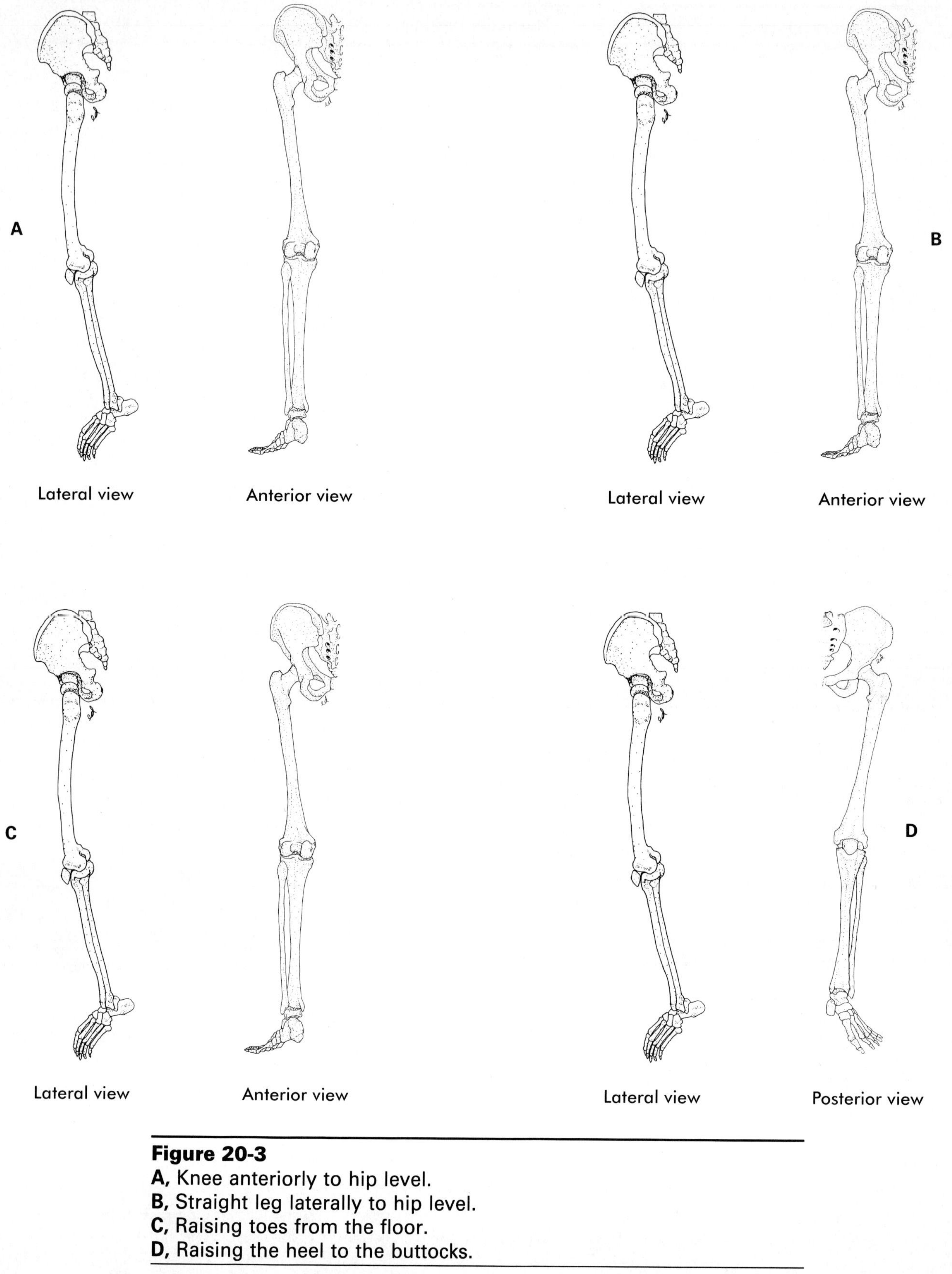

Figure 20-3
A, Knee anteriorly to hip level.
B, Straight leg laterally to hip level.
C, Raising toes from the floor.
D, Raising the heel to the buttocks.

Return to the preserved, fresh, and model joints at the demonstration table. Try to determine where muscles should be attached to move the articulating bones through the full range of motion. As you explore these materials, also examine your own joints and muscles. You should be able to feel most of your own muscles tighten as they contract.

Figure 20-3 shows diagrams of leg, hip, and foot bones from the lateral and anterior or posterior views. Draw the muscles required to bring the designated body part to each position described below. (If you have questions about the muscles, feel your muscles as you move in the described ways.)

A. Bring the knee up in front of you to hip level.
B. Lift a straightened leg out to the side to hip level.
C. Raise the toes from the floor.
D. Bring the heel to the buttocks.

CONCEPTS/TERMS

Bone and muscle interactions

To move one bone closer to another bone, muscles must be anchored to both bones. The attachment of the muscle to the stationary bone is called the **origin** of the muscle. The attachment of the muscle to the moving bone is called the **insertion** of the muscle. The particular movement of a bone depends on (1) the exact placement of the origin and insertion in respect to the location of the joint between the bones, (2) the structure of the joint, and (3) the amount of change in the length of the muscle.

Some terms associated with particular types of movements are the following:

1. **Flexion:** decreasing the angle of a joint
2. **Extension:** increasing the angle of a joint
3. **Abduction:** moving the distal end of a part away from the saggital midline (the plane separating right and left) of the body
4. **Adduction:** moving the distal end of a part toward the saggital midline of the body

APPLYING

Maintenance of upright posture

Muscle activity in the knee-jerk response

Return to the knee-jerk reflex explored in Laboratory 18. When nerve impulses traveled through the efferent (motor) neuron of the knee-jerk reflex arc to the muscles of the anterior thigh, the muscles contracted strongly, extending the lower leg rapidly. When this occurred, what do you think happened in the flexor muscles located in the posterior thigh?

15 ____

Muscles of the body are usually maintained in a partially contracted state. That is, within each muscle some of the muscle fibers contract while other muscle fibers relax. This gives the muscle the characteristic of **tone.**

The flexor muscles of the lower leg in the posterior thigh are also in this partially contracted state at the initiation of the knee-jerk reflex. When the extensor muscles contract fully in the knee-jerk reflex, they work against the partially contracted fibers in the flexor muscles. The nervous system does not just transmit impulses through the sensory and motor branches of the knee-jerk reflex arc. When the stretch receptor's message arrives in the central nervous system, the sensory neuron sends the message to the primary motor neuron *and* to interneurons. These interneurons initiate nerve impulses through other motor neurons to the flexor muscles, causing a decrease in the contracted state of the flexors. The neuronal activity that decreases the state of muscle contraction is called **inhibition.** The muscles that move bones in opposite directions are called **antagonistic muscles.** The flexor and extensor muscles of the knee joint are antagonistic muscles. Describe the location and action of three other sets of antagonistic muscles in your body.

16 ____

____

____

Experiment: does antagonistic muscle action help maintain upright posture?

Forming hypotheses. Put together all the information you now have about bones, joints, muscles, and nervous system activities to form hypotheses and predictions about the maintenance of a standing position. Think about the activity of joints and specific sets of muscles in the lower and upper leg as you develop your hypotheses. What are your hypotheses and predictions?

17 ____

____

____

On what information are you basing your hypotheses?

18 ____

____

Techniques for recording muscle contractions. One technique for determining whether a skeletal muscle is contracting or relaxing is simply to feel the muscle. You can compare the contraction/relaxation of two muscles by putting a hand on each simultaneously. However, this method does not pick up small or deep muscle contractions.

Another more refined method uses physiological measuring equipment to pick up, amplify, and record the electrical activity associated with muscle contraction. (Review Appendix H.) This is the same technique used to produce the electrocardiogram (ECG), the record of the electrical activity of heart muscle. The record of the electrical activity of skeletal muscle is called an **electromyogram (EMG).** If the available equipment has multiple channels to include information from two muscles, you can record activity of two antagonistic muscles simultaneously. Otherwise, record from one muscle and then record from the other muscle as the subject repeats the activity.

To record skeletal muscle activity, strap or tape two flat-plate electrodes (with salty electrode gel) to the skin over the opposite ends of the muscle. With cables, connect the electrodes to the recording equipment as directed by your teacher.

CAUTION
Keep safety of subjects in mind as you design the experiment.

Experimental design, execution, and written communication. Working with a small group or as designated by your teacher, plan the experimental procedure. Present your design to the teacher for suggestions.

Conduct the experiment, analyze the data, and draw your conclusion in regard to your hypotheses. Attach a written report of your experiment to this manual.

Interacting systems in balance

Complete the partial diagram in Figure 20-4 of the interacting components as you maintain balance when you lean forward or backward. Add the appropriate muscles to the bones of the leg. Also add sensory and motor neurons in the proposed pathways to the eyes and equilibrium receptors and to the parts of the brain involved with balance and unconscious muscle coordination. This

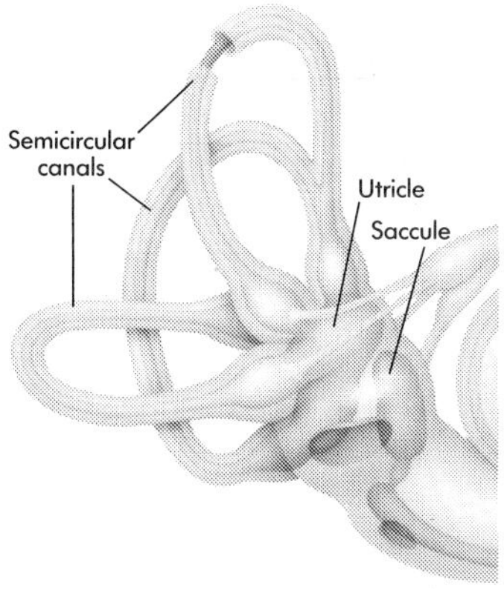

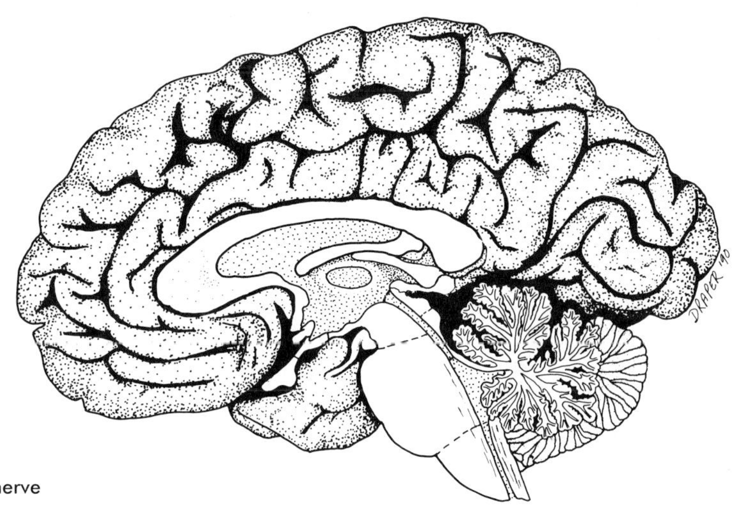

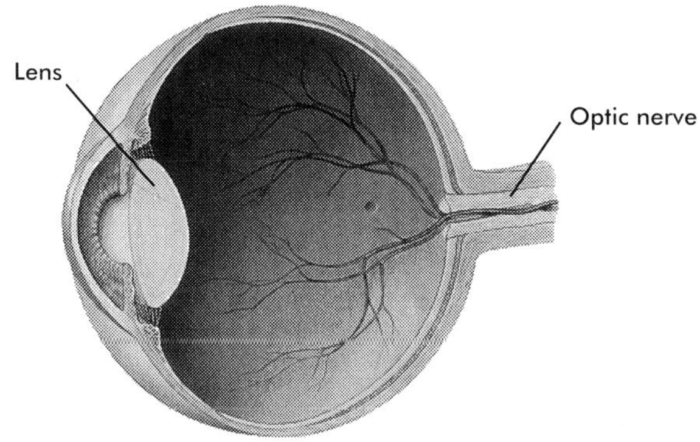

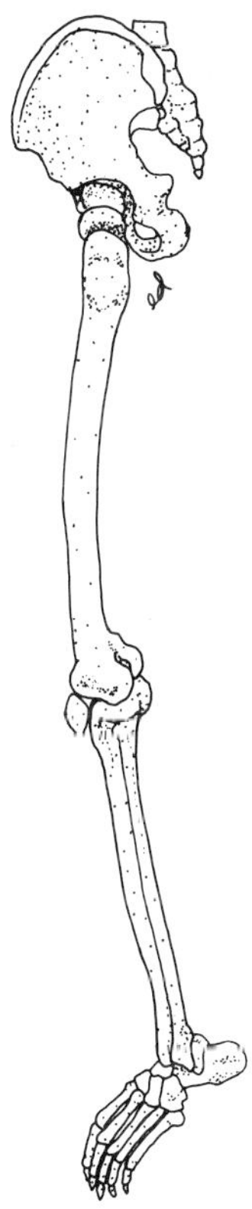

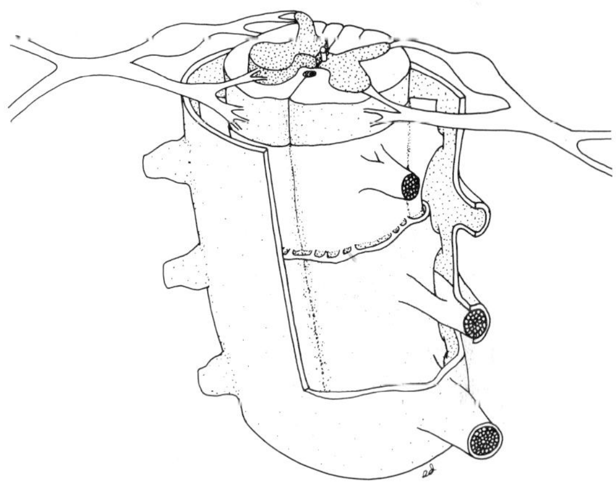

is a very complex interaction to diagram. You are not expected to complete the diagram as a neurosurgeon would. This is an opportunity to demonstrate your understanding of the homeostatic interaction of systems as the body maintains its steady state. Think about your explorations in this laboratory and those of reflexes in Laboratory 18.

Figure 20-4
Integration of systems in balance reflexes.

Abnormal conditions of the musculoskeletal system

Many factors can affect the condition of the articular cartilage in synovial joints. Among these are long-term mechanical stress from activities such as sports and dance, nerve problems associated with diabetes and hemophilia, joint instability resulting from sprains and dislocations, and drugs such as steroids. These factors increase the rate of cartilage loss. What will be the effect on the individual's ability to walk if steroids are used for a long period of time?

19 ____

Osteoporosis is a very common disease in which bone mass decreases. This process occurs in both men and women starting at about 40 years of age and continues throughout life. Spongy bone is reabsorbed more rapidly than compact bone. What are possible symptoms of this condition?

20 ______________________________

Osteomyelitis is a bone infection caused by bacteria. Based on bone structure, what are several reasons this infection is very difficult to treat?

21 ______________________________

____

How would a tear in the tendon (a strain) of the flexor muscle of the forearm affect your ability to lift a book?

22 ______________________________

Explain your answer.

23 ______________________________

____

LABORATORY 21

Reproductive system

Production of gametes and offspring

Role of reproduction

Endangered species are dying faster than they are replacing themselves with offspring. Other species, such as humans, are very adept at reproducing themselves and are producing offspring more rapidly than other members of the species are dying. For a species to survive, each generation must produce the next generation.

In most organisms, whether a simple alga, a pine tree, or a human, two mature organisms contribute genetic material that contains instructions for producing organisms that are very similar to (but not exactly like) themselves. This kind of reproduction is called **sexual reproduction.**

The genetic information contributed by each adult is contained within a specialized cell called a *gamete*. (See Laboratory 6.) In most sexually reproducing species, there are two types of gametes, the egg and the sperm. One egg and one sperm unite, forming the first cell of a new individual. The egg is a large and immobile cell. The egg contributes not only the genetic material from one of the adults but also food for the very early stages of the new organism. The sperm is a small and mobile cell. It contributes genetic material from the second adult.

In some organisms, eggs are retained within the adult for a period of time. While there, each egg may unite with a sperm and the developing organisms are protected and provided with additional nourishment.

Today you will focus on the system in mammals that contributes the genetic information, packages it into the gametes, supplies nourishment, and provides protection for the beginning stages of the next generation. This is the **reproductive system.**

You may want to review the process of meiosis in Laboratory 6 before beginning these explorations.

EXPLORING

Reproductive systems

Explore the reproductive structures of your fetal pig (female—pp. 234 to 240; male—pp. 240 to 243). Later, exchange your pig for one of the opposite sex with another laboratory group and explore its structures. In the fetal pig, the structures of the reproductive system are immature. Nevertheless, you can gain an understanding of the arrangement of the reproductive organs with respect to each other and can observe many of their general features.

An immature female reproductive system

Working with a partner, obtain the following materials:

Dissecting tools
A magnifying lens or dissecting microscope
A female fetal pig

Table 21-1 Reproductive organs of the female pig

Number/name	Description	Possible function

In the area between the urinary bladder and the posterior portion of the digestive tract, locate a Y-shaped tube. Using a magnifying lens or dissecting microscope as needed, trace each arm of the Y anteriorly until you find a small lump at the end of each tube. These lumps are anchored in mesenteries in the dorsal portion of the abdominal cavity posterior to the kidneys. Return to the Y-shaped tube where you began and trace posteriorly from the base of the Y to the opening of this system at the external surface. (If the ventral side of the pelvic girdle is not opened, open it now. Your teacher will assist you.) What is the structural relationship between the urinary system and the reproductive system in the female fetal pig?

1 ____

With fine-pointed scissors, open the reproductive tube along one side. Beginning at its most posterior (caudal) end, cut anteriorly toward the base of the Y.

Return to the lumps at the most anterior end of the tubes, and proceeding posteriorly, conduct a thorough exploration. As you work, fill in Table 21-1. Give the structures a number for now; you can find names for them later. Describe the tubes and associated organs in their anatomical order and include their locations. Describe any openings into them and give their size relative to other structures of the immature reproductive system. Leave space between each description of the fetal structures to add information from later explorations of mature structures.

Look for blood vessels associated with the reproductive structures. Describe the blood supply to each area of the tube and to the anterior lumps.

2 ____

____

Which of these structures seems the most likely to produce the gametes?

3 ____

Explain your reasoning.

4 ____

After additional exploration of more mature structures, you can add the function of each structure to the third column of Table 21-1.

CONCEPTS/TERMS

Organs of the female reproductive system

In most organisms, the female produces eggs. The lumps at the anterior end of the reproductive system on the right and left sides are the **ovaries.** The remainder of the system is essentially a muscular, tubular passageway with some expanded, specialized portions. The two tiny tubes in contact with the ovaries on each side are the **uterine tubes.** They are also called the **oviducts** or **fallopian tubes.** The larger-diameter, more posterior portions of the tubes that form the arms of the Y are the **uterine horns.** The base of the Y is the **uterus.** The posterior portion of the uterus with its inward folds is the **uterine cervix.** The cervix leads to the more posterior **vagina.** The vagina joins the urethra (urinary system), forming the **urogenital sinus.** In the fetus, the urinary and reproductive systems share the urogenital sinus, which opens to the external surface. The opening of the urogenital sinus is easily identified by a protrusion called the **urogenital papilla,** which is an identifying structure of the female. On the ventral wall just anterior to the urogenital papilla is a small mound of sensitive erectile tissue, the **clitoris.** In adult mammals, the urogenital sinus shrinks as the urethra and vagina extend posteriorly. In humans, the vagina and urethra are separate; each opens to the external surface.

EXPLORING

Mature female organs

Microscopic examination of organs

As you observe sections of two organs of the female reproductive system, continue to fill in Table 21-1. Obtain your microscope and the slides of mammalian ovary designated by your teacher. Some prepared slides may have all ovarian structures of interest on one slide; others will have these same structures on two or three slides.

Look at the ovary section before placing it on the microscope stage. Explore the section using the lowest magnification. After getting a general idea of its structure, examine the cellular details using each of the higher magnifications. In Figure 21-1, *A,* draw an ovary section that includes all the structures that you find in the slides. Include written notes if they help convey your observations. In Figure 21-1, *B,* construct a more detailed drawing of one of the circular structures that contains cells protruding into a fluid-filled space. These structures are called **ovarian follicles.** On your drawing in Figure 21-1, *A,* indicate the location of the follicle. Using high magnification, compare the details of the large cell in the center of the protruded group of cells to similar cells in other follicles in the ovary. If your slide does not have other fluid-filled follicles, examine your neighbor's ovary slide. Also explore follicles similar to the one you have just drawn except they lack the fluid-filled space. What is the

Figure 21-1
A, Ovary.
B, Ovarian follicle.

size relationship between the follicles lacking the space and the ones with the space?

5 ______________________________

Describe some of the differences, particularly regarding the nuclei, of the large cell within these follicles.

6 ______________________________

____

Describe any relationships you can see between the state of these nuclei and the phases of meiosis.

7 ______________________________

____

Look at the large, somewhat circular structures (which may be collapsed) that have thick walls and that seem to be filled with a homogeneous material. What is the possible relationship between these structures and the ovarian follicles?

8 ______________________________

Obtain a microscope slide of a section of the mature uterus. Explore this section under both low and high magnification. Describe the tissues of this portion of the reproductive tube.

9 ______________________________

____

Does the uterus contain many blood vessels?

10 ______________________________

Describe the evidence for your answer.

11 ______________________________

____

Reproductive organs of a pregnant pig

With your classmates, examine the reproductive structures of a pregnant pig. How many fetuses are in this uterus?

12 ______________________________

How many membranes surround each fetus?

13 ______________________________

Examine the surface of the outer fetal membrane with a magnifying lens. Describe the surface of this membrane and the surface of the uterine lining to which it was attached.

14

Describe the structural connection between each fetus and the uterus.

15

Examine the surface of the ovary. Which of the structures that you microscopically observed in the ovary section can you link with the lumps at the surface of the ovary of the pregnant pig?

16

What is the spatial relationship between the ovary and the open end of the uterine tube?

17

What are the most obvious changes in size relationships of the tubular reproductive system of the pregnant pig compared to those in the fetal pig?

18

Examine the umbilical cord of your fetal pig. (You may want to make a fresh cut across the cord.) What systems of the body are represented by tubes that run through the umbilical cord?

19

Which umbilical circulatory vessel carries blood with the highest oxygen content?

20

Which vessels carry blood with the highest concentration of waste products?

21

What information did you use to answer the last two questions?

22

CONCEPTS/TERMS

Activity of the female reproductive system

The female produces eggs within follicles of the ovaries (Figure 21-2 and Plate 11). Certain cells within an ovary enlarge by storing nutrient materials. These large cells are called **primary oocytes.** The primary oocytes divide by meiosis. As meiosis proceeds, the follicle (the future egg and its "helper" cells that produce the hormone estrogen) enlarges in size and develops the fluid-filled cavity (antrum). About once a month in women, a swollen, mature follicle breaks through the ovarian wall and the egg is released from the ovary. In most other

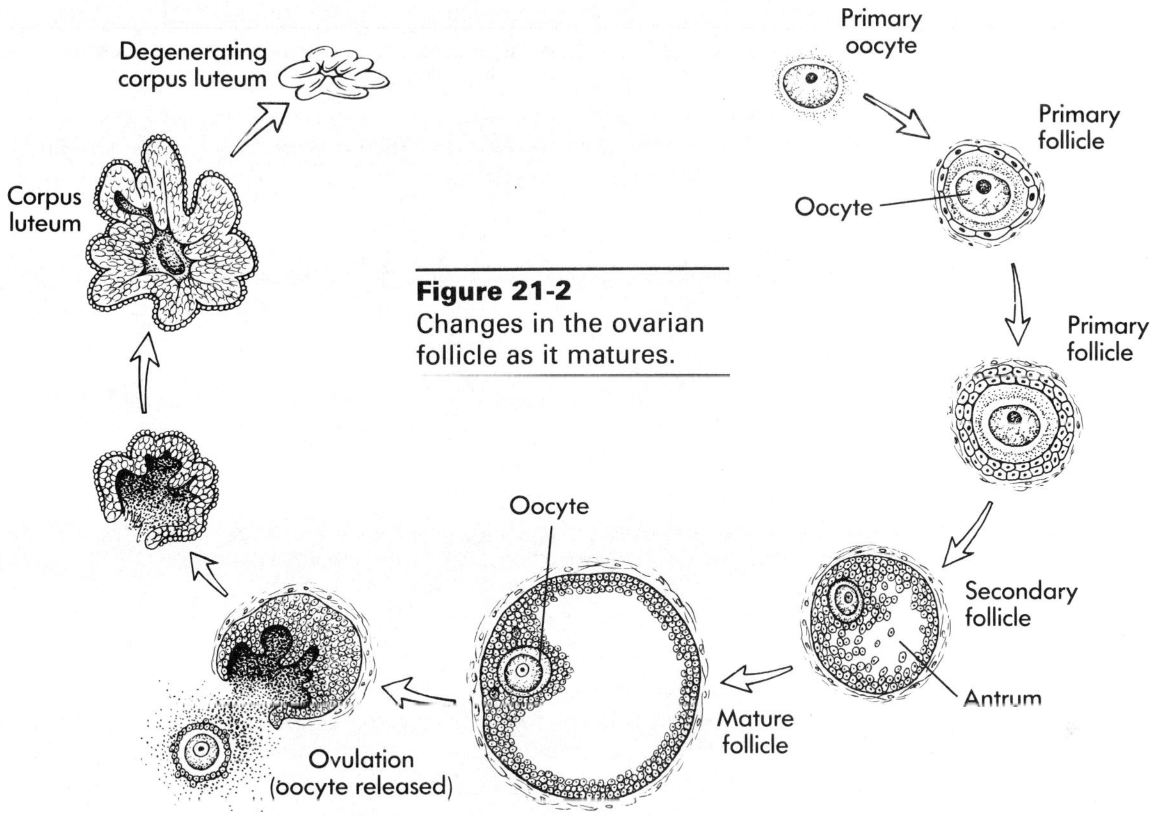

Figure 21-2
Changes in the ovarian follicle as it matures.

mammals, several eggs are released at one time. This rupture and release is called **ovulation.** Some of the helper cells that remain in the collapsed ovarian follicle enlarge and produce relatively large amounts of the hormone progesterone. After ovulation, the remains of the follicle is called the **corpus luteum** ("yellow body").

After ovulation, the ciliated cells lining the uterine tube set up currents that sweep the egg (or eggs) into the tube. If sperm are present in the uterine tube, one sperm and an egg may unite, forming the first cell of the new organism (the zygote). Continued action of the ciliated epithelial cells and contractions of the smooth muscle of the uterine tube propel the egg to the uterus.

If the egg is not fertilized, it continues moving through the tubular reproductive organs and is eventually expelled through the vaginal opening with the uterine lining. This sloughing (shedding) of the uterine lining is called **menstruation.**

If the egg is fertilized, it divides by mitosis and nestles into the thickened lining **(endometrium)** when it reaches the uterus. Uterine cells and cells of embryonic origin near the developing organism **(embryo)** interact, forming the **placenta.** The mother's blood vessels transport nutrients to the placenta. The nutrients diffuse through the placental membranes into the embryo's placental blood vessels. The umbilical vein carries the nutrients into the body of the embryo. The umbilical arteries deliver the waste products from the embryo's metabolism to the placenta. The waste products diffuse across the placenta into the mother's blood vessels. In this way, the female adult supplies the developing organism while it is protected within the uterus.

When the new organism has developed sufficiently for survival without the protection of the uterus, it is expelled. The smooth muscles of the uterine wall contract, pushing the fetus out of the uterus and through the vagina. These contractions also push the surrounding membranes and placenta through the birth canal, and the thick uterine lining begins sloughing off.

The entire process of developing a female gamete and preparing the uterus to nourish the embryo is highly regulated by hormones. These hormones are produced by the hypothalamus, pituitary gland, and ovaries.

Identify the names and functions of all structures in Table 21-1. If more information is needed, check reference materials available in the laboratory.

EXPLORING

An immature male reproductive system

Working with a partner, obtain the following materials:

Dissecting tools
A magnifying lens or dissecting microscope
A male fetal pig.

Remove the skin from one of the two swollen areas between the hind legs. In the exposed area, locate a sac that is anchored by thin tubes and cords that disappear into a tunnel (canal) through muscles of the abdominal wall. With a pair of fine-pointed scissors, cut the abdominal wall ventral to a canal on one side of the pig. This cut should expose the entire length of the tubes and cords. Identify the systems to which these tubes and cords belong.

23

Open the sac. With a magnifying lens or microscope, examine the contents. Turn the large oval organ over and examine all surfaces. In some specimens, the organ has not descended into the sac and you can find it in the abdominal cavity or canal. Trace the muscular tube (that does not carry blood) from the oval organ until it and a similar tube from the other side join a larger, midline tube that is also associated with the urinary system (the urethra). (If the ventral pelvic bones are not cut, probe for the ventral bones and cut them with guidance from your teacher. Avoid cutting a tube that is posterior and ventral to the ventral pelvic bones.) Follow the urethra until it opens to the exterior surface of the pig. Note any glands that may be associated with the urethra. Particularly look for small bumps on the dorsal side of the cranial end of the urethra where the small tubes from each sac attach. Remove the skin posterior to the base of the umbilical cord and examine the tube that encloses the urethra.

Return to the sac on the dissected side of the body. Starting with the large oval structure, give each structure a number and list each in its anatomical order in Table 21-2. In the second column, describe each structure and include the location and the size relative to other immature reproductive structures of the pig. Leave space after the description of the oval structure for additional information. Fill in as much of the third column as you can.

CONCEPTS/TERMS

Organs of the male reproductive system

The oval organ within the **scrotal sac** in each swollen area between the hind legs of the male pig is the **testis.** The testes (plural of testis) produce **sperm.** The remainder of the reproductive system consists of glands that contribute nourishing and lubricating fluids to the sperm and a system of tubes that stores the sperm. In addition, the muscular tubular system moves the mature sperm to the external opening of the urethra at the tip of the penis.

A group of tiny, coiled tubes attached to the surface of each testis collects and stores sperm from the testis. This storage portion of the tube system is called the **epididymis.** From it, the sperm are propelled into an uncoiled tube called the **vas deferens** (or ductus deferens). It extends (with blood vessels and nerves) through the **inguinal canal** in the abdominal wall and joins the **urethra** near the base of the urinary bladder. On the dorsal side of the urethra near this junction are immature glands, two small **seminal vesicles** and a smaller **prostate.** When mature, these glands add fluids to the sperm. To-

Table 21-2 Reproductive organs of the male pig

Number/name	Description	Possible function

gether, the sperm and fluids are called **semen.** The semen moves down the urethra within the ring of the pelvic bones. In the pelvis, the **bulbourethral glands** (also called **Cowper's glands**) add more fluid. From the posterior end of the pig, the urethra conducts the semen anteriorly to a surface opening near the base of the umbilical cord. In this area, the urethra is surrounded by the erectile **penis.**

EXPLORING

A mature testis and sperm

Obtain the following materials:

Compound light microscope
Prepared slide of testis, x.s.
Prepared slide of human sperm, w.m.

Examine the testis slide under low power. Describe the general construction of this organ.

24 ____

Thoroughly examine several of the circular structures (cross sections of **seminiferous tubules**) using high magnification. One tubule cross section may not contain all types of cells produced during the meiotic process. Particularly note sizes of cells and whether they have tails. In Figure 21-3, *A,* draw a cross section of a tubule including its contents. If all types of cells are not within one tubule's cross section, draw another that includes other cell types. In Figure 21-3, *B,* enlarge a pie-shaped section from each of your cross sections. Draw

Figure 21-3
A, Seminiferous tubule, cross section.
B, Details of the cells in the seminiferous tubule.
C, Human sperm.

the details of each type of cell in the pie sections. If you find cells between tubules, add them to Figure 21-3, *A*. Examine the human sperm slide with high magnification and draw one sperm in Figure 21-3, *C*.

CONCEPTS/TERMS

Activity of the male reproductive system

The testes are composed of many seminiferous tubules. Within these tubules and near their periphery are large cells—**primary spermatocytes**—that divide by meiosis. As meiosis proceeds, the products are pushed into more interior regions of the tubules. The sperm are the small cells with tails in the center of the tubules.

The production of sperm is highly regulated by interaction of hormones from the pituitary gland, the **interstitial cells** between the seminiferous tubules, and the hypothalamus.

After their production, sperm move into the tubules of the epididymis where they mature and are temporarily stored. During sexual arousal and intercourse, muscular contractions move the sperm into the vas deferens and urethra and ejaculate them into the female vagina. From the vagina, the sperm travel through the uterus to the uterine tubes, where one may unite with an egg. Complete Table 21-2. If additional information is needed, consult reference materials available in the laboratory.

APPLYING

Comparison of pig and human reproductive systems

Explore the models or charts of the human female and male reproductive systems available in the laboratory. List the major similarities and differences between the pig and human reproductive structures.

25 Female similarities ______________________________

____

Differences ______________________________

____

Male similarities ______________________________

____

Differences ______________________________

____

Reproductive information to aid and prevent pregnancy

Promotion of pregnancy

Some couples have difficulty in conceiving a child. Under these circumstances, knowing when the woman ovulates is helpful. The egg can unite with a sperm for only about 24 hours after ovulation. Changes in a woman's temperature can be used to help determine when ovulation occurs. A slight increase in temperature (about 0.6° F) occurs near the time of ovulation. Because an individual's temperature also cycles with normal activity during the day, the woman must take her temperature when at complete rest. Usually, she takes her temperature on awakening each day for several months to establish her normal temperature pattern. In this way she may be able to detect the temperature change caused by ovulation. In addition, the woman can check for changes in the cervical mucus that occur at the time of ovulation.

On a fetal pig, attach paper labels (1 and 2) with straight pins to (1) the structure that ovulates and (2) the structure that is checked for mucus changes.

Prevention of pregnancy

Many couples conceive children unintentionally even though a number of methods can prevent conception. Only one method is 100% effective. Pregnancy prevention methods usually involve the female more than the male. Most birth control methods are designed to prevent the two healthy gametes from being in the same place at the same time, thus preventing their union.

Several of the following contraceptive methods require a doctor to measure dimensions of the organs, perform surgical procedures, or prescribe hormones. Other methods use materials readily available from a variety of stores.

A few of the contraceptive methods provide some protection from sexually transmitted diseases.

Abstinence. If no sperm are deposited in or near the vagina, no conception can occur. Therefore abstinence (not having sex) is 100% effective in preventing pregnancy. In addition, sexually transmitted diseases (such as syphilis, gonorrhea, herpes infections, and AIDS) cannot be transferred from one individual to another without sexual contact. Abstinence stops the spread of these diseases. (Abstinence does not stop transfer of AIDS through blood and blood-contaminated materials.)

Consider the possible consequences of nonabstinence. On the fetal pig, label the most external portion of the reproductive system from which sperm, if deposited, could travel to an egg (3). Also label the reproductive organ in which a resulting fetus would develop (4).

Rhythm method. Because the egg can only be fertilized within about a day of ovulation, some couples practice abstinence during those fertile days. For this method to be effective, the time of ovulation must be known to determine when to abstain from sexual intercourse. Because sperm survive in some women's reproductive organs up to 5 days, abstinence for 5 days before and 2 days after ovulation is recommended. Many women, however, have irregular ovulation times and have difficulty in accurately predicting ovulation; this lowers the reliability of the rhythm method in prevention of pregnancies. The rhythm method does not protect against sexually transmitted diseases.

Condom. Placing a sheath of sperm-impenetrable material over the penis holds the released sperm (Figure 21-4, *A*). This blocks the sperm from reaching the egg and minimizes direct contact of reproductive organs of the partners. The quality and type of the condom material can affect its degree of protection. The correct use of a sound (without any holes or splits) latex condom helps prevent the transfer of sexually transmitted diseases.

The newer female condom lines the woman's vagina and the nearby external surfaces (Figure 21-4, *B*). This condom prevents the sperm from traveling through the female reproductive system to the egg. Because the female condom covers more surface area than the male condom, it is expected to give greater protection against the transfer of sexually transmitted diseases.

With small pieces of plastic wrap, cover the comparable fetal pig's organs that are covered by condoms in humans. Use pins or thread to anchor the plastic in the appropriate place.

Diaphragm and cervical cap. These flexible, sperm-impenetrable barrier devices fit over the cervix (Figures 21-4, *C* and *D*). The correct size is determined by a doctor. What is the most posterior part of the reproductive tube that these devices prevent sperm from entering?

26 ______________________________

Do you expect these devices to help prevent the spread of sexually transmitted diseases?

27 ______________________________

A

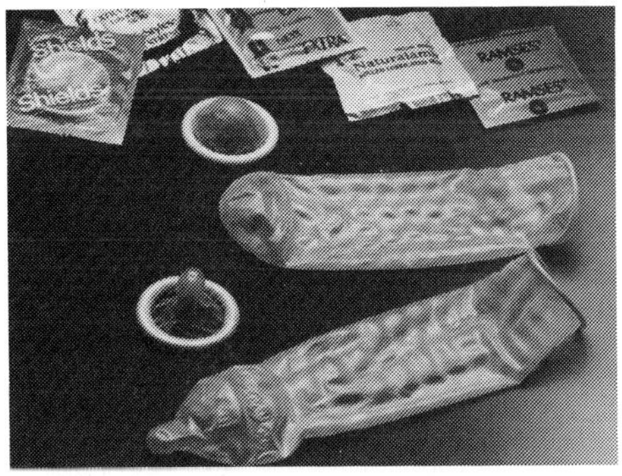

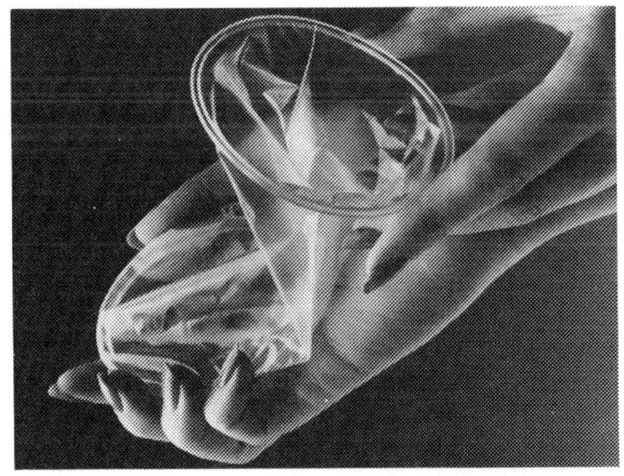

B

C

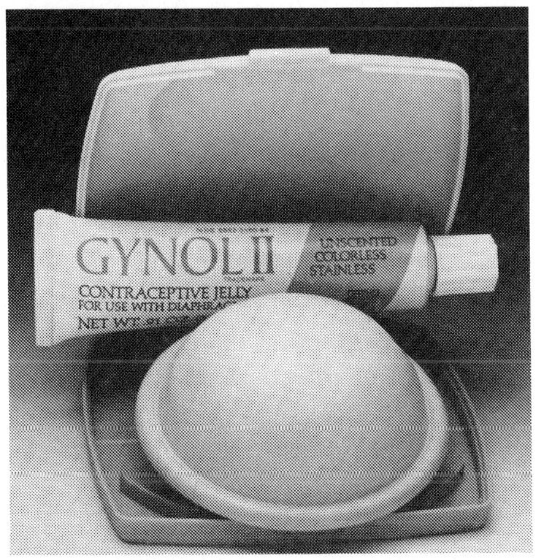

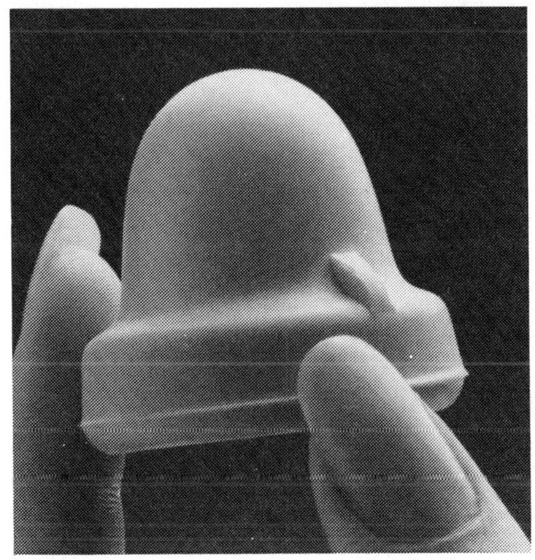

D

Figure 21-4
Contraceptive materials.
A, Male condom.
B, Female condom.
C, Diaphragm.
D, Cervical cap.
E, Spermicidal foam and inserts.
F, Vaginal sponge.
G, Oral contraceptives.
H, Hormone implants.

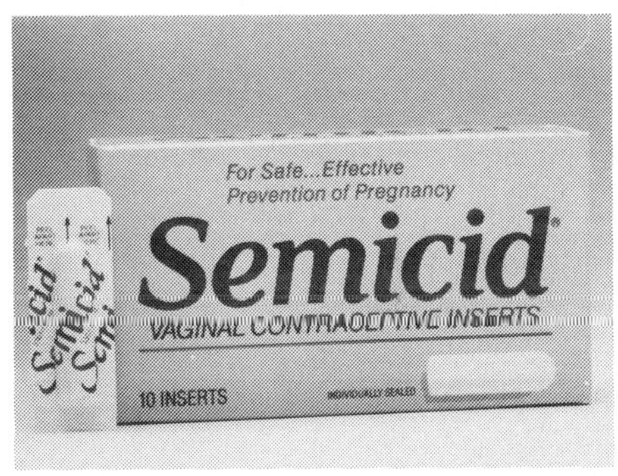

E

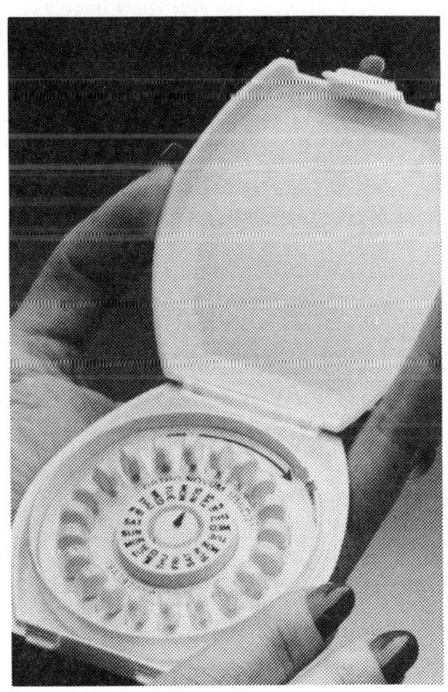

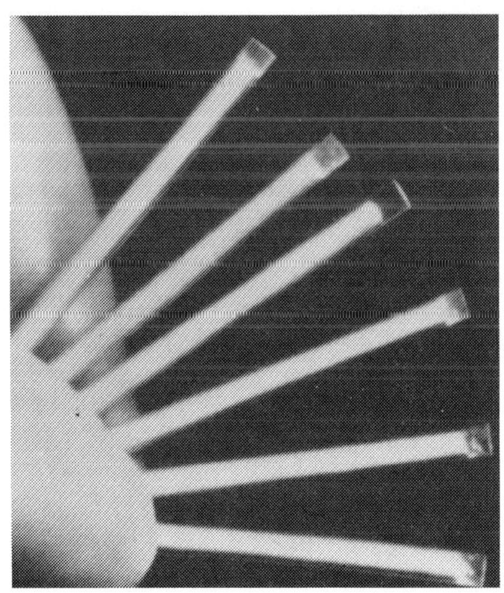

F

G

H

Explain your reasoning.

28 ______________________________

Pin a label (5) to the portion of the reproductive tube where sperm blockage occurs with the use of these pregnancy prevention methods.

Spermicides. In this method, foams or creams containing sperm-immobilizing chemicals are inserted in the vagina with an applicator (Figure 21-4, *E*). These chemicals lose their effectiveness after about a half hour; they need to be used just before intercourse.

Another spermicidal method is the vaginal sponge (Figure 21-4, *F*). The sponge is placed into the vagina where it releases small quantities of spermicide for several hours. The vaginal sponge is available in stores and is not individually fitted by a doctor. Considering the anatomy of the reproductive system, what is the most effective placement of the sponge in the vagina?

29 ______________________________

Spermicides are also used with the diaphragm and with the cervical cap to increase effectiveness.

Surgical blocks. Surgical procedures on small tubular pathways create barriers to gametes traveling through the reproductive system. Doctors remove portions of (or just cut) the small tubes and tie the ends closed. Vasectomy (*ectomy* meaning "cut out") is the surgical blockage of the vas deferens. Would you expect the scrotal sac portion of the vas deferens to be cut or its abdominal portion?

30 ______________________________

Explain your reasoning.

31 ______________________________

Tubal ligation is the surgical blockage of the uterine tube. It is usually done through small incisions in the abdominal wall with laparoscopic equipment. (With laparoscopic techniques, doctors use tubular equipment inserted through small openings to view, manipulate, and remove organs such as gall bladders and parts of uterine tubes.)

Which of these surgical blockage procedures—vasectomy or tubal ligation—do you think is the easiest to perform and the least involved for the patient?

32 ______________________________

Explain your reasoning.

33 ______________________________

Both of these surgical procedures should be considered as permanent methods of birth control. Sometimes delicate surgery can reconnect and reopen the tubes, but this is not always possible.

Perform a vasectomy and tubal ligation on one side of a male and a female fetal pig. Your neighbors can perform the same procedures on the other side of the same pigs. Make your cuts in the regions of the tubes that are most accessible to the surgeons. Explain your choice of cut locations.

34 ____

____

Label the vasectomy and tubal ligations with a 6 and 7, respectively.

Prevention of egg formation: oral contraceptives **("the pill") and** hormone implants **(Norplant).** The pill and Norplant supply extra estrogen and progesterone, which are hormones produced by the ovaries. At these higher hormone levels, eggs do not mature and are retained in the ovary. This is similar to the suppression of egg production that occurs during pregnancy.

Women take birth control pills on a doctor-prescribed daily schedule (Figure 21-4, *G*).

Norplant is a newer method of hormone delivery (Figure 21-4, *H*). Six matchstick-sized tubes of the hormones are surgically slipped under the skin of the upper arm. These tubes deliver hormones in egg-suppressing quantities for up to 5 years.

Label the organs of the pig that have their activity mimicked by this conception prevention method (8). What effect do you think this method has on the transfer of sexually transmitted diseases?

35 ____

LABORATORY 22

Animal development

From gametes to an organism

Development

Developmental biology is concerned with the natural, progressive changes occurring in a living organism throughout its lifetime. It includes all the events initiating the formation of an organism (the production and union of gametes) through the aging process that eventually results in death of the organism. In this laboratory period, you will explore some of the earliest developmental stages of three animals.

You are familiar with the nutritive value contained in the chicken egg. However, do you realize that when this female gamete unites with a sperm, the fertilized egg has the potential to become a baby chick? Certainly the egg does not look like a baby chick, so a very complex set of developmental processes must transform the fertilized egg into a down-covered chick. There is no known case of animals of one species begetting new organisms of a different species. So the egg and sperm of the parents must contain information that precisely controls the developmental events of the new organism or it would not always resemble its parents.

EXPLORING

Fertilization in living sea urchins

To begin the explorations, follow the development of one member of the echinoderm phylum, the pin-cushion-like sea urchins. Living sea urchins, which may produce eggs and sperm when stimulated, or microscope slides that have a number of early developmental stages of sea urchins are available in the laboratory. Use the sections of the manual that correspond to your type of sea urchin material. (See p. 254 if you have prepared whole mounts.)

Obtaining gametes (a class activity)

Examine the living sea urchin with and without a magnifying lens. Where is its mouth?

1 ______________________________

Describe the structures on its surfaces.

2 ______________________________

What kind of symmetry does it have?

3 ______________________________

What other organisms, if any, are living with or on the sea urchins?

4 ______________________________

Sea urchins are fertile at specific times of the year. Different species vary in their times of fertility. To obtain the eggs and sperm of sea urchins, inject fertile adults with a solution of potassium chloride. Your teacher will demonstrate the injection process. After injection, place the urchins, oral side down, on paper towels and wait a few minutes for the release of gametes.

If yellow or purple material is released, immediately place the sea urchin upside down on a small beaker completely filled with sea water. If white or cream-colored material is released, place the sea urchin upside down on a dry watch glass or petri plate. Try to determine which type of gamete each urchin is releasing. What characteristics are you using to identify the gametes?

5 ______________________________

If you cannot identify the gametes with your naked eye, observe a wet mount of each type of material with a compound light microscope. Compare your observations with what you know about the size of chicken eggs and perhaps caviar (fish eggs) to identify each type of sea urchin gamete. What color is each type of gamete?

6 ______________________________

Wash eggs free of the potassium chloride by changing the sea water several times. Your teacher may give you instructions to dilute the sperm fluid with additional sea water.

Developmental biologists estimate that a single "ripe" male sea urchin can hold about 1×10^{11} (100 billion) sperm (see Appendix D). Biologists estimate that the human male produces about 1.2×10^{8} sperm per milliliter of semen. A ripe female urchin contains about 8×10^{6} eggs. What is the usual number of mature eggs produced by the human female at one time?

7 ______________________________

Only one sperm is needed to fertilize each egg. When many sperm are around an egg, chances that more than one sperm will enter the egg at the same time increase. This condition is known as *polyspermy*. It is detrimental to the development of the new offspring.

CAUTION
Gametes and embryos are extremely sensitive to detergents and many other types of chemicals. Therefore all equipment should be very clean and detergent free. Rinse all glassware with distilled water before using. Embryos are also sensitive to heat. When you are not observing the developing eggs, turn off your microscope light.

The events of fertilization

Working with your partner or as designated by your teacher, add a dropper of washed eggs to a fingerbowl with sea water about 1-cm deep. Transfer three or four of these eggs to a seawater-filled depression slide and add a cover slip. Set the depression slide of eggs aside for a moment while you fertilize the eggs remaining in the bowl.

Fertilize the eggs in the fingerbowl by adding a minute amount of undiluted sperm on the end of a toothpick or by adding two drops of diluted sperm suspension. Label the bowl with your names, the laboratory number, the time of sperm addition, and the date. About 20 to 30 minutes after adding the sperm to the eggs, carefully pour or pipet off the old seawater and cover the fertilized eggs with about a 2-cm depth of clean sea water.

Return to the depression slide and observe the unfertilized eggs with the scanning and low power objectives of the microscope. (The depression of the slide has a curved bottom surface; as you examine different eggs, you may need to adjust the focus with your microscope's coarse-adjustment knob. Note that eggs are large cells and probably will not be completely in focus in one depth of field.) With a drawing and written notes, record your observations of an unfertilized egg in the second column of Table 22-1. (Table 22-1 consists of four manual pages. Remove pp. 256 and 257 from the manual and staple them to the bottom of pp. 252 and 253 respectively, to form the complete table.)

After you examine and record information about the sperm-free eggs in the depression slide, add a tiny bit of sperm fluid at the edge of the cover slip. Observe with the low and high power objectives. What are the very small, moving dots?

8 ______________________________

(If you cannot see the moving dots, reduce the light intensity. If you still have trouble, review the microscope information in Appendix A or ask for assistance from your teacher.) Carefully observe the interactions between the eggs and sperm for several minutes. Describe all the events occurring in and around the eggs.

9 ______________________________

____

____

In the second column of Table 22-1, draw one of the fertilized eggs from those on the slide. Add written comments and labels to your drawing. In the first column, record the amount of time (minutes, hours, or days) that has elapsed from the addition of sperm.

CONCEPTS/TERMS

The events of fertilization

Fertilization is the union of two gametes. The sequence of events you observed as sea urchins' sperm and eggs interacted on the depression slide occurs with minor variations in most animals during the process of fertilization. The small mobile sperm swims to the large immobile egg. The sperm sticks to the egg's surface and releases enzymes that dissolve the covering of the egg at the sperm attachment site. The nucleus in the head of the sperm penetrates the egg. This causes changes in the egg that prepare its nucleus for union and "raises" a **fertilization membrane** from its surface. This membrane prevents other sperm from entering the egg and identifies a **fertilized egg.** Inside the egg, the sperm nucleus swells as it moves to the center, where it fuses with the egg nucleus. This union of the two nuclei produces one diploid nucleus within the cell. It is the first cell of the new generation and is called a **zygote.**

EXPLORING

The early developmental stages of sea urchins

Living sea urchins (a week-long exploration)

Check the sea urchins' development at least once every 20 minutes for the remainder of this laboratory period. Whenever the developing sea urchin has a new shape or structure, draw the new developmental stage in Table 22-1. If previously fertilized eggs are available, observe their developmental progress. You can follow the developmental events more easily if you record your observations in order of the time elapsed from fertilization. If you record older embryo observations between those of your own developing sea urchins, leave space in Table 22-1 for the intervening stages. Note the time of fertilization and add the elapsed time to the first column of Table 22-1.

As you observe the sea urchins' progress, consider the following questions: (1) As the fertilized egg divides, how does the size of the resulting cells compare to the original egg? (2) How does the overall size of the organism compare to the egg? (3) How many cells are in the organism? (4) What are the positions of cells with respect to one another? Include your answers in your drawings in Table 22-1.

Table 22-1 Animal development

Time from fertilization	Sea urchin stage (diagram, notes, and name of stage)	Comparison of sea urchin and frog
0 min	Unfertilized egg	
	Fertilized egg	

Stages of frog (diagram and notes)	Comparison of mollusk to frog and sea urchin	Stages of mollusk (diagram and notes)

Continued.

At the end of the laboratory period, follow the directions of your teacher for storing the fingerbowl. Return to the laboratory each day to observe the progress of the sea urchins.

Whole mounts and films of sea urchins development*

Obtain a microscope slide of the stained and preserved whole mounts of early sea urchin developmental stages. Using low power, explore the mixture of early developmental stages on the slides. Carefully examine the number of cells in the different stages, the cells' nuclear structure (if nuclei are stained), the size of the individual cells, and the overall size of the entire developmental stage. In the multicellular stages, note how cells are arranged. Are cells stacked directly above or below each other or are the cells of "one row" nestled into the grooves between the cells of the next row?

10 ______________________________

Try to determine the order of stages in the developmental sequence. (Hint: After two gametes fuse, the newly formed zygote divides repeatedly by mitosis.) Briefly describe each stage in its proper developmental order.

11 ______________________________

____

____

____

____

Discuss your reconstruction of the developmental sequence with your partner and other classmates. With your classmates, observe the film of sea urchin fertilization and early development.

Return to the slide of sea urchin developmental stages. Draw an unfertilized egg and a fertilized egg in the second column of Table 22-1. Add written notes and labels to your drawings. Using observations from your slide and the developmental sequence shown in the film, draw the other stages of sea urchin development in their normal sequence. Draw all stages through the formation of a stage that has a digestive tube in its interior (larval stage). Add notes to your drawings. Leave the first column of Table 22-1 blank.

CONCEPTS/TERMS

Stages of early development

From your observations of the sea urchin, in what ways do you agree and disagree with the statement: "Development of the sea urchin is a gradual process of change rather than an abrupt movement from one stage to the next." Support your answer with specific evidence from your sea urchin observations.

12 ______________________________

____

____

*Use this section if live sea urchins are not available.

Staple to p. 252.

Staple to p. 253.

Discuss your evidence with your partner and teacher.

Although developmental biologists recognize that development is a gradual process of change, they designate the formation of certain structures as identifying stages of development. Breaking the continual process into stages provides an easy way for biologists to refer to a particular interval in the developmental process. Figure 22-1 presents the sequence of developmental stages.

After fertilization, the zygote divides by mitosis and forms two cells with the same genetic information as the original fertilized egg. Each of the two cells divide, forming a four-celled stage. Cell division continues, forming an increasing number of smaller cells. This series of mitotic divisions in the early stages of development is called **cleavage.** As the cells cleave in sea urchins and some other animals, the mitotic spindle is positioned either perpendicular or parallel to the axis of the egg. When cytokinesis occurs, the resulting daughter cells are either stacked one above the other or horizontally to each other. The cleavage pattern that leads to the stacked arrangement of cells is called **radial cleavage.** In other animals, the mitotic spindle is diagonal to the axis of the egg. This second cleavage pattern results in daughter cells nestling into the grooves between the cells just below or above them. This type of cleavage pattern is called **spiral cleavage.**

After each set of early divisions the resulting stage is identified by the number of cells that make up its structure (2-cell, 4-cell, 8-cell, and so on). The overall size of each of these stages is about the same as the original egg. (What explains the lack of increase in overall size?)

When many cells are produced, a cavity forms in the middle of the mass of cells. This hollow ball of cells is the blastula. The layer of cells that forms the blastula encloses a space called the **blastocoel.**

The cells in the blastula rearrange their positions with respect to one another by moving in complex patterns that vary with the type of organism. These movements result in the formation of a three-layer structure, the **gastrula.** The movements of cells resulting in the formation of the gastrula is called **gastrulation.**

Figure 22-2 shows the gastrulation process of sea urchins. Cells on one side of the blastula change shape (the inner parts of these cells expand and the outer parts contract), forming wedge-shaped cells. This shape change causes this portion of the layer of cells to fold into the cavity of the **blastula.** This infolding forms a double-layer structure similar to a tennis ball with one side pushed in. The outer cell layer is the **ectoderm.** The inner cell layer is the **endoderm.** The new cavity formed by the infolding process is the **primitive gut.** The opening into the primitive gut is called the **blastopore.** The complete tube-like structure of the digestive tract forms as the primitive gut continues to elongate in the interior of the gastrula. When the cells of the future digestive tube touch the cells on the opposite side of the gastrula, a second opening de-

Sperm ↘

Unfertilized Egg → Fertilized Egg → (cleavage begins) → 2-Cell Stage → 4-Cell Stage → 8-Cell Stage

→ 16-Cell Stage → 32-Cell Stage → → → → → → (cleavage continues) Blastula → → → (gastrulation)

Gastrula → → → (organogenesis) Larva → → → (maturation) Adult

Figure 22-1
Sequence of early developmental stages.

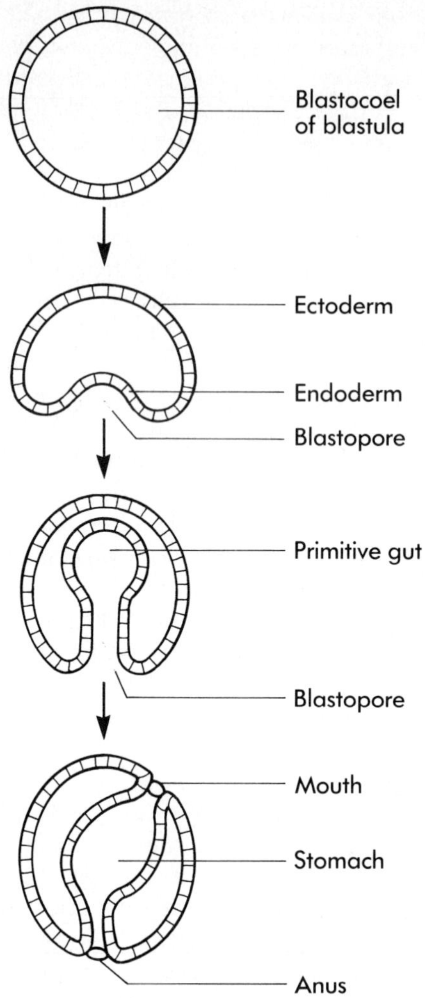

Figure 22-2
Gastrulation in the sea urchin.

velops. This digestive tube forms the basic **tube-within-a-tube** body plan of many animals. In sea urchins as well as some other animals, this new opening becomes the **mouth** and the blastopore becomes the **anus.** These animals are called **deuterostomes** (DOO-ter-o-stomes), meaning "second mouth" (from the Greek *deuteros* and *stome*). In other animals the blastopore forms the mouth, and the new opening forms the anus. These animals are called **protostomes** (PRO-toe-stomes), meaning "first mouth" (from the Greek protos).

The deuterostomes and protostomes also differ in their types of cleavage. Deuterostomes have radial cleavage, and protostomes have spiral cleavage.

In all except the most primitive animals, a third layer of cells called the **mesoderm** forms between the ectoderm and endoderm layers. How this middle layer forms differs between deuterostomes and protostomes. The mesoderm of deuterostomes forms from cells that bulge out (outpocketing) and pinch off from the digestive tube. This pocket of mesoderm has a central cavity. The mesoderm of protostomes forms from a mass of cells that migrates from near the blastopore to the space between the ectoderm and endoderm. A cavity forms within the mesoderm of the protostomes as the mass develops an internal split.

These developmental-pattern differences (the fate of the blastopore, type of cleavage, and the mesoderm formation) are very important in determining relationships among animals. They are used to divide the family tree of more complex animals into two major branches, the protostomes and the deuterostomes. Complete Table 22-2.

The organs of animals develop from the three gastrula layers **(germ layers).** Each layer forms the following organs of the animal: (1) Ectoderm forms the skin and the nervous system. (2) Endoderm forms the inner layer of the digestive system. (3) Mesoderm forms muscles, bones, and the urinary and circula-

Table 22-2 Developmental characteristics of protostomes and deuterostomes

Characteristic	Protostomes	Deuterostomes
Type of cleavage		
Fate of blastopore		
Origin of mesoderm		

tory systems. It also interacts with endodermal cells that form the liver and respiratory system. The process of **organogenesis** involves complex interactions among neighboring cells, groups of cells, and cell layers. Eventually the cells acquire specialized structures and functions. When this occurs, the cells are said to be *specialized,* or **differentiated.** When organs form, most developing organisms are called **larvae** (singular, larva).

EXPLORING

Frog development

At the demonstration table, find a series of small dishes containing the preserved, early stages of developing frogs. Using dissecting microscopes, examine each frog stage. Draw and add notes about these frog stages in the fourth column of Table 22-1. Place the frog information at the same level in the forth column as the corresponding stage of the sea urchin in the second column.

In the third column of Table 22-1, describe the ways in which the development of each stage of these two organisms are similar and different.

Are frogs protostomes or deuterostomes?

13 ______________________________

Do frogs have radial or spiral cleavage?

14 ______________________________

CONCEPTS/TERMS

Thinking about relationships

After comparing the development stages of the frog and sea urchin, do you agree or disagree with the statement: "Frogs and sea urchins exhibit the same developmental patterns."

15 ______________________________

Cite at least four pieces of evidence from your explorations to support your answer.

16 ______________________________

____

____

Do some of your observations of the stages—zygote through larva—refute your answer to the sea urchin and frog development question?

17 ______________________________

If so, what is the evidence?

18 __________

Write a summarizing statement about development in both the sea urchin and frog that includes *all* observations from zygote through larval stages.

19 __________

EXPLORING

Mollusk development

Explore slides of the development of a mollusk, either the slipper snail or a bivalve such as a clam. As you examine each mollusk stage, consider the same types of questions as you did for the sea urchin and frog (for example, the number, size, and position of the cells with respect to one another).

In the sixth column of Table 22-1, draw the developmental stages of the mollusk across from the corresponding stages of the sea urchin and frog.

In the fifth column, compare the sea urchin, frog, and mollusk by describing the ways in which early development is similar and different.

Is the development of the sea urchin more similar to the mollusk or to the frog?

20 __________

Support your answer by giving at least three of the most important pieces of evidence.

21 __________

CONCEPTS/TERMS

Phylogenetic relationships

Ancestral relationships

Are you more closely related to your sibling (sister and brother), if you have one, or to your cousin?

22 __________

Who is the closest **common ancestor** of you and your sibling?

23 __________

Who is the closest common ancestor of your and your cousin?

24 __________

Do you and your sibling or do you and your cousin have a more recent ancestor in common?

25 __________

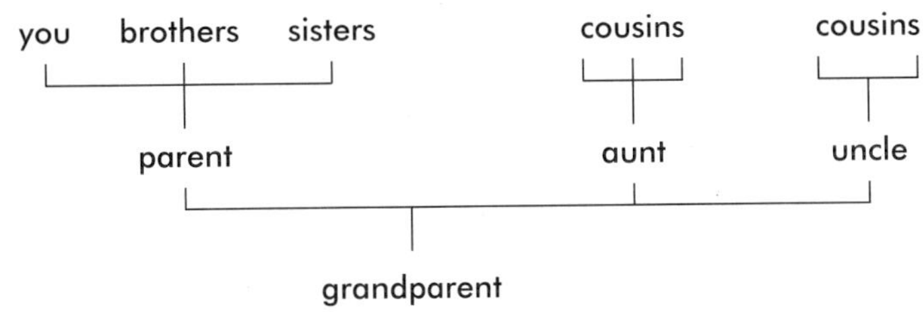

Figure 22-3
A family tree.

In what ways do you agree and disagree with the statement: "The more closely related you are to a person, the more recently you have had an ancestor in common."

26 ______________________________

____

Concept of phylogeny

Phylogeny is the evolutionary history of types of organisms. You can think of phylogeny as the ancestral history of species. It is very much like your own ancestral history except that phylogeny deals with ancestral relationships among groups of organisms rather than individuals.

The ancestral relationships among groups of organisms drawn in the form of a tree (similar to the family tree in Figure 22-3) is known as a **phylogenetic tree.**

APPLYING

The phylogenetic relationships to sea urchin, frog, and mollusk

Based on their developmental patterns, which two of these three organisms are the most closely related?

27 ______________________________

On what observations did you base this answer?

28 ______________________________

____

____

Which two of these three organisms had a common ancestor in the most recent past? (Hint: Think in geological time.)

29 ______________________________

Do you think that the closest common ancestor of all three of these organisms existed before, after, or at the same time as the closest common ancestor of the two organisms chosen?

30 ______________________________

Think back through the reasoning you used to answer these questions. What assumption did you make as you developed answers for the last two?

31 ______________________________

In Figure 22-4, draw a phylogenetic tree of these three types of organisms. Indicate on the tree the location of the most recent common ancestor of all

Figure 22-4
Phylogenetic tree: ancestral relationships of sea urchin, frog, and mollusk.

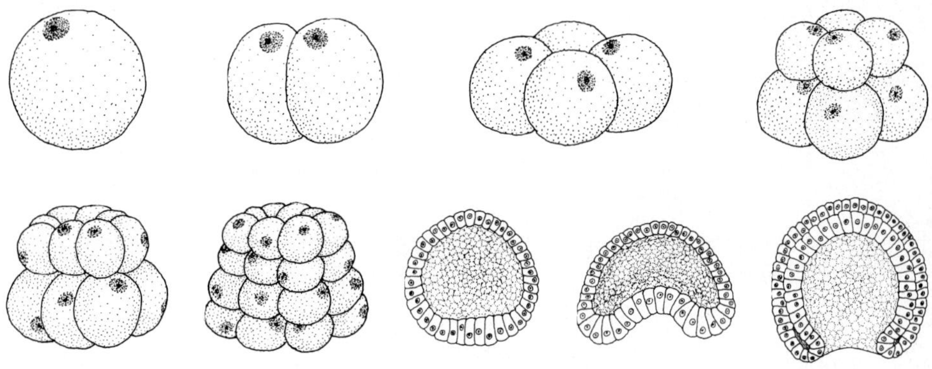

Figure 22-5
Early developmental stages of an unknown animal.

three of these organisms. Also indicate the location of the most recent common ancestor of the two organisms that have the most similar developmental patterns.

Where does the unknown animal fit?

Figure 22-5 shows the early developmental stages of a fourth animal. Which of the three animals that you examined in the laboratory is the most closely related to this unknown animal?

32 ______________________________

Describe the developmental characteristics of the unknown animal that you matched to the other three organisms to determine relatedness.

33 ______________________________

____

Where would you place this animal on the phylogenetic tree in Figure 22-5?

34 ______________________________

LABORATORY 23

Monerans, protists, and fungi

The five kingdoms

Scientists estimate that approximately 4500 kinds of mammals and several million kinds of insects exist on earth. Some scientists estimate that perhaps more than 30 million species (kinds) of organisms have yet to be identified in just the tropical rainforests alone. To deal with this tremendous diversity, biologists group organisms, placing them with other organisms that share certain characteristics. They also form a hierarchy of organismal groups just as we commonly do with things in our lives. For example, we associate rocking chairs, dining chairs, highchairs, folding chairs, ladder-back chairs, and recliners in a group that we call *chairs*. These chairs share structural features such as a flat surface and a "back" that can support our bodies. Then we associate chairs with tables, beds, chests, and other groups in a more inclusive group called *furniture*.

Initially, biologists form groups of organisms that share a vast number of characteristics. If one characteristic shared among all group members is successful production of offspring, then biologists call this group level a **species.** They place several species that share other characteristics (other than the ability to interbreed) into a more inclusive group called a **genus** (plural, genera). All members of a genus exhibit fewer similarities and more differences than the members of a species. Biologists continue this process of forming increasingly more inclusive groups of organisms until they form the most inclusive groups of living organisms, which they call **kingdoms.** Currently, most biologists recognize five kingdoms: **Monerans, Protists, Plants, Fungi,** and **Animals.**

In this laboratory period, you will examine organisms from three kingdoms: monerans, protists, and fungi. Before beginning, review Laboratory 4, which explores differences between prokaryotic and eukaryotic cells. Your teacher may direct you to move around the laboratory to different stations or areas of the room to prepare wet mounts and observe materials or may direct you to obtain material from the supply table and use them at your own desk.

PART ONE Kingdom Monera

EXPLORING

Monerans

At the demonstration table or at the designated station, observe prepared slides or wet mounts of four monerans. A sign at each microscope gives an identifying number for the organism and the diameter of the field of view for the magnifying system. For the two monerans designated by your teacher, determine the length of the cells (in micrometers) by dividing the estimated number of cells that can fit across the diameter of the field of view into the diameter of the field of view. (You may recall using this method of estimating cell size in Laboratory 4.) In Table 23-1, diagram the overall structure of each organism and show how its cells are arranged. If some of the organisms have only one cell, show the arrangement (if any) between the organisms. Describe each type of cell in each moneran and include the estimated cell size for the two designated monerans.

Examine the electron micrographs in Figure 23-1. Match these two micrographs with two of the four monerans you observed. Place the letter of each electron micrograph in the first column of Table 23-1 by the number of the most similar moneran. Do not go by the overall shape of the organism but by the possible contents of the cells. What type of cells do these monerans have?

1 ______________________________

CAUTION

You may examine some of the slides with oil immersion objectives. Use only the fine adjustment knob to sharpen focus with the oil objective.

Which, if any, of the observed monerans are multicellular organisms?

2 ______________________________

What evidence supports your answer?

3 ______________________________

What is your definition of multicellular?

4 ______________________________

Figure 23-1
Electron micrographs of two monerans.

A

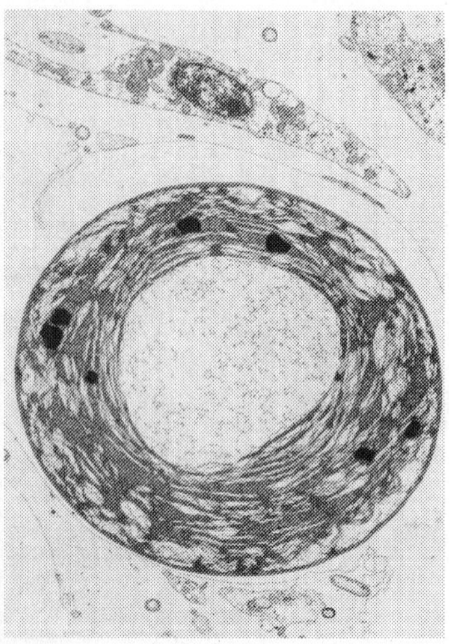

Where can you find monerans?

Locate a series of clear, flat, covered dishes. These dishes are called *petri dishes* or *plates*. Microbiologists frequently grow bacteria in plates filled with a semisolid, nutrient medium. (A nonnutritive substance called **agar** [produced by a red seaweed] solidifies the medium to resemble gelatin. The microbiologist can vary the nutrients mixed into the agar.)

Observe the nutrient agar plates. The surface of each plate has been exposed to the environment in a specific way. Note the characteristics (such as color, texture, and type of surface) of all growths. If the different types of growths are separated, note the contours of their edges. Describe each type of growth in Table 23-2. Based on your observations, which environments on Earth are free of microorganisms?

5 ______________________________

B

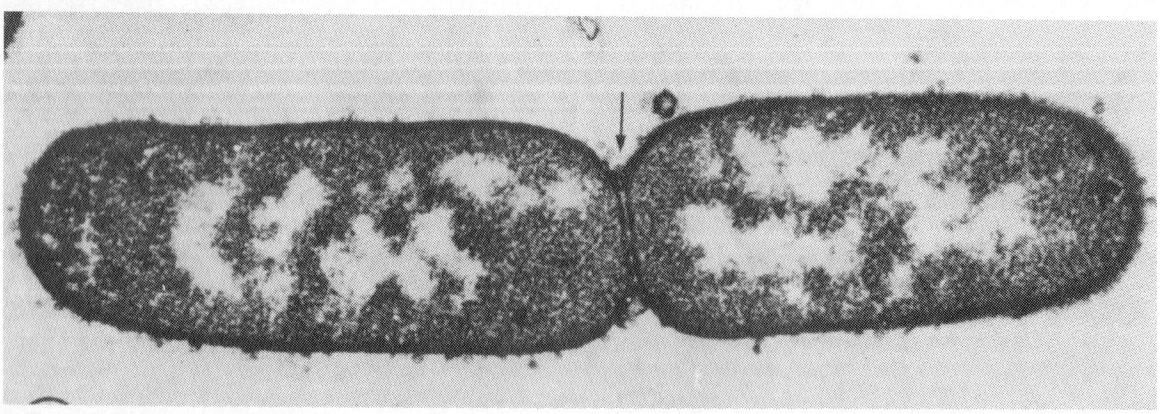

Exposing nutrient agar plates to microorganisms

Working in small groups or as designated by your teacher, obtain four sterile (without living organisms) nutrient agar plates. (Your teacher may substitute plates with different kinds of nutrients.) Expose each plate to a different environment or smear some material from your environment (for example, plaque from the base of your teeth, scrapings from beneath your fingernails, fluid from your drinking cup, an eating utensil, or a baby's pacifer). To apply liquids or semisolids to the agar surface, dip a sterile swab or toothpick into the material and lightly rub it back and forth over the agar's surface. Be careful not to gouge the agar. From your observations in Table 23-2, how do you think you should hold the lid of the petri dish to prevent possible contamination from the air as you introduce the materials?

6 ______________________________

Table 23-1 Representative monerans

Number	Diagram of organism showing relationship among cells	Description of cell types	Size	Name
1.				
2.				
3.				
4.				

Table 23-2 Growths on nutrient agar

Treatment of agar surface	Description of growths
Wiped with paper towel used on laboratory desk	
Touched with rim of beaker	
Exposure to laboratory air for 15 minutes	
Thumb smear	

CAUTION

Dispose of all materials that have come in contact with body fluids in the biohazard containers as directed by your teacher.

Using two small pieces of tape, fasten the lid of the petri dish to its bottom. Label the bottom of each plate with your names, date, laboratory section, and its treatment. Tape all of your plates together, turn them upside down, and place them in the warm place designated by your teacher. Incubate these plates for about 2 days (or as indicated by your teacher). Observe them; record the types of observations you made previously on a table that you construct. Attach your table to this manual. When observing your plates, leave the lids taped because some of the growing organisms may be pathogenic (disease producing). After recording your observations, follow your teacher's directions to dispose of your plates.

What materials or environments did you find to be free of microorganisms that can grow on the nutrients in the plate?

7 ______________________________

If different kinds of nutrients were supplied in the agar, do you think different types of organisms might have grown on the plates?

8 ______________________________

Explain your answer.

9 ______________________________

____

CONCEPTS/TERMS

Unity and diversity in the kingdom Monera

All monerans are prokaryotes, and all prokaryotes are monerans. Having a prokaryotic cell structure is a characteristic that separates the members of the kingdom Monera, the **bacteria,** from all other living organisms. (Remember that prokaryotic cells are small compared to eukaryotic cells and lack membrane-bound organelles such as nuclei, mitochondria, and chloroplasts.)

Monerans are ubiquitous; they are found almost everywhere on and in the surface layers of Earth and in the atmosphere. Even though some bacteria stick together after division, forming particular spatial patterns (arrangements), and may form simple interacting groups, monerans are essentially one-celled organisms. Very few moneran cells are specialized within the groups.

Although all monerans are prokaryotes, this kingdom of organisms is quite diverse. For example, the type of molecules they use for nutrients (their mode of nutrition) and the way they obtain energy varies.

One group, the **cyanobacteria** (blue-green bacteria), uses simple inorganic substances (such as carbon dioxide and water) and energy from the sun to make their own food molecules. These are **photoautotrophs,** which means "light self-feeders." They have a bacterial chlorophyll that absorbs light energy during photosynthesis. Under certain environmental conditions, some cyanobacteria develop a specialized cell called a **heterocyst.** A heterocyst can fix nitrogen gas, which means that it can build a nitrogen-containing organic compound from nitrogen gas. *Oscillatoria* (one of the four observed monerans) is a cyanobacterium. *Anabena* and *Gleocapsa,* explored in Laboratory 4, are also cyanobacteria. Cyanobacteria are the largest of the prokaryotes.

A second group of bacteria are **chemoautotrophs,** which means "chemical self-feeders." These bacteria use simple inorganic substances (such as carbon dioxide) as nutrient building blocks (as all autotrophs). However, they build these molecules using energy from the breakdown of other inorganic molecules such as hydrogen sulfide and ammonia.

Bacteria in the third group cannot build their own nutrient molecules. These bacteria are called **heterotrophs,** which means "other feeders." Heterotrophic bacteria obtain their energy by breaking down organic materials produced by other organisms. Some of these bacteria are pathogenic, causing diseases in plants and animals. Many heterotrophic bacteria are **decomposers,** breaking down dead organic materials (such as dead leaves) into simpler molecules. These molecules are then available for another cycle through living organisms. An example of a heterotrophic, ecologically important bacterium is *Rhizobium,* a symbiotic (meaning "living together") bacterium that lives inside the nodule cells on the roots of legumes such as soybeans. It and a few other bacteria convert nitrogen gas into ammonia. The bacteria are the initial suppliers of nitrogen to plants. This fixed nitrogen eventually cycles through *all* organisms.

The heterotrophic bacteria are further characterized by the shape of their individual cells and colonies. The shape of bacterial cells may be round (coccus), rod like (bacillus), or spiral (spirillom). Colonies are visible masses of bacterial cells on agar media. Some of the ways in which bacterial colonies differ are in color, texture, shape, and edge contours.

APPLYING

Grouping monerans by their mode of nutrition

Using your observations recorded in Table 23-1, place each of these representative monerans into one of the following groups: photoautotrophs, chemoautotrophs, or heterotrophs. Place the group name in the right margin, next to the organism's name in Table 23-1.

PART TWO Kingdom Protista

EXPLORING

Protists

Examine three different sets of protists. In Table 23-3, describe each organism. Some questions or suggestions to guide your exploration are the following:

1. What is the overall form of the organism?
2. What is the color of the organism's cell(s)?
3. Is the organism a heterotroph or an autotroph? How do you know?
4. What is the size(s) of the organism's cell(s)?
5. What is the number of cells in the organism?
6. What cell organelles can you identify?
7. If the organism moves, what is its mechanism of movement?
8. Does the organism have specialized cells?
9. Is the organism multicellular?

For the first set, prepare and explore wet mounts from cultures of *Chlamydomonas* (CLAM-me-doe-MON-as), *Pandorina, Volvox,* and dinoflagellates. (Use a depression slide for the *Volvox* and the dinoflagellates.) Also examine a prepared slide of *Oedogonium* (EE-doe-GON-i-um) and a preserved kelp.

For the second set, prepare wet mounts of *Amoeba* and *Stentor*. (You can dislodge amoeba from the bottom of the culture dish with a jet of water from the transferring dropper.) Add a small amount of a water mixture of yeast and red dye particles to each wet mount. As you observe these two organisms, decide whether these protists take in particles from their environments. Also examine a prepared slide of *Giardia.*

For the third set, explore a water mold and a slime mold with a dissecting microscope. Examine a dish with the water mold growing on a hemp seed or dead insect. Are the strands of this organism divided into individual cells? Do any elongated tips of the strands encase a number of small cells (spores) or do other round tips hold eggs? If so, include these in your diagram and description of water molds in Table 23-3.

Table 23-3 Representative protists

Organism	Diagram and description
Chlamydomonas	
Pandorina	
Volvox	
Oedogonium	
Dinoflagellate	
Kelp	
Amoeba	
Stentor	
Giardia	
Water mold	
Slime mold	

With a dissecting microscope, observe the slime mold growing on the surface of an agar-filled petri dish. How does a slime mold move? Does this slime mold have a specific shape?

10 ______________________________

Describe its shape.

11 ______________________________

CONCEPTS/TERMS

Unity and diversity in the kingdom Protista

All protists have eukaryotic cells, as do members of all kingdoms except Monera. The protists are a very diverse group. Many are one-celled organisms. Some are relatively simple colonies of cells, and others are multicellular organisms with specialized structures. Both sexual and asexual reproduction occurs in most species but the details vary. Gamete sizes and the life cycles also vary.

In general, protists can be grouped by their mode of nutrition. One group consists of plant-like photoautotrophs. A second group consists of animal-like heterotrophs. A third group consists of fungus-like heterotrophs.

Plant-like protists

The photoautotrophic protists (sometimes called *algae*) vary in the kinds of pigments they contain and the type of food they produce from their photosynthetic activity. Green algae—which include *Chlamydomonas, Pandorina, Volvox, Oedogonium, Spirogyra* (see Laboratory 4), and *Haematococcus* (Laboratory 4)—have some characteristics in common with plants. They use the same chlorophylls and carotenoid pigments for photosynthesis and produce starch.

Dinoflagellates and large brown algae, the kelps, are also photosynthesizing protists. These plant-like algae have different pigments (except for chlorophyll *a*) and photosynthetic food products than the green algae and plants.

The plant-like protists vary in the number of cells and the complexity of the organism's structure from unicellular *Chlamydomonas* and dinoflagellates to *Volvox* and the kelps, which may be 100 meters in length. (Some biologists include the green, brown, and red algae in the plant kingdom because some of their members are multicellular and have several types of specialized cells.)

Animal-like protists

Animal-like protists such as *Amoeba, Stentor,* and *Giardia* are frequently called **protozoans.** These heterotrophs acquire their food in ways that are frequently associated with their method of locomotion.

In amoeba-like protozoans, part of the protoplasm temporarily flows in a particular direction, pushing out cellular extensions called **pseudopods.** You observed this amoeboid movement as the *Amoeba* engulfed the yeast. The pseudopods surround the yeast and form a food vacuole by phagocytosis. The amoebas add enzymes to the vacuole that digest the food. The products of digestion diffuse or are actively transported across the membrane of the vacuole. The digestive products nourish the amoeba. The undigested materials are eliminated from the organism by exocytosis.

Ciliates are protozoans that use cilia for locomotion and feeding. The coordinated activity of the short cilia move these protozoans rapidly. Some cilia that are located around a funnel-shaped gullet sweep food into the gullet. Most ciliates form food vacuoles at the bottom of the gullet. The food vacuoles pinch off and circle through the cytoplasm as the food is digested. The undigested materials are ejected from the organism by exocytosis. The *Stentor* and the *Paramecium* (see Laboratory 4) are members of this set of protists.

Heterotrophic **flagellates** use one or more long flagella for locomotion. The whipping rotation of the flagella propel the flagellates in a jerky motion. The protozoan flagellates absorb food from their environment through the cell membrane. *Giardia* is a flagellate.

Fungus-like protists

Fungus-like protists are heterotrophs that secrete digestive enzymes into their environments and absorb the digested materials. The slime molds and water molds are examples of this type of protist.

Slime molds spend part of their lives as amoebas (or flagellated cells), feeding on bacteria in compost materials. In one type, acellular slime molds, two amoebas fuse nuclei, forming a zygote. The zygote nucleus divides by mitosis without cytoplasmic division, forming a multinucleated, soft mass called a **slug.** The slug moves slowly and engulfs food particles by extending pseudopodia. Under adverse environmental conditions, the slug erects an upright stalk with a ball-shaped tip, the **sporangium.** Cells within the sporangium divide by meiosis, forming haploid **spores.** After release from the sporangium, the spores become amoebas and the cycle of life forms repeats. Another type of slime mold, the cellular slime molds, forms slugs from clumping haploid amoeba without fusion of nuclei.

Water molds belong to a group of protists that look like but have fundamental differences with members of the kingdom Fungi. The body of water molds consists of diploid, multinucleated filaments called **hyphae.** A collection of hyphae forms the cottony mass, the **mycelium,** of the water molds and the body of many members of kingdom Fungi. You may have observed structures for sexual and asexual reproduction at the tips of the hyphae. Meiosis occurs in sexual structures that include large, egg-producing tips called **oogonia** and smaller, sperm-producing tips called **antheridia.** Asexual motile spores are produced in hyphal tips called **sporangia** by the process of mitosis.

PART THREE Kingdom Fungi

EXPLORING Fungi

Using a dissecting scope or hand lens, examine the petri dish of the common black bread mold. Obtain a clean glass slide and a 2-inch piece of double-side transparent tape. Hold each end of the tape between a thumb and index finger. As your partner momentarily raises the lid of the plate, touch the tape to the fungal growth. Tape the opposite (unused) side of the tape to the glass slide. Examine the tape mount with a microscope. Diagram and describe this fungus in Table 23-4. Are the strands of the fungus divided into cells?

12 ______________________________

Look at a prepared slide that shows where strands of two different forms of the black bread mold have grown together. Note any structures developing at the contact points between short extensions from two hyphae. Add these observations to Table 23-4. What structures on the petri plate or on your slide appear to have either a sexual or an asexual reproductive function? (Hint: It takes two to mate in sexual reproduction; only one is needed for asexual reproduction.)

13 ______________________________

____

Now explore the slide of a cup-shaped structure of either *Peziza* or *Patella*. Particularly note long sacs that hold a row of spores. How many spores are in each row?

14 ______________________________

Draw and describe your observations in Table 23-4.

Table 23-4 Representative fungi

Organism	Diagram and description
Black bread mold	
Peziza or *Patella*	
Mushroom	

Examine a mushroom with the naked eye and a dissecting scope, noting its parts. Remove small portions of its stalk and its cap. (Tease these pieces apart with two dissecting needles.) Prepare wet mounts of each, and examine them with all magnifications levels of a microscope. What type of microscopic structures form the body of the mushroom?

15 ______________________________

Examine the structures on the lower side of the cap. These layered, thin structures are called **gills.** Obtain a prepared slide of a section through a mushroom cap or prepare a wet mount of a small piece of a mushroom gill. Examine the slide on high power, particularly noting small, club-shaped structures covering the surface of the gills. What is attached to some of the club-shaped structures?

16 ______________________________

What is the maximum number of these attachments on any one of the clubs?

17 ______________________________

Describe your observations in Table 23-4.

CONCEPTS/TERMS

Unity and diversity in the kingdom Fungi

The fungi differ from the fungus-like protists. The body form, or morphology, of both fungi and water molds consists of strands called hyphae, which interact forming a cottony mat called a *mycelium*. However, the walls of the hyphae are formed from different chemicals, and the fungal hyphae are sometimes incompletely divided into cells. In addition, the hyphae nuclei of the water molds are diploid; fungi hyphal nuclei are haploid.

At one time, fungi were placed in the plant kingdom, but fungi are quite different from plants. Both have eukaryotic cells with cell walls surrounding them, but their cell walls are composed of different materials. Fungi lack chloroplasts and are heterotrophs that absorb food from their environment such as dead trees and foods kept too long in the refrigerator. In contrast, plants have chloroplasts and are photoautotrophs. The hypha and mycelium body of fungi are very different from plant structures.

In sexual reproduction, two hyphae of different mating types (− and +) join, and the different types of nuclei fuse, forming zygotes. In some groups of fungi, the two types of nuclei may exist together in the same hypha for a time before fusions occur. The zygotes divide by meiosis and produce haploid spores that are distributed to new areas by air currents or animals. These spores form new hyphae and mycelia.

Mycologists (biologists who study fungi) divide fungi into three groups based on their reproductive structures. In fungi such as the black bread mold, short extensions of the + and − hyphae touch and the nuclei from each fuse, forming a zygote. The wall around the zygote thickens and hardens. Because of the formation of this well-protected zygote case, this group of fungi is called the **zygote-forming fungi.**

A second group of fungi has small, elongated, sac-like reproductive structures that form on hyphae containing two different mating types of nuclei. A hypha with two types of nuclei is called a **dikaryotic** ("two nuclei") **hypha.** It forms when a − type hypha contributes its nuclei to the + type hypha. In the reproductive sac of the dikaryotic hypha, the two different types of nuclei fuse, producing a zygote. The zygote divides by meiosis, producing four haploid spores. In some of these fungi, each of the four spores divides once, forming eight spores. When released, the spores produce new hyphae. In some of these

Figure 23-2
Life cycle of the fungus
______________________.

fungi, the reproductive sacs line a cup-like structure that forms from intermingled hyphae. These fungi are called the **sac-fungi.** Examples include *Peziza, Patella,* and the gourmet delicacies of morels and truffles.

The third group of fungi has small, club-shaped reproductive structures, the **basidia** (singular, basidium), that form on dikaryotic hyphae. The basidia are housed in larger structures such as mushrooms that are formed from intermingled hyphae. The basidia cover the surfaces of the mushroom gills. Within a basidium, two nuclei of different mating types fuse and produce a zygote. The zygote divides by meiosis, producing four haploid nuclei. These nuclei move to the surface of the basidium and push out as spores. Air currents distribute spores to new areas, where the they develop into haploid hyphae. Haploid hypha of the two mating types intermingle their contents, forming dikaryotic hyphae, and the life forms repeat. These fungi are called the **club fungi.**

Fungi are ecologically important as decomposers.

APPLYING

Organisms in a hay infusion

Construct a table that consists of the identifying characteristics of each kingdom. After using these characteristics to identify organisms in a culture, attach your table to this manual.

Prepare wet mounts of the microscopic organisms living in a container of water that has been enriched with hay or other organic material. This type of culture is called a *hay infusion.* You may want to add methyl cellulose as you prepare your wet mounts to slow any fast-moving organisms.

Draw the different organisms and identify their kingdom using your knowledge of each kingdom's characteristics. Attach your drawings to this manual.

Constructing the life cycle of a fungus

Using your observations of one of the three fungal groups and information on pp. 273-274, reconstruct the life forms in the order they are produced by the fungus. Because this sequence is repeated, this sequence of life forms is called a **life cycle.** Diagram and label the life cycle in Figure 23-2. Some of the terms that you may want to use are *basidium, diploid, haploid, hypha, meiosis, mitosis, mycelium, sac,* and *zygote.* Also name the fungus you are drawing.

A

B

C

Plate 1

A, A brook trout. **B,** A large rainbow trout. **C,** Brown trout.

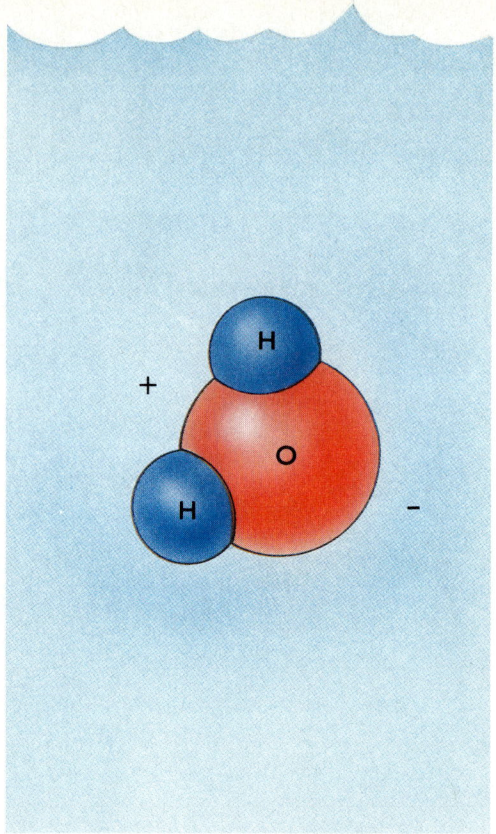

A

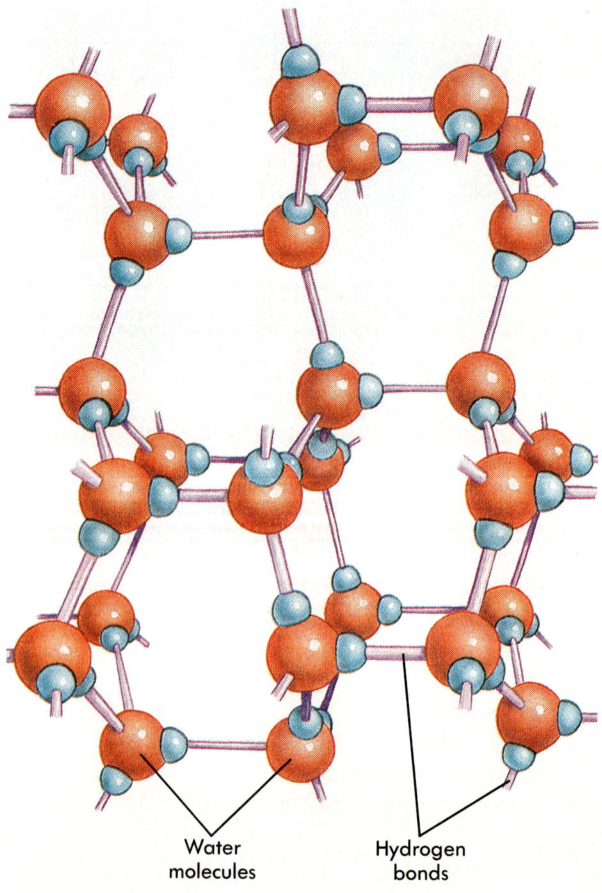

B

Plate 2
Models of the structure of water. **A,** A partial positive charge is on the hydrogen side of the water molecule. A partial negative charge is on the oxygen side of the molecule. **B,** Water molecules "hang together" as the positive charges at the hydrogens and the negative charges at the oxygen of neighboring water molecules attract each other.

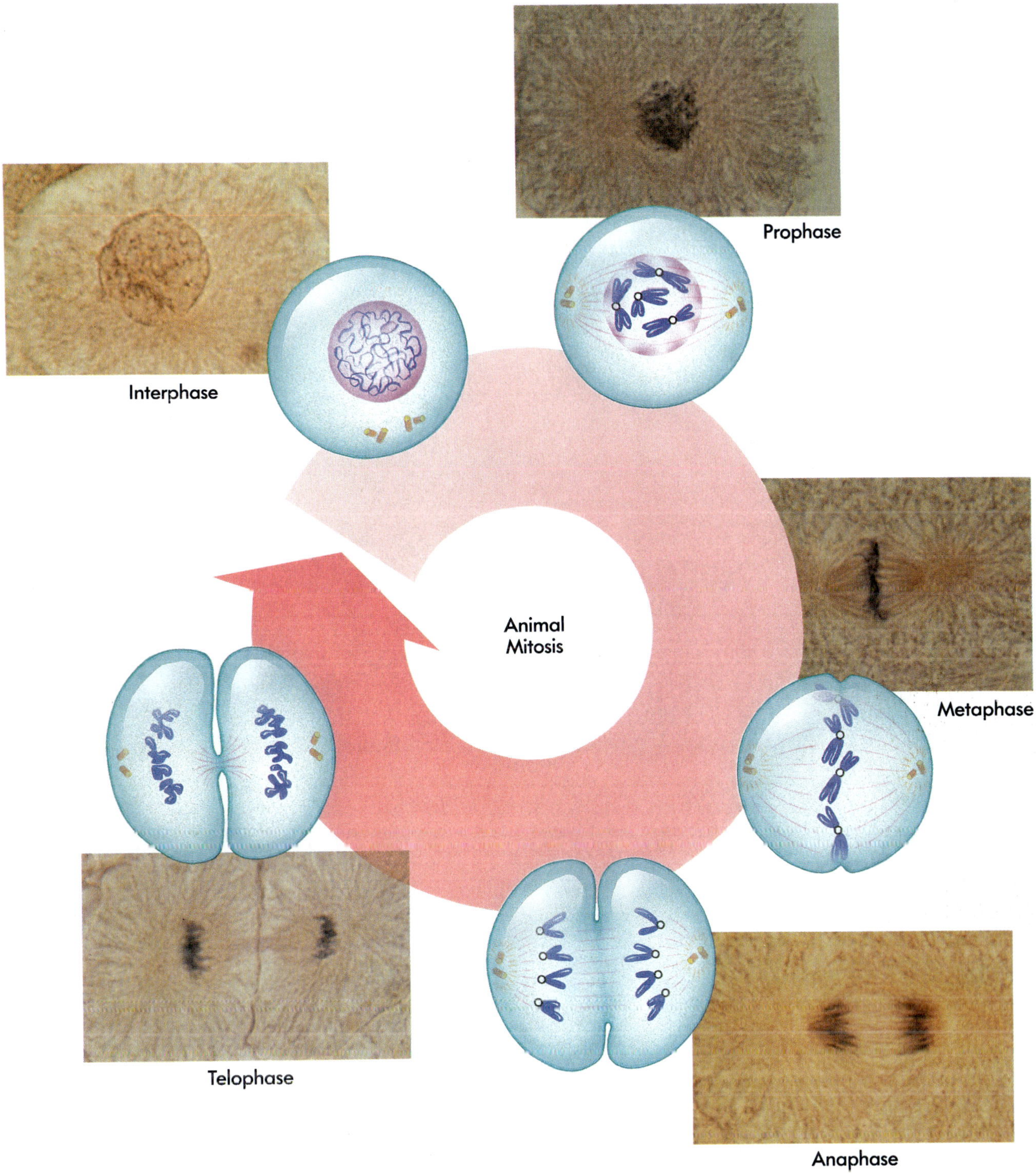

Plate 3
Mitotic cell division in whitefish. The photographs and drawings show the four phases of mitosis and interphase, which is the nondividing phase of the cell.

A

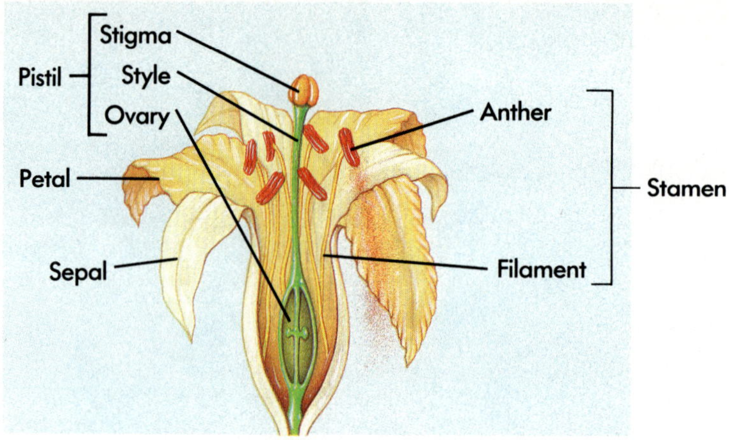

B

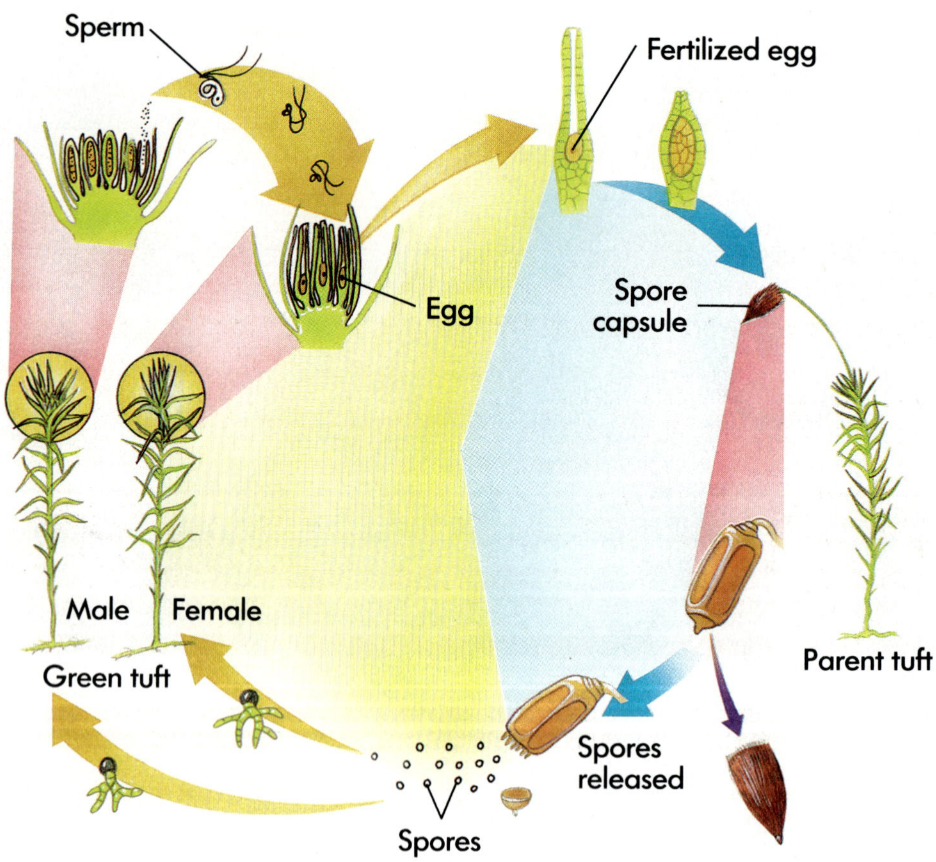

Plate 4

A, The structure of the flower of a lily. **B,** The life cycle of a moss.

A

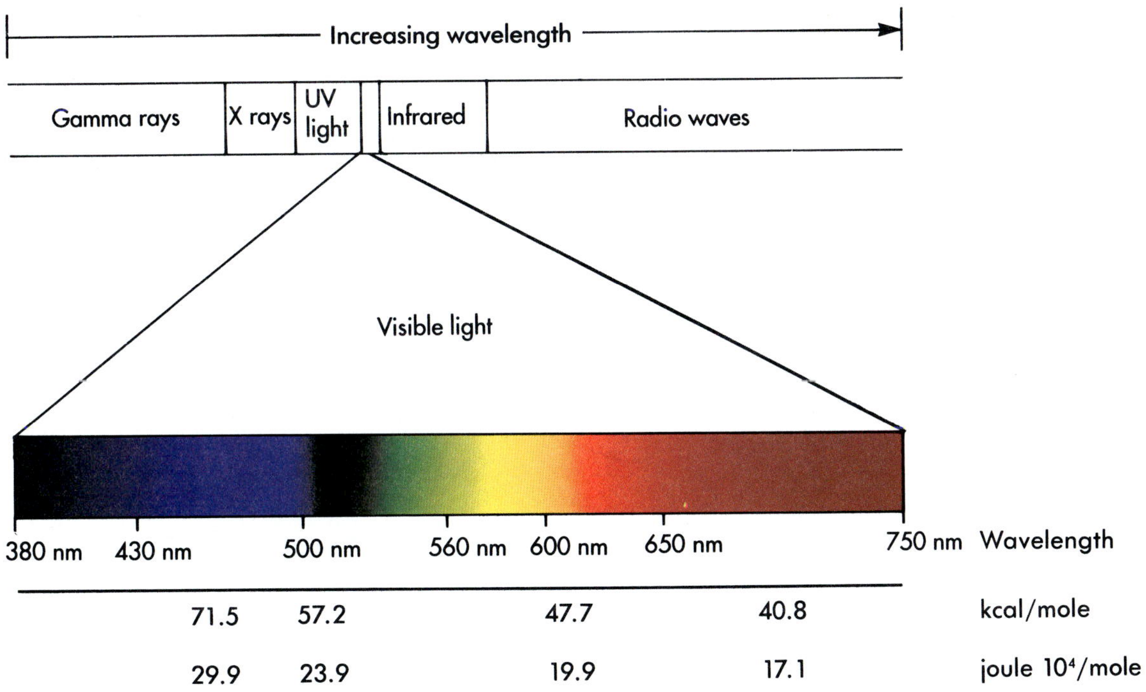

B

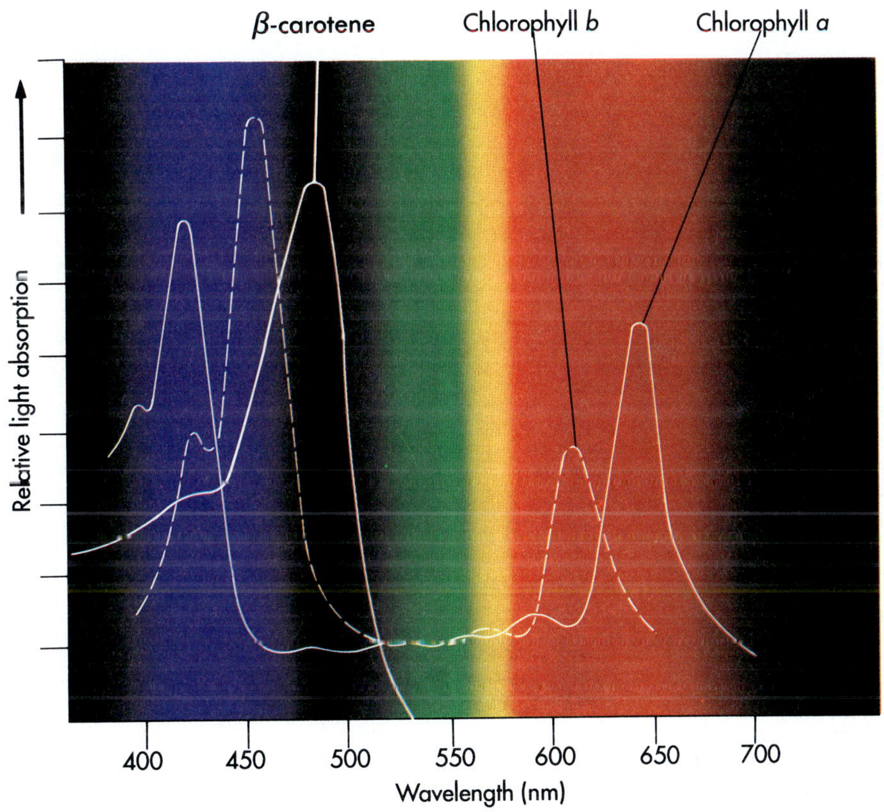

Plate 5
A, The electromagnetic spectrum with the visible light portion enlarged. The wavelengths and photon energy levels of visible light are included. **B,** Absorbance spectra of some common pigments.

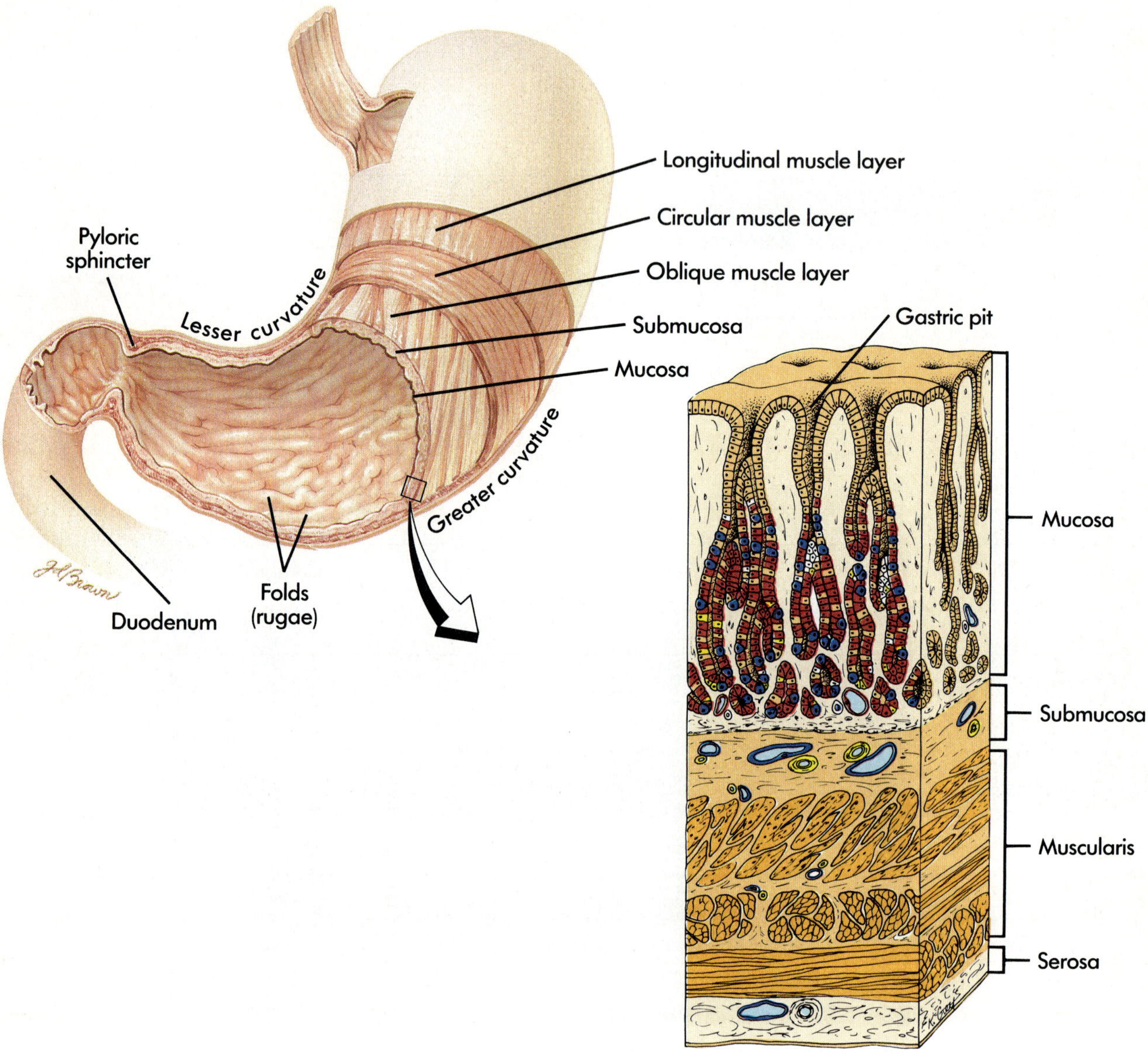

Plate 6

The structure of the stomach. Removal of the outer layers reveals tissues that include three layers of muscles. The insert shows additional details of the tissues in the stomach wall.

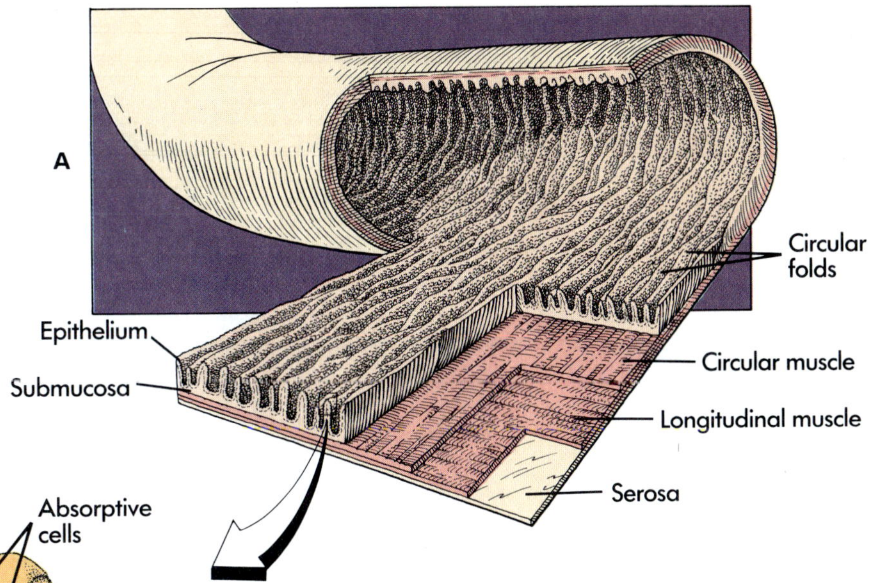

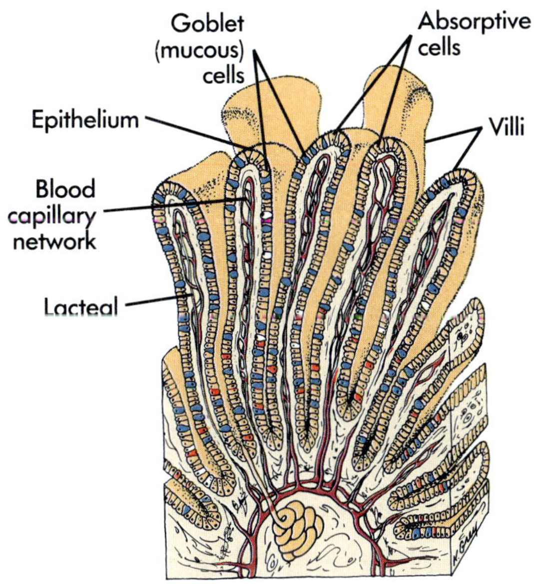

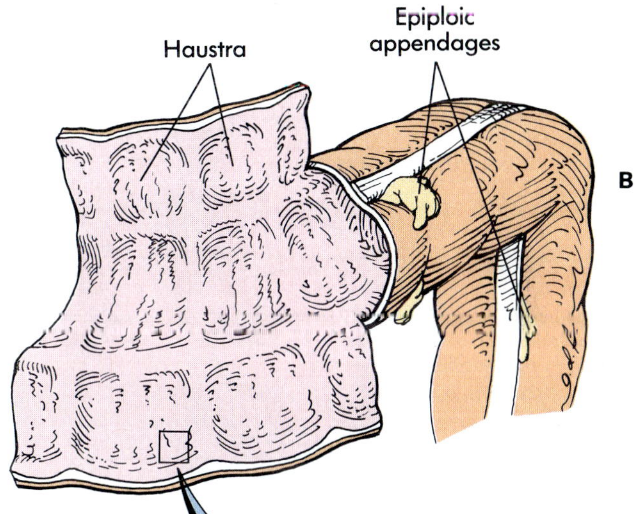

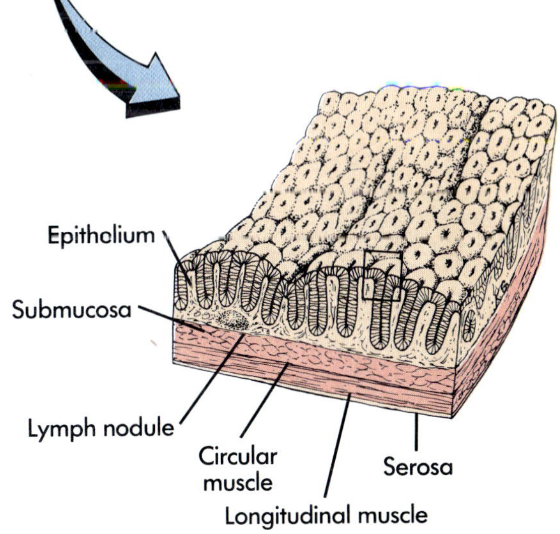

Plate 7
A, The structure of the small intestine (duodenum). The insert shows tissues of the villi, which line the cavity of the small intestine. **B,** The structure of the large intestine. The insert shows additional details of the tissues in the wall of the large intestine.

A

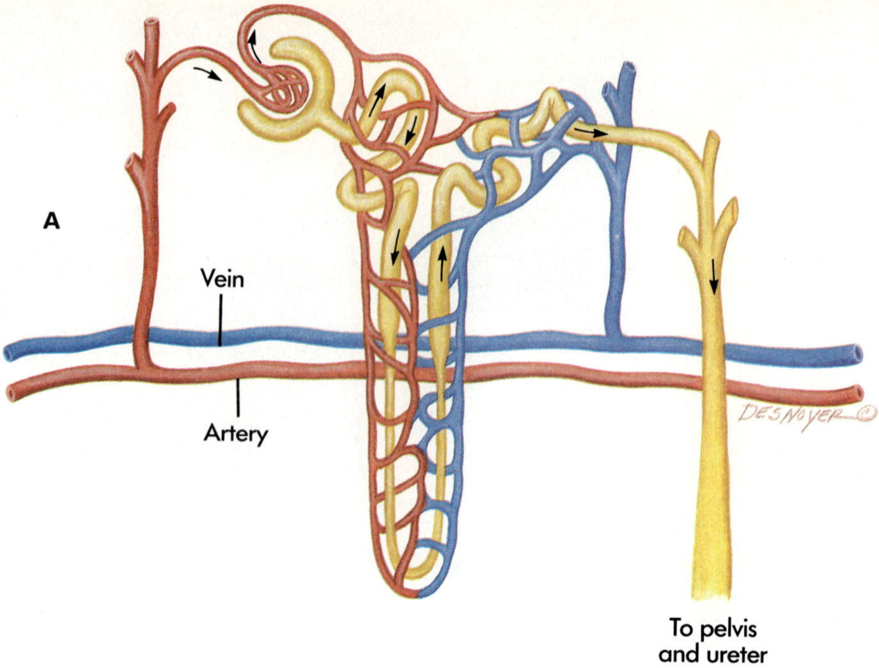

B

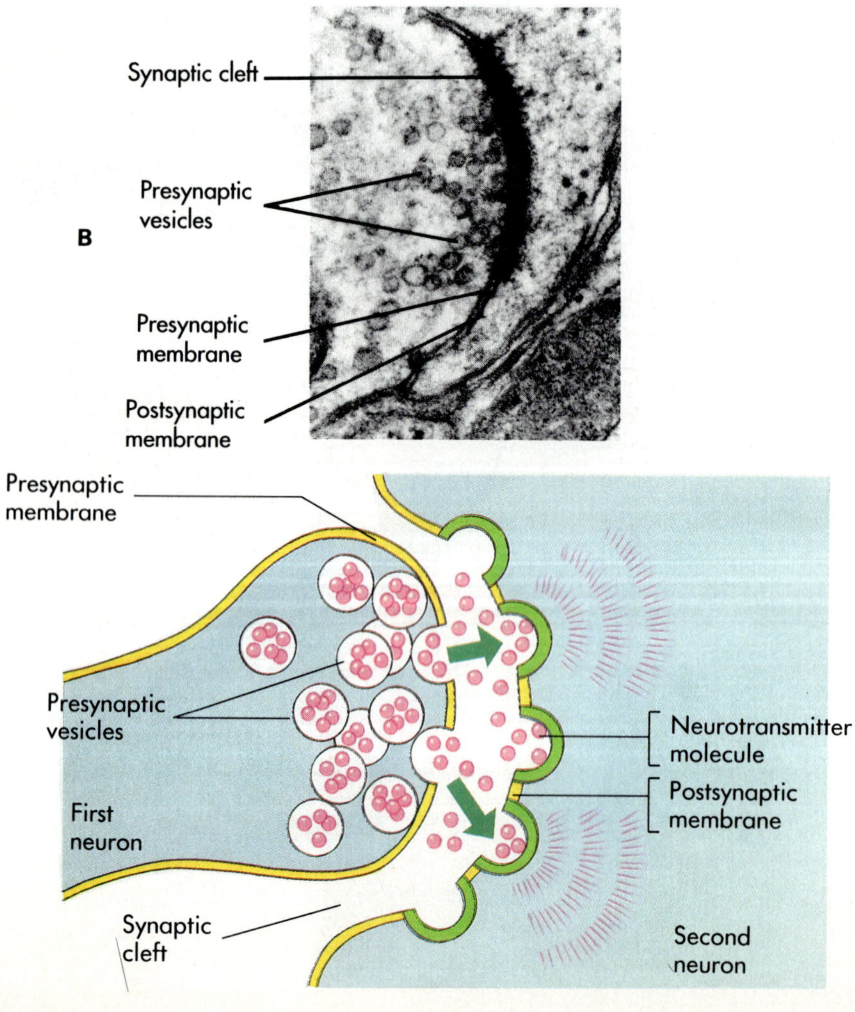

Plate 8
A, The structure of a nephron, the functional unit of the kidney. The fluid is altered as it flows through the nephron tubule. **B,** An electronmicrograph *(top)* of a synapse between two neurons. A drawing *(bottom)* of a synapse showing the release of transmitter molecules into the space (cleft) of the synapse.

A

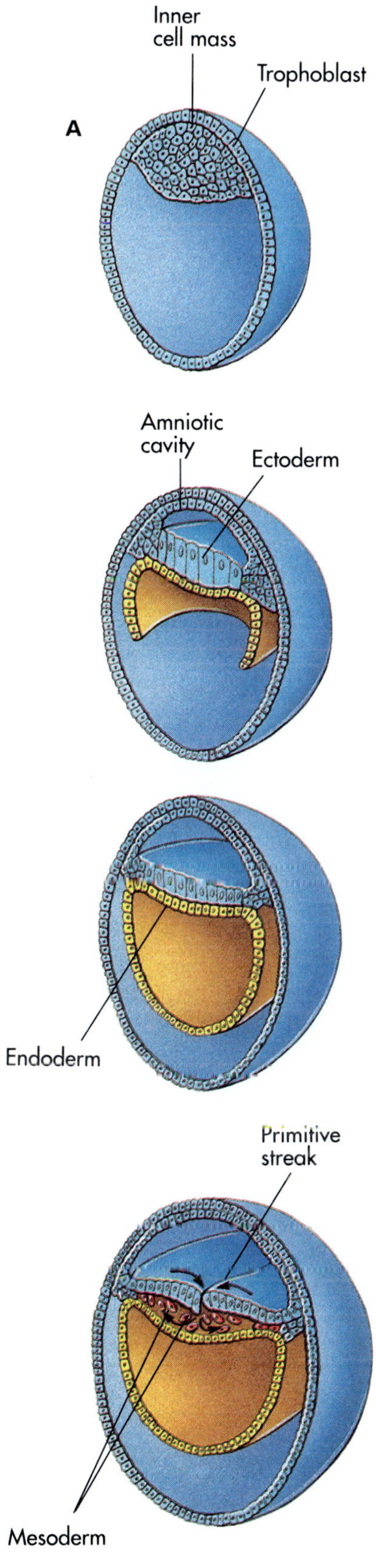

B

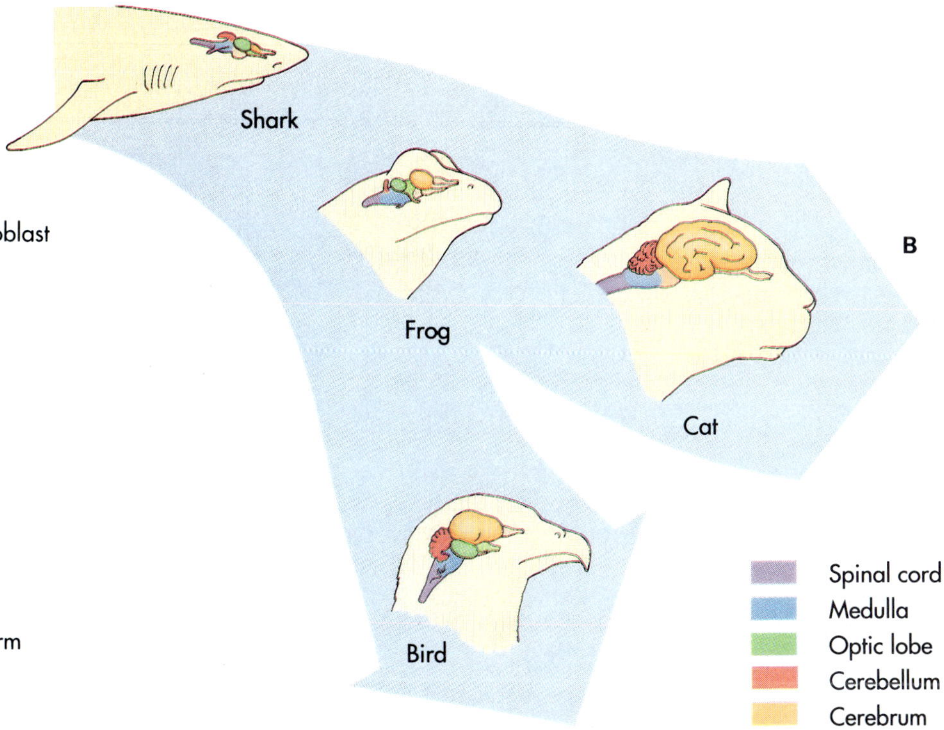

C

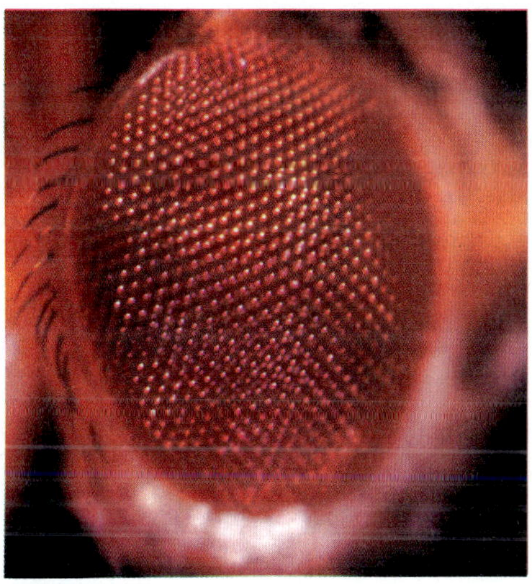

Plate 9

A, Human gastrulation. The human gastrula forms from a inner mass of cells. The ectoderm and endoderm form as the cells move and differentiate into two layers. The mesoderm forms as cells move between the ectoderm and endoderm from the area near the primitive streak. **B,** Vertebrate brains. As the vertebrates evolved, the structure of their brains changed. **C,** The compound eye of the fruitfly, *Drosophila melanogaster.*

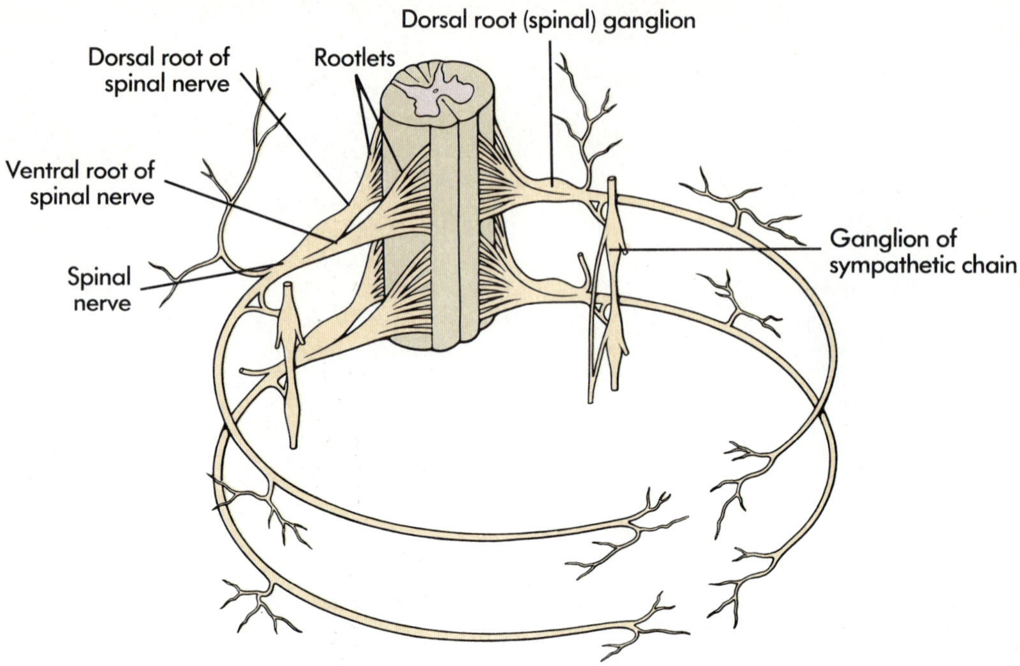

A

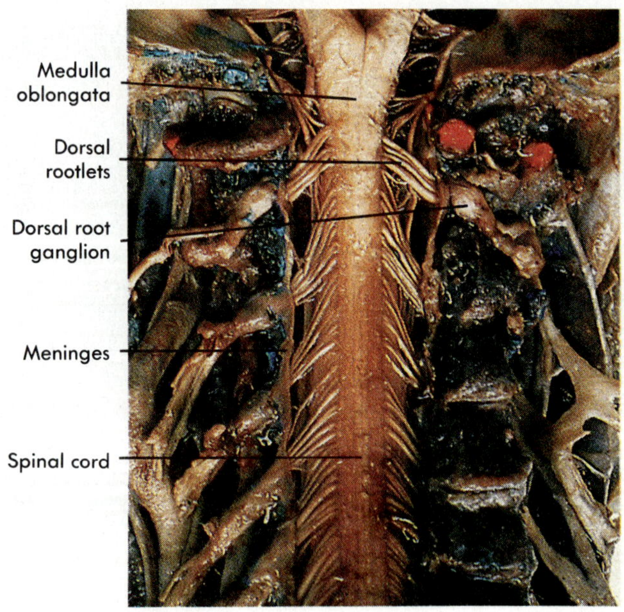

B

Plate 10

A, Typical spinal nerves in the thorax. The two small vertical structures represent the ganglionic chains of the sympathetic nervous system. **B,** The dorsal roots of the spinal nerves entering the spinal cord of a pig.

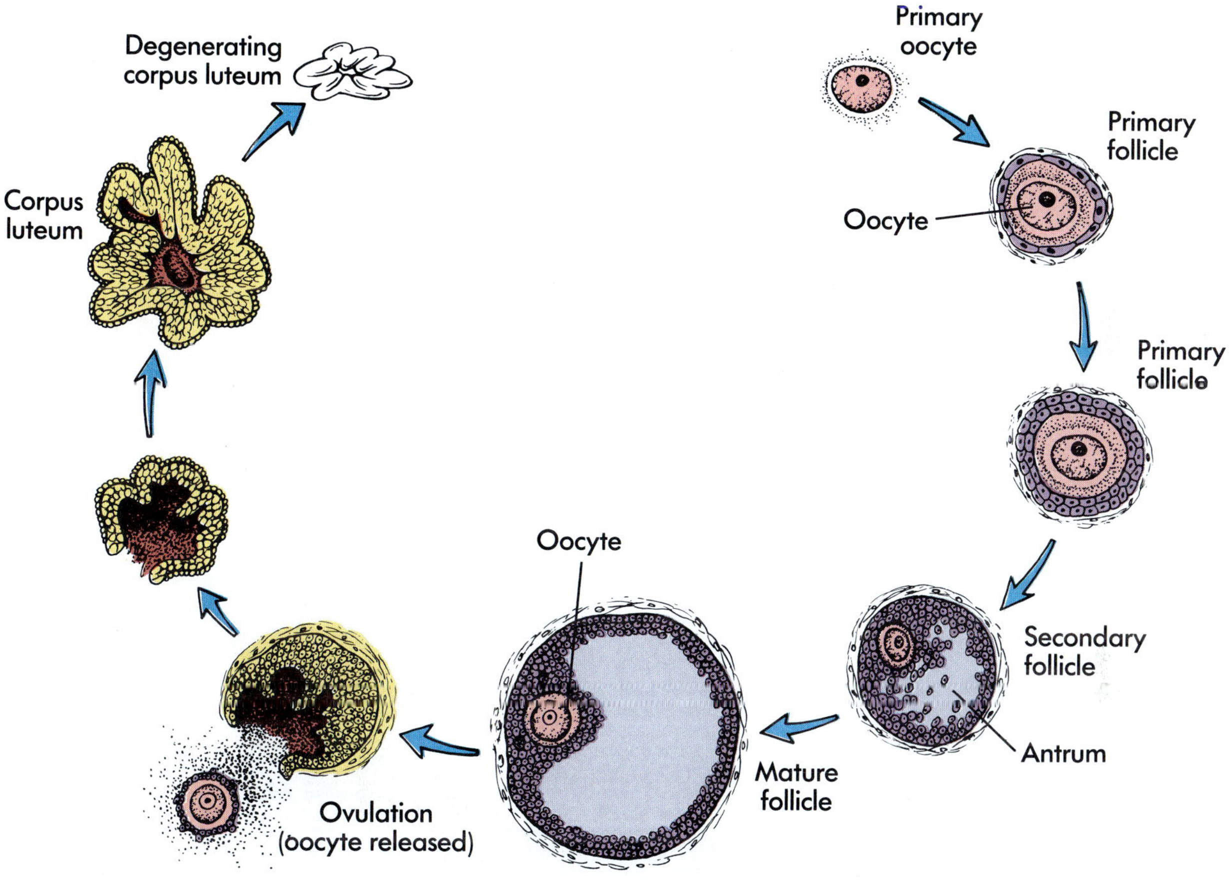

Plate 11
Changes in the ovarian follicle as it matures.

Plate 12
A, Monocots and dicots. In most monocots *(top),* the roots are fibrous, the seeds have one cotyledon, the leaves have parallel veins, and the flower parts are in threes or multiples of three. In most dicots *(bottom),* the roots are taproots, the seeds have two cotyledons, the leaves have net-like veins, and the flower parts are in fours or fives or their multiples. **B,** Types of roots. Taproots in a dandelion *(left).* Adventitious roots in corn *(center).* Fibrous roots in grass *(right).*

A

B

A

Simple squamous epithelium

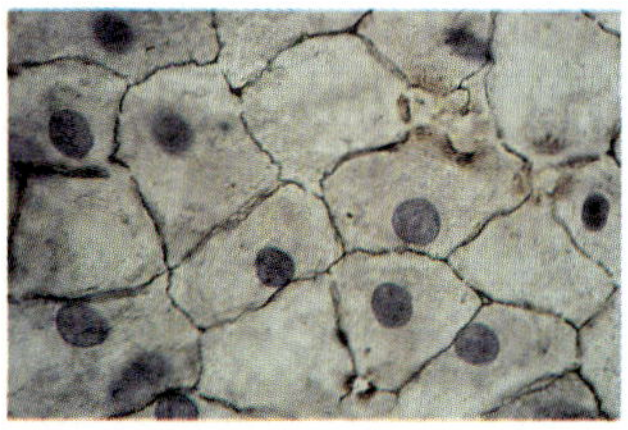

Cuboidal epithelium

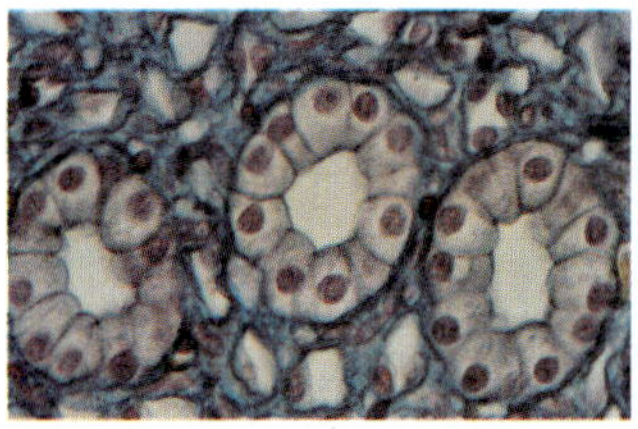

Stratified squamous epithelium

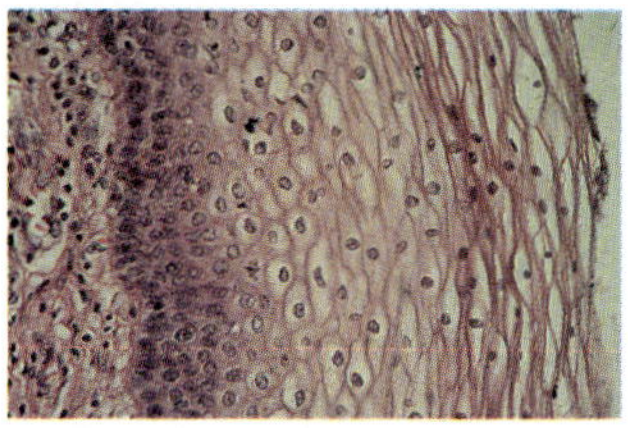

Columnar epithelium

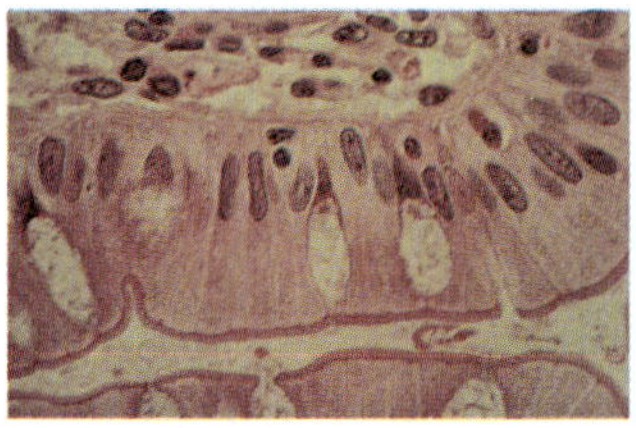

Pseudostratified epithelium

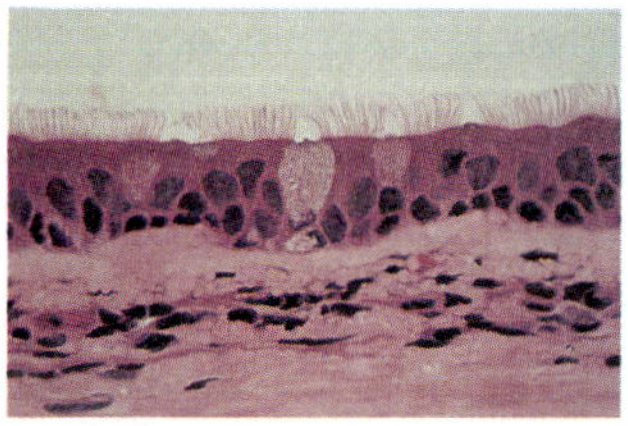

Transitional epithelium

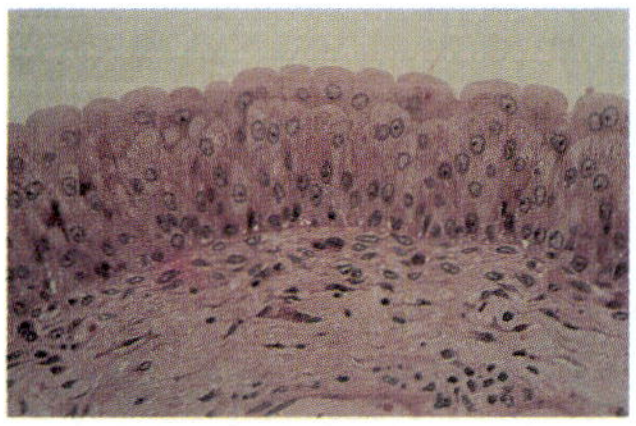

B

Skeletal muscle

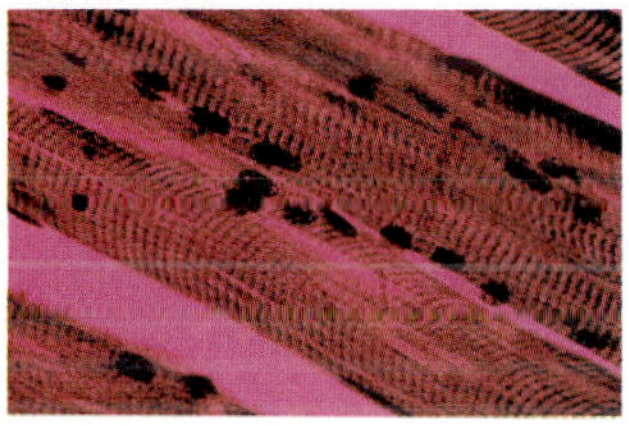

Smooth muscle

Cardiac muscle

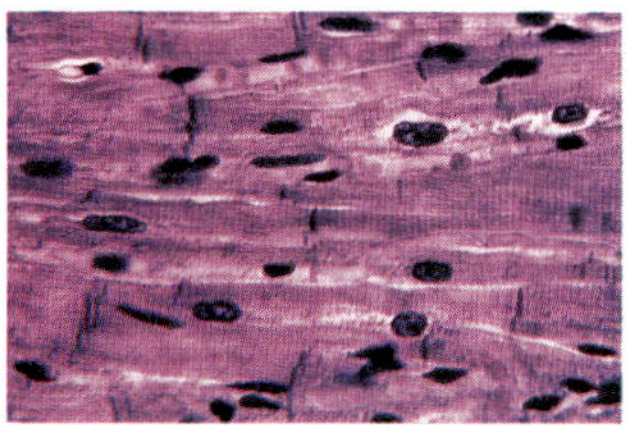

C

Nervous tissue

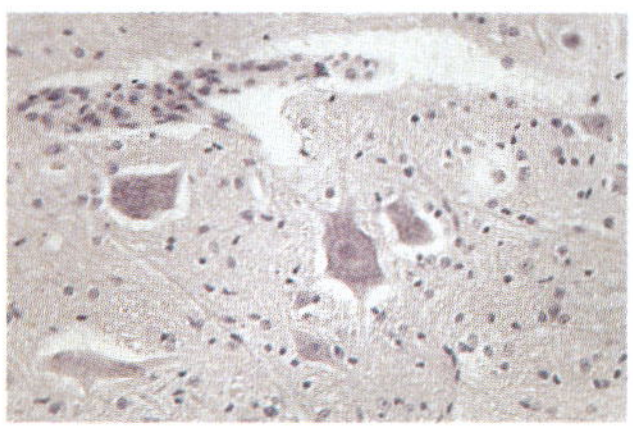

Plate 13

A, Epithelial tissues. Epithelial tissues cover organs and line cavities. **B,** Muscular tissues. The elongated cells that form muscular tissues move attached structures or decrease the size of organs as they shorten. **C,** Nervous tissue. Nervous tissue is made up of the large message-transmitting neurons and small helper cells.

Dense fibrous connective tissue

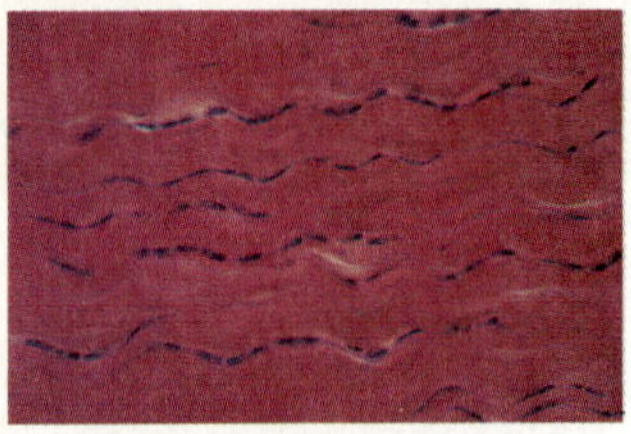

Reticular connective tissue

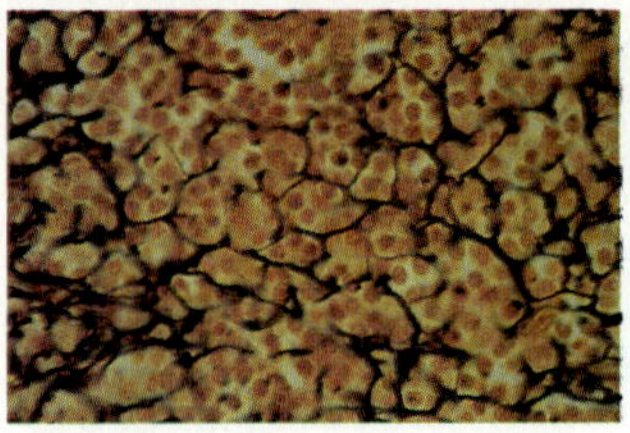

Loose connective tissue

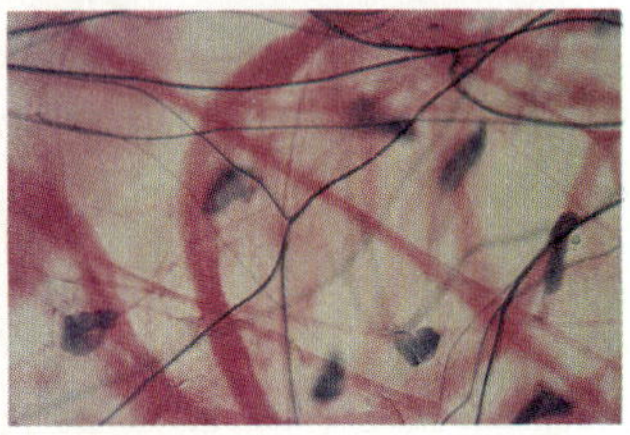

Fibrocartilage

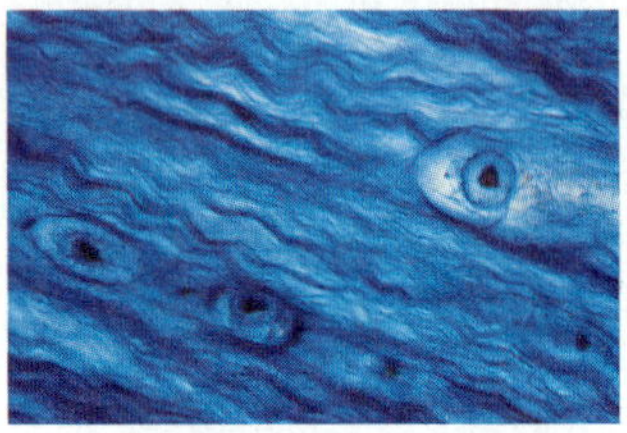

Elastic connective tissue

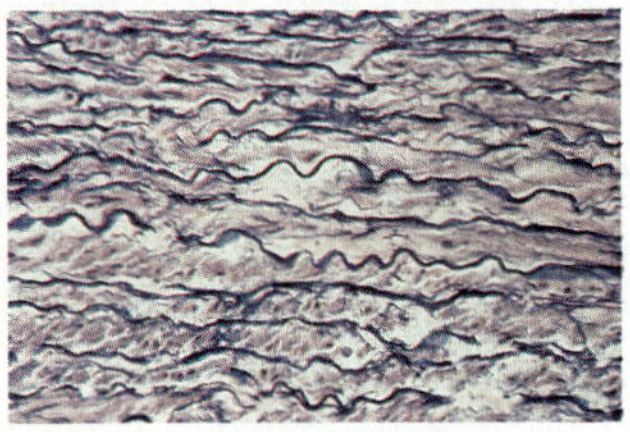

Blood

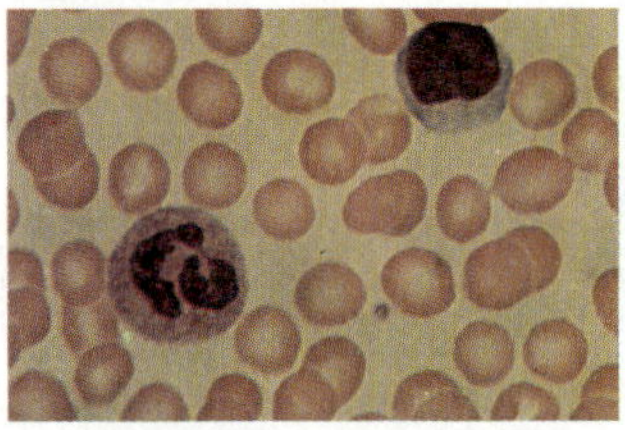

Elastic cartilage

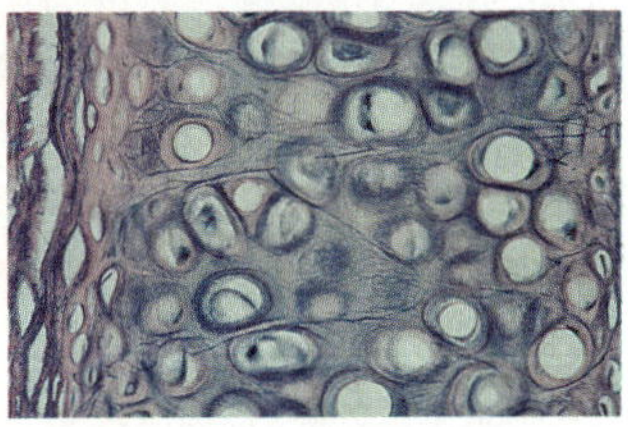

Bone

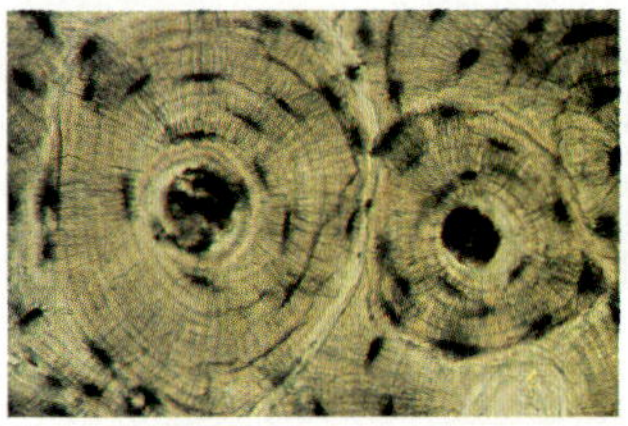

Hyaline cartilage

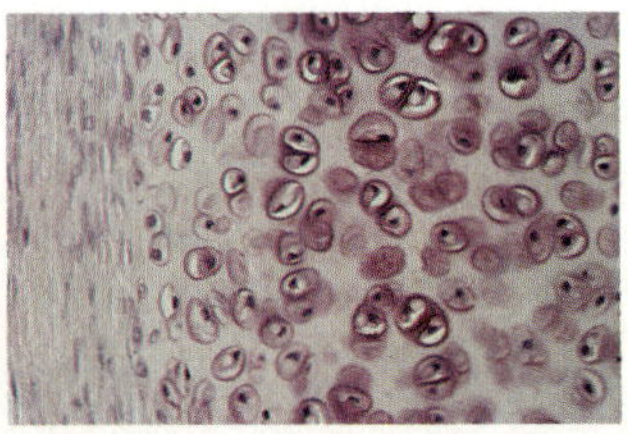

Adipose connective tissue

Plate 14
Connective tissues. The cells of connective tissues are separated from each other by a nonliving matrix. The matrix is important in the function of each connective tissue.

A

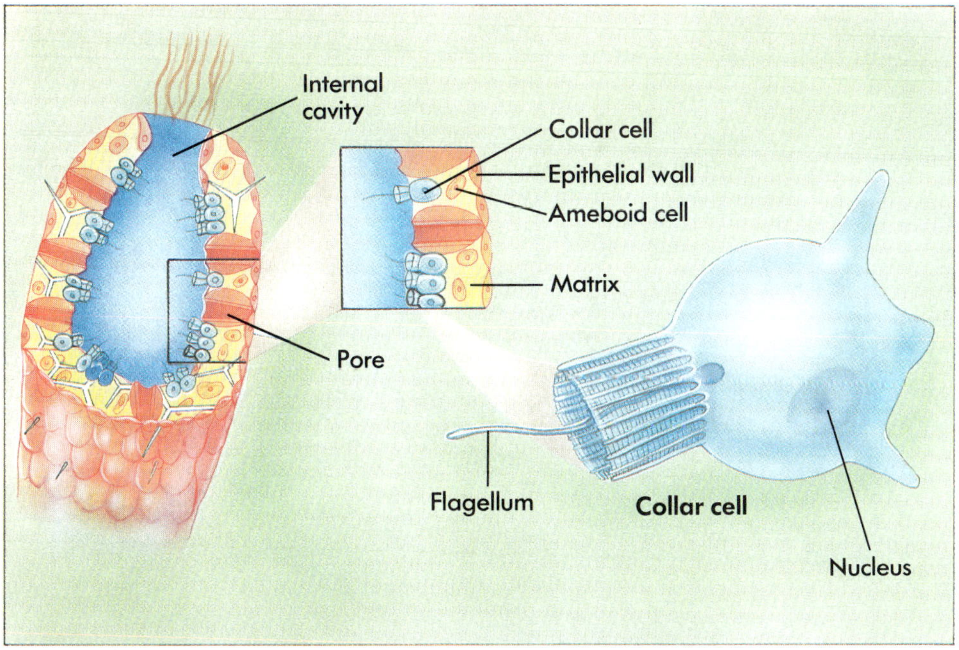

B

C

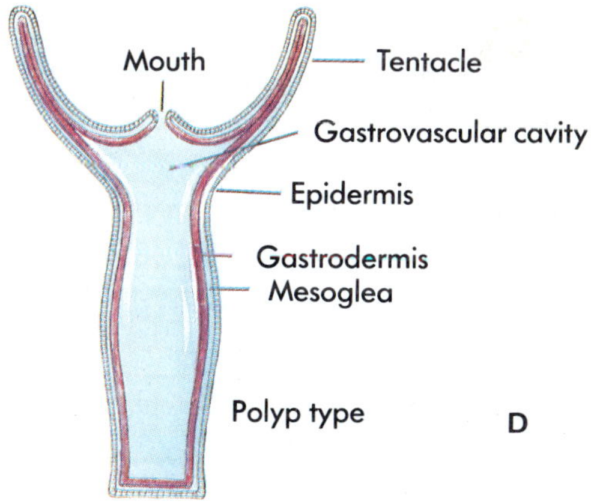

D

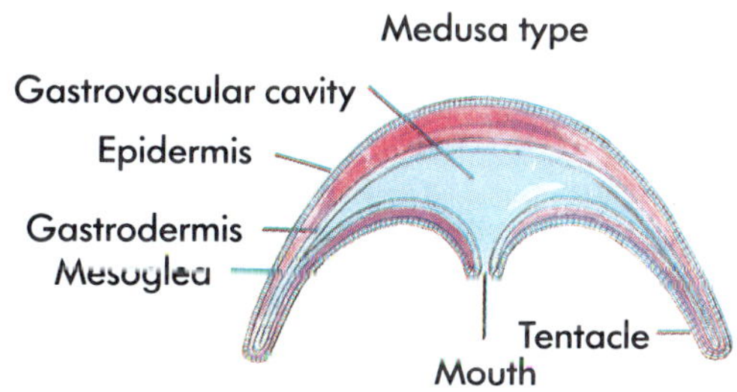

Plate 15
The sponge. **A,** Structure of a sponge with inserts showing addition details of the wall and a collar cell. **B,** Barrel sponges. **C,** A red boring sponge. **D,** The polyp and medusa forms of cnidarians.

A

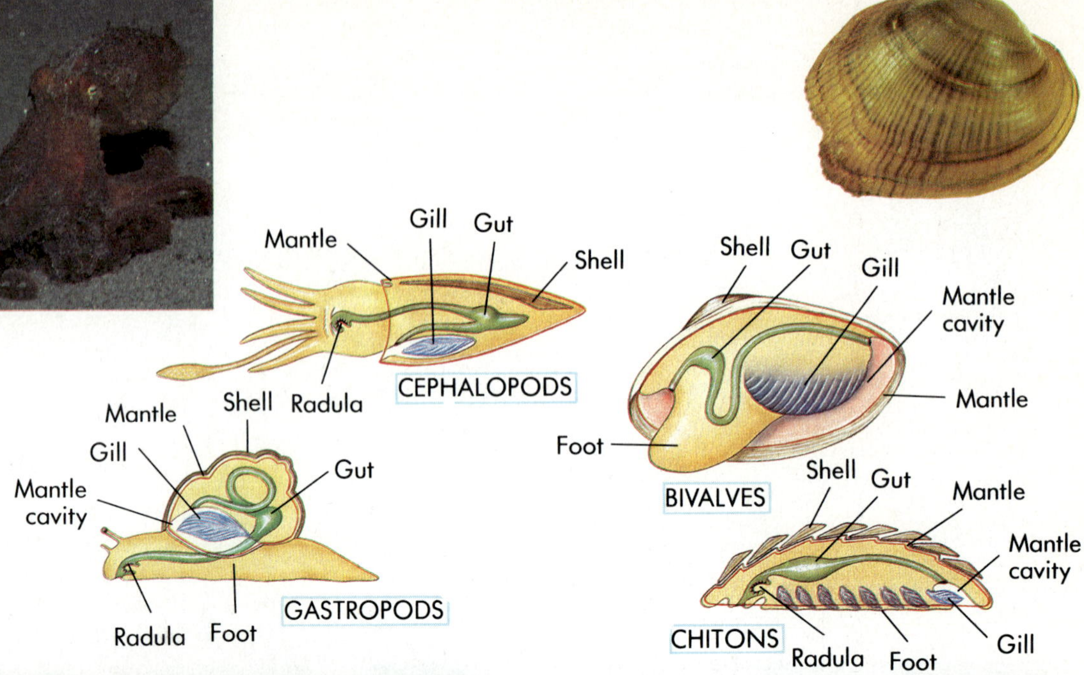

Plate 16
A, Diversity in the mollusks. **B,** Two common poisonous spiders. The female black widow spider *(left)* is identified by the red hour glass on its ventral abdomen. The brown recluse spider *(right)* is identified by the dark brown violin on its dorsal cephalothorax.

B

LABORATORY

24

Plants

Alternation of generations

Alternation of phases in cycles of life

When biologists step back and think about all forms of life, they realize that sexually reproducing organisms alternate between a diploid phase and a haploid phase. The processes of **meiosis** and **fertilization** transfer organisms between these phases. Organisms move from the diploid to the haploid phase when some of their cells divide by meiosis. They return to the diploid from the haploid phase when their gametes unite by fertilization. One sequence that includes meiosis, a haploid phase, fertilization, and a diploid phase is called a **life cycle.**

The prominence of the haploid phase compared to the diploid phase varies among groups of organisms. Organismal groups differ in (1) the amount of time they spend in each phase, (2) the size of the haploid forms compared to diploid forms, and (3) the complexity of the forms. An important key in distinguishing among the somewhat confusing life cycles of organisms is noting whether mitosis occurs in the phases. Two contrasting ones are the animal (for example, human) and fungus life cycles.

In humans, the diploid fertilized egg **(zygote)** divides many times by mitosis and the resulting cells differentiate. Through these processes, the zygote develops into the specialized, multicellular, diploid adult human. Cells in mature human ovaries and testes divide by meiosis forming the gametes—the inconspicuous, brief haploid phase. The gametes do not divide by mitosis. Instead, they unite in the process of fertilization, forming the zygote, and the diploid phase of humans begins again. This sequence of events and organismal forms repeats in cycles of human life. Figure 24-1 illustrates the relationship between diploid (2n) and haploid (1n) phases in humans.

In fungi, the alternating phases differ from those in humans. The fungal diploid phase consists of one cell (the zygote), and the haploid phase is multicellular and specialized. The zygote divides not by mitosis, as do animals, but by meiosis. This initiates the haploid phase. The haploid spores (products of zygote meiosis) divide repeatedly by mitosis, producing the multicellular (or multinucleated) haploid hyphae of the mycelium. When the hyphal nuclei of the two mating types (+ and −) fuse, the fungus returns to the 1-cell diploid phase. Figure 24-2 illustrates the relationship between the haploid and diploid phases in fungi.

Some of the photosynthesizing protists also alternate phases in the same way as the fungi. In other protists, however, both phases are equally prominent or the diploid phase is more prominent.

Plants have a sequence of multicellular forms in both phases. Botanists (biologists who study plants) call all forms of an organism that possess the same genetic complement (either diploid or haploid) a **generation.** The cycling between haploid and diploid generations is called the **alternation of generations**

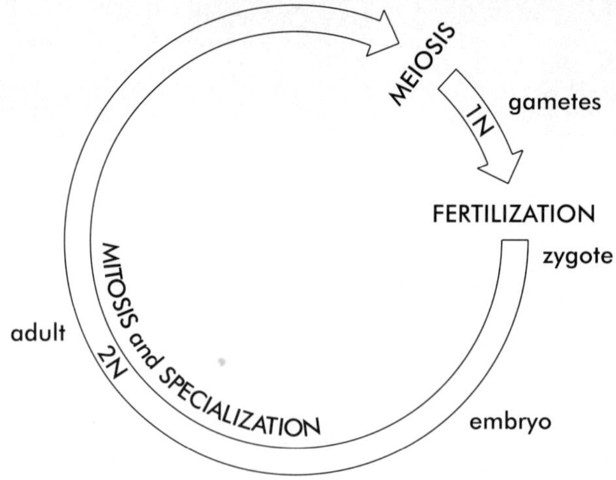

Figure 24-1
Human life cycle.

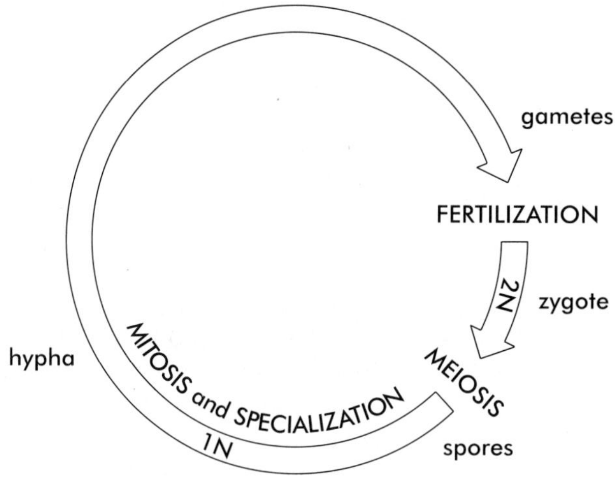

Figure 24-2
Fungus life cycle.

life cycle. The generation that is most conspicuous (exists for the longer time, is the larger, and is the more complex) is called the **dominant generation** of the plant's life cycle.

Today you will explore the alternation of generations life cycle in four major plant **divisions** (the level of organismal grouping comparable to phyla in animals).

PART ONE Bryophytes (mosses, liverworts, and hornworts)

EXPLORING The forms of a moss

Obtain the following materials:
- Preserved or fresh moss specimens
- A microscope slide set of smaller moss structures
- A compound light microscope
- A dissecting scope or hand lens

Begin with the preserved or fresh tuft of green moss that has a slender stalk with a container-like capsule, or **sporangium,** at its tip. Particularly observe the color, size, and relative complexity of its parts as well as the arrangements of each part to the others. Which parts, if any, grow into the soil?

1 ______________________________

In mosses, the cells of the green tufts are haploid and the cells of the slender stalk and sporangium are diploid. From this information, which of these generations is the dominant generation in mosses?

2 ______________________________

Describe the evidence on which you based your answer.

3 ______________________________

____

What is the major role of the haploid green tuft in the alternation of generations life cycle?

4 ______________________________

What is the major role of the diploid sporangium in the alternation of generations?

5 ______________________________

With your naked eye or the scanning objective, examine two microscope slides, one labeled *archegonia* and the other labeled *antheridia.* Determine what part of the preserved or fresh moss is on these two slides. Archegonia is from

6 ______________________________

Antheridia is from

7 ______________________________

Examine each of these slides with the low power and then high power objectives. Using the archegonia slide, locate one or more elongated, flask-shaped (or vase-shaped) structures at the tip of the specimen. Using the antheridia slide, locate relatively large, oval containers. Compare the cell in the center of the expanded portion of the vase-shaped archegonium to the cells inside the oval antheridium. Describe these cells.

8 Archegonium cell ______________________________

____

Antheridium cells ______________________________

____

At this point, what do you think is inside the archegonium?

9 ______________________________

What do you think is inside the antheridium?

10 ______________________________

On what evidence did you base your answers?

11 ______________________________

____

With your naked eye or the scanning objective, examine two slides labeled *young capsule* and *mature capsule*. What part of the preserved or fresh material is on these slides?

12 ______________________________

Examine the two capsule slides with high magnification. Compare the cells in the interior of the young capsule (sporangium) to those in the mature capsule. Based on your observations, what process occurs inside the capsule?

13 ______________________________

On what evidence did you base your answer?

14 ______________________________

____

What role do the cells in the interior of the sporangium have in the life cycle of moss?

15 ______________________________

Examine a slide labeled *protonema*. Describe this moss form.

16 ______________________________

CONCEPTS/TERMS

The life cycle of moss

Mosses have an alternation of generations life cycle with a dominant haploid generation. The haploid green tufts are what you probably recognize as "moss." The haploid generation is called the **gametophyte,** which simply means "the plant that produces the gametes." In the tip of one type of green tuft, the vase-shaped **archegonium** produces a haploid, nutrient-rich cell, the **egg.** The egg is not produced by meiotic cell division but by the mitotic division and differentiation of the haploid cells that make up the gametophyte generation. On the tip of another green tuft, the daughter cells of other dividing haploid cells within the oval **antheridium** differentiate into **sperm.** When mature, the antheridium ruptures and releases its motile sperm. Water such as splashing raindrops transfers the sperm to the female (egg-producing) gametophyte. The sperm swim down the open neck of the archegonium to the egg, and the two gametes fuse, forming a zygote. The resulting zygote is the first cell of the diploid **sporophyte** ("the plant that produces spores"). The zygote divides by mitosis and forms the stalk and capsule of the multicellular sporophyte. The sporophyte remains attached to and receives most of its food, water, and minerals from the gametophyte. The sporophyte completes its role in the moss life cycle when the cells within the sporangium divide by meiosis and produce haploid **spores.** When environmental humidity levels change, tooth-like structures at the end of the sporangium fling these spores into the air. If they land in a moist environment, the spores divide by mitosis and form a simple filamentous stage, the **protonema.** Buds on the filaments develop into the green tufts of the mature gametophyte. Then the cycle repeats.

APPLYING

The life cycle of moss

Draw the life cycle of moss in Figure 24-3. Label the important structures of each generation. Terms that you may want to use are antheridium, archegonium, diploid (or 2n), egg, fertilization, filaments (or protomena), gametophyte, haploid (or 1n), meiosis, mitosis, sperm, sporangium (or capsule), spores, sporophyte, and zygote. Using the metric system, give the approximate size of these structures.

Compare your moss life cycle with your neighbors'. Is the ordering of the moss forms different? Are the terms *meiosis, fertilization,* and *mitosis* in the same place? Discuss any differences and describe your reasons for drawing the life cycle as you did.

What evidence supports botanists' contention that the haploid generation is the dominant generation in mosses?

17 ____

____

CONCEPTS/TERMS

The alternation of generations life cycle in plants

All plants have an alternation of generation life cycle in which both generations have multicellular forms. The haploid forms are called the **gametophyte generation,** and the diploid forms are called the **sporophyte generation.**

As in moss, a few cells of the haploid multicellular gametophyte differentiate into **gametes.** The gametes unite, producing the first cell of the diploid sporophyte generation, the zygote. The zygote divides by mitosis, and the

Figure 24-3
Moss life cycle.

daughter cells differentiate into a mature multicellular sporophyte. A few cells of the mature sporophyte divide by meiosis and produce the first cell of the gametophyte generation, the spore. The plant completes one life cycle when the spore divides by mitosis and the daughter cells differentiate into the mature gametophyte.

Plants vary in the dominance of the two generations in the life cycle. Systematists (biologists who study relationships among organisms) base the major branching point in the family tree of plants in part on whether the dominant generation is the gametophyte or sporophyte.

Next you will explore another major group of the plants, the ferns, to understand their version of the alternation of generations life cycle.

PART TWO Pterophytes (the ferns)

EXPLORING

The forms of a fern

Obtain a set of microscope slides of the smaller structures and either fresh or preserved specimens of the larger forms of ferns.

Explore the plant that you know as a fern. The cells of this form are diploid. Some suggestions and questions to initiate your explorations include the following:

1. Which parts of the plant are leaves, roots, and stem?
2. Which of these organs are photosynthetic?
3. If very young green leaves are present on the fern, what is their shape?

In Figure 24-4, draw this form of the fern and include written notes.

If the fern you have been exploring has many small, round structures on the underside of its leaves, remove one of these by slipping under it with a sharp razor blade. Place this structure on a dry microscope slide. If the fern

Figure 24-4
Fern structures.

lacks these structures, obtain a prepared slide labeled *sori* (singular, sorus) or *sporangia.* Using your microscope, explore the sorus and describe its contents.

18 ______________________________

Are all containers within the sorus filled with small dense structures, or are some open and empty?

19 ______________________________

If you are observing fresh material, watch for one of the containers to pop open and eject its contents as it dries. Based on your knowledge at this point, what is the probable role of the containers and their contents in the life cycle of ferns?

20 ______________________________

Using your knowledge of moss, what should these "containers" be called?

21 ______________________________

Obtain slides labeled *mature fern prothallium* and *old fern prothallium.* Using the light microscope, explore the small, heart-shaped structure on the mature fern prothallium slide. Try locating small, dense, dome-like structures on the underside of the heart in the area surrounding the notch. Also try locating small, ball-like, less dense structures in areas away from the notch. The cells of the fern prothallium are haploid.

Explore the old fern prothallium with your naked eye. Then examine it with your microscope. Describe the differences between the old prothallium and the mature prothallium.

22 ______________________________

Are any prothallia growing on the soil surface of potted ferns in the laboratory?

23 ______________________________

What generation of the fern does the prothallium represent?

24 ______________________________

What is the role of the prothallium in the life cycle of ferns?

25 ______________________________

What is the new structure growing from the old prothallium?

26 ______________________________

Is the prothallium or the first fern form that you examined the most dominant form in the fern life cycle?

27 ______________________________

Figure 24-5
Fern stem.

On what evidence did you base this answer?

28 ______________________________

Examine a cross section of a fern stem with the microscope. Particularly note the areas of different cell types, the size of the cells, and the cell wall thicknesses. In Figure 24-5, sketch the cross section and outline the areas of different cell types. Draw the details of one cell of each type.

CONCEPTS/TERMS

The life cycle of a fern

The sporophyte generation of the fern (the plant with the leaves, horizontal stems, and roots) dominates the fern life cycle. In fact, the leaves are so large and complex that biologists give them the special name of **fronds.** Stems called **rhizomes** store food and run horizontally in the soil.

The ferns cycle from the diploid to the haploid generation within the containers, the sporangia, of the sorus. The diploid spore mother cells within the sporangia divide by meiosis and produce haploid spores. In drying environments, the sporangia pop open, flinging out the spores. If they land in a favorable environment, the spores produce the small, heart-shaped gametophytes. The gametophyte (the **prothallium**) produces eggs in its archegonia near the notch. The ripe eggs are fertilized in the archegonia by the sperm that swim in a film of water from the antheridia (the less dense structures away from the notch). The zygote divides by mitosis and forms the young sporophyte. The gametophyte soon disintegrates, and the young sporophyte grows into the large fern.

In addition to having a dominant sporophyte generation, the ferns differ from the bryophytes in possessing specialized fluid-conducting tissues called **vascular tissues.** Two types of conducting tissues form **vascular bundles.** The empty, stiff, long, thick-walled cells of the **xylem** transport water with its dissolved minerals. The softer, thinner-walled cells with adjacent companion cells of the **phloem** conduct food from photosynthetic areas of the plant. These transport systems were instrumental in the plants successfully invading land environments.

In the plant kingdom, bryophytes are the only major group, or division, of organisms that has a dominant gametophyte generation. The ferns and other divisions of the plant kingdom have a dominant sporophyte generation. In addition, plants with the dominant sporophyte generation also have vascular tissues; in general, bryophytes lack these tissues.

Figure 24-6
Fern life cycle.

APPLYING

The life cycle of a fern

Draw the life cycle of fern in Figure 24-6. Include the important structures of each generation and label them. Add terms such as *antheridia, archegonia, diploid* (or *2n*), *egg, fertilization, gametophyte* (or *prothallium*), *haploid* (or *1n*), *meiosis, mitosis, sperm, sporangium, spores, sporophyte,* and *zygote*. Also give the approximate size of each structure.

PART THREE Gymnosperms

EXPLORING

The forms of a conifer

Explore either a fresh or preserved branch of a pine tree. Also examine a slide of a pine stem cross section. What kinds of cells are in this stem that are also in the fern stem?

29 ______________________________

What organs does a pine tree possess?

30 ______________________________

Either on the branch or in nearby dishes, locate cones of various sizes. Arrange the cones in order of their size. If you find small, somewhat softer cones that are covered with a powder or softer cones without the powder, set these cones in a separate row. (For easy reference, call these softer, flimsier cones *type I cones.*) Cut a powdery and a nonpowdery type I cone longitudinally with a sharp, single-edge razor blade. Closely examine these sectioned cones with a dissecting scope or magnifying lens. If a prepared slide of a longitudinal section of the cone is available, examine it with the light microscope. In Figure 24-7, sketch the structure of these cones.

Figure 24-7
Pine cone (type I).

When you examined the type I cone, did you locate oval containers inside the scales (modified leaves extending from the cone's central axis)?

31 ______________________________

If not, examine the sectioned cone again. If you are unsure, ask your teacher for assistance. Add these structures to Figure 24-7. Describe these containers and their contents.

32 ______________________________

____

What part of the life cycle of pine is represented by the containers' contents? (The contents could be individual cells or the cells could have developed into simple structures.)

33 ______________________________

What term also used for fern and moss can be given to the containers' contents?

34 ______________________________

When the contents are released from the container, they are called a term that is familiar to most people (particularly those who suffer from hayfever). What term do you think is associated with these structures?

35 ______________________________

Return to your row of the sturdier cones (type II cones). With a single-edge razor blade or large knife, cut the cones longitudinally. Using a dissecting scope or hand lens, explore the cut surfaces. If prepared slides of longitudinal sections of the small cones are available, examine them with your light microscope. As you observe these materials, particularly note the changes in the

Figure 24-8
Pine cone (type II).

scales as these cones grow and mature. In Figure 24-8, sketch both a small and a large cone. With your drawing and written notes, describe the changes occurring in the scales.

What common term (also a biological term) refers to the structures at the base of the scales in the largest cones?

36 ______________________________

Where do you think these oval structures fit into the pine life cycle?

37 ______________________________

How do you think they are distributed to new environments?

38 ______________________________

What generation of the pine life cycle does the big pine tree represent?

39 ______________________________

On what evidence and information did you base your answer?

40 ______________________________

CONCEPTS/TERMS

The life cycle of a conifer

Gymnosperms such as conifers are vascular plants with a dominant sporophyte generation. This is the first plant division you have explored that produces **seeds.**

In conifers the gametophyte generation is reduced in size and is retained in the sporophyte generation, the pine tree. In fact, the female gametophyte grows entirely within the large, sturdy **cones** of the old sporophyte generation. In addition, the new sporophyte generation begins in the cone where a female gametetophyte produced an egg. The new sporophyte generation is the embryo within the seed. Seeds contain the sporophyte embryo and its food (the remaining female gametophyte tissue) encased in a protective seed coat. (One simple definition of a seed is a baby in a box with its lunch.) How can seeds increase the survival of these species?

41 ______________________________

In the gymnosperms, two plants are part of the gametophyte generation. One gametophyte produces the sperm nuclei, and the other produces the egg. These gametophytes develop in the scales of two different types of pine cones. The male (type I) cones develop on the pine trees in the spring. In the cone's containers, the **microsporangium** (meaning "small spore container"), spore mother cells divide by meiosis and produce small haploid cells called **microspores.** Each microspore divides a few times and forms a simple, multicellular male gametophyte called **pollen.** In the pollen grain is a nucleus that functions as a sperm. The other gamete, the egg, is produced in the scale of the type II cone in a container called an **ovule.** In the ovule, spore mother cells divide by meiosis and produce one big haploid spore called a **megaspore** and three useless nuclei. The megaspore divides by mitosis, filling the ovule with haploid cells of the female gametophyte. Over a period of months, the female gametophyte produces the egg.

In late spring, the pollen is released from the microsporangium. Pollen is transferred (in the process of **pollination**) by wind to the female cone even before the egg is formed. A tube from the pollen grain slowly grows through the female gametophyte as the egg forms. The sperm nucleus is moved

Figure 24-9
Pine life cycle.

through the pollen tube from the pollen grain and fertilizes the egg about 15 months after pollination. The zygote remains in the ovule and divides by mitosis, forming the embryo of the new sporophyte. The female gametophyte cells become the embryo's food. The ovule forms a thick coat around the embryo and its food and is now called a *seed.* A gymnosperm seed consists of three generations: (1) the old diploid tissue (the seed coat), (2) haploid tissue (the food), and (3) new diploid tissue (the embryo).

APPLYING

The life cycle of a conifer

Draw the life cycle of pine in Figure 24-9; include written notes. Label the diagram with the terms *archegonium, cone* (male and female), *diploid* (or *2n*), *egg, fertilization, gametophyte* (male and female), *haploid* (or *1n*), *meiosis, mitosis, ovule, pollen, sperm nucleus, microsporangium, seed,* and *sporophyte.*

PART FOUR Angiosperms

EXPLORING

The forms of a flowering plant

As before, work with either fresh or preserved specimens of the larger forms and with prepared slides of the small structures.

Examine the structures of a flowering plant. What organs can you find?

42 ______________________________

Carefully pick one flower from the stem, making sure to remove the entire flower. Beginning at the base of the flower, remove the lowest circle of leaf-like structures. These structures are usually green but have other colors in some flowers. Tape this first circle of structures in a row in the lowest space reserved for Figure 24-10. Then remove the next circle of structures, the **petals.** Tape these structures above the first row. Continue removing circles of structures and taping them in ascending order in Figure 24-10. Using your knowledge of the life forms in pine, add labels to the flower parts. Terms from the life cycle of pine you may want to use are *diploid* (or *2n*), *egg, gametophyte* (male and female), *haploid* (or *1n*), *ovule, pollen, sperm nucleus, microsporangium, seed, sporophyte,* and *zygote.* Other terms include *anther, ovary, petal, sepal, stigma,* and *style.* If you cannot apply all of these new terms now, apply what you can. Then complete your application of these terms after additional observations.

From another flower, obtain the centermost structure without damaging the remaining structures. Cut the expanded, lower portion of this structure, the **ovary,** in half longitudinally. Then cut one of the resulting halves again in cross section. Examine the ovary sections with the dissecting scope or hand lens. Get a prepared slide of a cross section through a flower ovary and examine it with your microscope. Describe the structure of the ovary.

43 ______________________________

____

Remove the circle of structures that surrounded the ovary in your second flower. Place one of the heads of these structures on a microscope slide and examine it with the dissecting microscope or a hand lens. Also examine a prepared slide of a section of this structure, the **anther,** with a light microscope. Describe the structure of the anther.

44 ______________________________

____

Figure 24-10
Parts of the flower.

What events do you think occur in the interior of the anther?

45 ________________________________

Describe the evidence on which you based your answer.

46 ________________________________

____

Examine seeds that have been in a damp environment for 1 to 4 days. Open the 1-day bean seed along its edges. Describe its parts.

47 ________________________________

Cut a corn seed longitudinally, separating its broadest surfaces in half. Examine the cut surfaces. Describe how the structure of the corn seed differs from the bean seed.

48 ________________________________

Describe the changes occurring in the bean or corn seeds over time.

49 ____

____

Return to the list of terms for Figure 24-10. Add all unused terms to Figure 24-10 that you can apply.

CONCEPTS/TERMS

The life cycle of a flowering plant

The angiosperms are vascular plants with a dominant sporophyte generation. They produce seeds in a unique structure called the **flower.** Biologists consider the flower to be an important difference between the flowering plants **(angiosperms)** and pine **(gymnosperms).** The angiosperm ovule, from its spore mother cell stage through the mature seed, is completely enclosed within the ovary of the flower. The pollen forms inside the anther similarly to pollen forming inside the male cone in conifers. Pollen is transferred by wind or animals to the sticky surface, the **stigma,** above the ovary. From the pollen grain, a pollen tube grows down the long cylinder, the **style,** to an ovule inside the ovary. Within the ovule, a cell has divided by meiosis and produced one functional, haploid megaspore. The megaspore divides three times by mitosis and produces eight nuclei. These eight muclei inside the ovule make up the entire female gametophyte. One of the nuclei serves as the egg nucleus and unites with a sperm nucleus from the pollen grain. Two of the remaining nuclei move to the center of the ovule and unite with a second sperm nucleus from the pollen grain, forming a triploid (3n) nucleus. This triploid nucleus divides many times, filling the ovule with triploid tissue called **endosperm.** The uniting of two sperm nuclei with female gametophyte cells is unique to flowering plants and is called **double fertilization.** The embryo, which develops from the fertilized egg, uses the endosperm for food. Together, the embryo, the endosperm, and the covering are the seed. The food in the seeds of conifers has a different origin than the food in the seeds of angiosperms. The pine seed's food is old gametophyte tissue and the flower seed's food is the triploid endosperm.

In many flowering plants **(dicots),** some of the endosperm nutrients are stockpiled in two, thick seed leaves called **cotyledons** (bean). In other flowering plants **(monocots),** the nutrients are transferred to only one cotyledon (corn).

The seed is an excellent form for surviving adverse conditions. During the last stages of seed formation, the seed dehydrates and the embryo's growth stops. Under unfavorable conditions, seeds remain dormant. Under favorable environmental conditions, seeds germinate. In **germination,** the seeds take in water and the embryo resumes growing. Some 10,000-year-old seeds found in animal burrows in the Arctic germinated when placed in a favorable environment.

APPLYING

The life cycle of a flowering plant

Draw the life cycle of flowering plants; add notes and labels in Figure 24-11.

Constructing a phylogenetic tree of the plant kingdom

Review the information you now have about plant life cycles. Compare the information from each division to the other three. Construct a phylogenetic tree of the four plant divisions. As you construct the tree, add terms showing new structures that developed in the plant divisions or terms giving major differences or similarities between divisions. For example, you could use the terms *nonvascular* and *vascular* to indicate a major difference between the branch leading to the bryophytes and the branch leading to the three other divisions. Some other terms you may want to use are *cones, dominant gametophyte, double fertilization, dominant sporophyte, enclosed seeds, flowers, naked seeds,* and *pollination.* Construct your tree in Figure 24-12.

Figure 24-11
Flowering plants life cycle.

Figure 24-12
Phylogenetic tree of the plant kingdom.

LABORATORY

25

Plants

Roots and stems

Plants and their environment

Most organisms within the plant kingdom are photosynthetic land dwellers. To produce sugars, immobile plants require raw materials in their immediate environment and the appropriate structures. This is similar to a sporting goods factory requiring raw materials and equipment to produce skates, soccer balls, and exercise machines. Because most plants are land dwellers, plants must be able to not only acquire the raw materials for photosynthesis on land but also grow and reproduce in the sometimes dry, windy, and varied temperatures of terrestrial ecosystems.

In this laboratory, you will focus on two of the three major, nonreproductive plant organs. It will be helpful to begin by briefly examining the complete plant.

EXPLORING

Plants as organisms

Using sight and touch, explore the plants that are on the demonstration table. Hand lenses are available to aid in gathering information. The common names of the plants and information about the environment in which they grow are given. Some have been unpotted and placed in water so that you can examine their underground parts. Keeping in mind that plants need nutrients and energy for sugar production, think about how the plants acquire them. Also think about how these plants survive in different environments.

In Table 25-1, list the nonreproductive organs that are present in most of the plants and describe their structure. First, describe the common characteristics of each organ. Second, describe how the organs differ among the plants. Some of the characteristics you may want to look at include the following:

1. Surface structure
2. Spatial arrangement of its parts
3. Types of visible tissues
4. Branching of structures within the organ
5. Flexibility, strength, and total surface area

What other characteristics can you examine?

1 ____________________

Do particular arrangements of the line-like structures in the leaves seem to be linked to particular types of root structures? If so, describe which forms of the organs that are linked.

2 ____________________

____

Table 25-1 Characteristics of plant organs

Organ	Description of organs	
	Common characteristics	Different characteristics

Table 25-2 Organs and acquisition

Requirement	Organ(s)	Evidence and reasons
Carbon dioxide		
Water		
Minerals		
Light		

Use your observations with whatever previous knowledge you have about plants to associate an organ or organs with energy and nutrient acquisition. The nutrients and energy requirements for photosynthesis are listed in the first column of Table 25-2. In the second column, add the organ (or organs) that take in each requirement. Also add organs that may move the requirement from the acquisition site to the site in which it is used. In the third column, state the evidence for your observations and your reasons for matching the organ (or organs) to the requirement.

Now consider some of the differences and modifications in each of the organs. Particularly think about the differences or modifications that may help the organism survive and produce healthy offspring in its normal environment. For each type of environmental extreme listed in Table 25-3, describe some of the specializations of organs that increase the survival of plants in that environment.

Which organ and what form of it do you think is most effective in preventing soil erosion?

3 ______________________________________

What do you think are the effects of forest removal on soils, their water content, and erosion?

4 ______________________________________

CONCEPTS/TERMS

Plant organs

The sporophytes of most vascular plants possess three nonreproductive organs: roots, stems, and leaves. **Roots** absorb water and minerals from the soil, transport water and dissolved minerals, and anchor the plant in the soil. Some roots, such as carrots', are also primary storage sites for starch produced from the plant's sugars. In general, plants exhibit three patterns of root development. A **taproot** consists of one large vertical root with many smaller roots extending from its sides. It grows deeply and anchors the plant firmly into the soil. **Fibrous roots** consist of numerous string-like roots that spread into the upper layer of the soil. They hold soil particles together, helping to prevent soil erosion. **Adventitious roots** develop from the stem or leaves and frequently anchor the plant. (The roots that anchor ivy to walls are adventitious roots.) Flowering plants can be divided into two groups partly based on their type of root system. Taproots are usually found in dicots (which have two cot-

Table 25-3 Adaptations of plants

Environment	Description of adaptations
Low and/or infrequent moisture	
High moisture	
High winds	
Low temperature	
High temperature	
Low light	
High light	

yledons or seed leaves) such as begonias, geraniums, carrots, oaks, and roses. Fibrous root systems are usually found in **monocots** (which have one cotyledon) such as grasses and lilies. (See Plate 12.)

Leaves are the photosynthetic manufacturing sites for sugars. They capture light, which provides energy for photosynthesis. Leaves also acquire carbon dioxide gas from the air and release the photosynthetic waste product, oxygen. Leaves are attached to the stem in particular spatial patterns at areas called **nodes.** In general, plant leaves are arranged in three patterns: (1) two leaves on opposite sides of the stem at each node; (2) one leaf, which alternates between sides of the stem, at each node; and (3) three or more leaves attached at one level in a whorl. Frequently, the type of leaf arrangement on the stem can help identify plants. In what ways can the spatial arrangement of leaves affect the acquisition of light energy?

5 ______________________________

Leaves are **blades** of photosynthesizing tissues that are attached to the stem by a thin stalk called the **petiole.** Sometimes the blade is separated into leaflets. The blade is permeated with **veins,** which contain the transporting tubes of xylem and phloem. The two groups of flowering plants can also be distinguished by the way veins are distributed in their organs. In dicot leaves, the veins branch, forming a net-like system. In monocots, they are generally parallel to one another.

Stems raise the leaves toward the light, support leaves against the pull of gravity, hold leaves against the wind, and help protect against foreign invaders. In addition, stems conduct water and its dissolved minerals from the roots to the leaves and conduct sugars from the sites of photosynthesis to the sites of active metabolism or to storage areas of the plants. In some plants, stems are also storage areas. Together, stems and leaves are called **shoots.**

EXPLORING

Roots

Obtain petri plates of germinating seeds (growing embryos) of a dicot and a monocot. Explore their root systems with a dissecting scope. Describe the structures that you think are most important to the absorption of water and minerals.

6 ______________________________

Explain your choice of absorption structures.

7 ______________________________

Describe the structural arrangement or structures that you think are most valuable to the root's function of anchoring the plant in the soil.

8 ______________________________

____

On what concepts did you base your choice of root structures?

9 ______________________________

In Figure 25-1, sketch the root system of a dicot and a monocot plant.

Using a microscope, explore a prepared slide of a longitudinal section of an onion root tip. Note areas of cells that differ in cell shape and thickness of walls, and look for areas of dividing cells. Carefully examine cells that seem to have corkscrew-like walls as well as other cells in the same region. Also examine the outer covering, the epidermis. Do you find any modifications of the epidermal cells? If so, where along the root are these modifications located?

10 ______________________________

In Figure 25-2, sketch the longitudinally sectioned onion root. Draw examples of the different cell types.

Figure 25-1
Dicot and monocot root systems.

Figure 25-2
Onion root.

Figure 25-3
Cross sections of dicot and monocot roots.

Compare prepared slides of cross sections of a young dicot root and a young monocot. First observe these slides at low power and then at high magnification. In Figure 25-3, sketch these two cross sections, showing areas of different cell types. Enlarge a pie-shaped portion of each cross section to show the details of the cells. Add notes that point out differences between dicots and monocots.

CONCEPTS/TERMS

The structure and function of roots

A root consists of several areas of different tissues. The tip of the root is covered with cells that protect the remainder of the root as the root pushes between soil particles. This group of cells is the **root cap.** Next to the root cap is an area of undifferentiated cells that divide by mitosis. This tissue is the **apical meristem** (which refers to dividing cells at the *apex* or *tip*). The cells produced at the meristem/root cap boundary replace root cap cells that wear away as the root grows through the soil. The cells produced at the proximal (closest to the onion bulb) meristem boundary become specialized cells. These cells begin their differentiation by elongating, which increases the root's length. The area of elongating cells is called the **zone of elongation.** The cells on the proximal boundary of the zone of elongation differentiate into the specialized cells of the mature root. These differentiated cells make up the **zone of cell differentiation.**

The cross sections of mature roots have three visible areas. Starting at the periphery and moving inward, the areas are a narrow epidermis, a wide cortex, and a central stele. The **epidermis** absorbs water and minerals. The projections from the epidermal cells, the **root hairs,** increase the surface area of the root. The epidermal cells actively pump minerals internally, and water follows by osmosis. The thin-walled parenchymal cells of the **cortex** store food. The innermost layer of the cortex, the **endodermis,** surrounds the vascular tissues of the stele. The edges of the single layer of endodermal cells are cemented together, preventing movement of materials between the endodermal cells. This structural block requires that substances move from the cortex into the stele only through the regulating endodermal cells. The central cylinder of the root, the **stele,** contains vascular tissues. In the center of the stele of dicots, the thick-walled, water-transporting **xylem** tubes form a star-shaped arrangement and the food-conducting **phloem** occupy the areas between the arms of xylem. In contrast to the stele of dicot roots, monocots have the vascular xylem and phloem surrounding a central **pith** of parenchymal storage cells. The outermost layer of the stele is the **pericycle.** The pericycle's cells can divide, forming lateral roots.

APPLYING

Terms to drawings

Return to your drawings in Figures 25-1 through 25-3. Label the areas of the roots with the appropriate terms.

EXPLORING

Roots and soils

At the demonstration table are examples of several types of soils: clay (fine particles), sand (coarse particles), peat moss (partially decomposed organic materials), and a mix of the three soil types. Feel these soils. Also examine samples of these soils to which an equal amount of water was added 2 days ago. Considering the functions and structures of roots, predict how the root structures are modified in those plants that can grow well in soils with a high amount of one of the soil types. Also predict whether fibrous or taproots would be the best anchoring roots in each type of soil. Finally, explain your reasoning.

11 Clay soils

Root structures ______________________________

____

Anchoring root type ______________________________

Reasoning ______________________________

____

Sandy soils

Root structures ______________________________

____

Anchoring root type ______________________________

Reasoning ______________________________

____

Organic soils

Root structures ______________________________

____

Anchoring root type ______________________________

Reasoning ______________________________

____

Growth of roots

At the demonstration table, examine the roots of young plants growing in petri dishes. Each plate of new plants has been treated differently. Plates A, B, and C have been kept in a stationary position with the same edge down throughout the time period. Plates D, E, and F have been rotated 90 degrees

Table 25-4 Directional growth of roots

	Light conditions		
Orientation	**Dark**	**Light from above**	**Light from below**
No rotation	(Plate A)	(B)	(C)
90-degree rotation every 24 hours	(D)	(E)	(F)

clockwise every day. In addition, the plates have been exposed to the following light conditions:

Plates A and D—in the dark
B and E—lighted from above
C and F—lighted from below

In Table 25-4, describe the direction of root growth in each of these plates.

How does light affect the direction of root growth?

12 ______________________________

Explain your reasoning and give the data you used in forming your conclusion.

13 ______________________________

____

____

Is it possible that this demonstration of root growth has a second independent variable?

14 ______________________________

If so, what is the variable and how does it affect the direction of root growth?

15 ______________________________

____

____

CONCEPTS/TERMS

Tropisms and root growth

Plant growth toward or away from an environmental stimulus is called a **tropism** (meaning "turn") Growing toward the stimulus is termed a **positive response** and growing away is a **negative response.** The roots turn when cells on one side of the root elongate more than cells on the other side. Root cells on the lighted side elongate more than those on the dark side, and the growing root tip turns away from the light. This growth response of the root away from the light source is called **negative phototropism.** Cells on the side away from the force of gravity elongate more than those on the side with a stronger gravitational force and the new growth turns downward. This growth response toward the gravitational force is called **positive gravitropism.**

EXPLORING

Stems

Examine stems of the soft (herbaceous) and stiff (woody) dicots, a monocot, and a conifer. Explore their surfaces. Some questions for you to consider are the following:

1. What is the texture of the surface?
2. Does it have a coating or protective layer?
3. Does it have openings in the surface layer?
4. Does photosynthesis occur in the surface layers?

Examine the attachment of leaves at the nodes.

1. Are there areas along the stem where leaves were formerly attached? If so, what observations lead you to make that judgement?
2. Are there small buds at the nodes? If so, do they have a specific position in regard to the leaf?

3. If the stem has branches, what relationship do the branches have to the position of the leaves and the buds?

Examine the tip or apex of the stem.

1. What is the structure of the stem's tip?
2. As you proceed from the stem tip to its base, does the distance between nodes, the **internodes,** change? If so, in what way?

What other types of observations can you make?

16 ______________________________

____

Compare these structural characteristics among the herbaceous and woody dicots, the monocots, and the conifers. In Table 25-5, record the characteristics that are common to all of the stems, then record those that vary.

Using prepared slides of herbaceous dicot and monocot cross sections, compare their stem structures. What are their similarities?

17 ______________________________

____

What are their differences?

18 ______________________________

____

What types of tissues are found in both stems and roots?

19 ______________________________

If there are tissues common to both root and stem, describe any differences in their arrangement in the organs.

Table 25-5 Structure of stems

	Description	
Structure	**Common characteristics**	**Different characteristics**
Surface		
Nodes		
Apex		
Internodes		
Others		

20 ____

____

Compare stem cross sections of a herbaceous dicot and a woody dicot. What are their similarities?

21 ____

____

What are their differences?

22 ____

____

Examine a pie-shaped block cut from a deciduous tree (one that loses leaves in winter or dry seasons). Relate its structure to the microscopic cross section of the young woody dicot. What type of tissue forms the "rings" observed on the cut cross-sectional surface of the tree?

23 ____

What tissues strengthen the tree sufficiently to support limbs and vast numbers of leaves against the pull of gravity and high winds?

24 ____

What are the locations of the active conducting tissues in the tree?

25 ____

CONCEPTS/TERMS

The structure and function of stems

Stems have tissues that transport minerals, water, and sugar solutions; provide strength to withstand winds and gravity; store materials; provide cells for growth; and protect internal tissues. In addition, some stems also form the valuable economic product wood.

Stems have xylem and phloem but have them in different arrangements than roots. In monocot stems, the bundles of xylem and phloem tubes, the **vascular bundles,** are scattered throughout the internal part of the stems and are surrounded by parenchymal storage tissues. In herbaceous dicot stems, the vascular tissues frequently form a ring of vascular bundles surrounding the central pith of parenchyma. Between the ring of vascular bundles and the epidermis, additional parenchyma and other tissues form the cortex. Some of the tissues in the cortex provide strength with the thickened corners of their cell walls.

In woody dicot stems, a layer of dividing meristem cells between the inner xylem and outer phloem adds new cells to the vascular tissues. This meristem is called the **vascular cambium.** New cells on the inner boundary of the meristem add to the xylem, and new cells on its outer boundary add to the phloem. These new cells increase the diameter of stems. The old xylem in the central portions of the stem ceases to transport water but continues to strengthen the dicot and is harvested by the lumber industry. Only the newer xylem, closest to the periphery, functions as water-transporting tubes.

Xylem growth is affected by weather conditions. The larger xylem cells form in the spring when water is available and other growing conditions are good.

The smaller cells form in the summer when growing conditions are less favorable. In woody dicots, this alternating cell size is visible as the **annual growth rings.** These rings have been used by scientists to determine the ages of trees and identify past weather conditions.

The epidermis of stems protects underlying tissues. Frequently, the epidermal cells are covered by a waxy coating, the **cuticle,** that decreases water loss. In some plants, the epidermal surface has small air spaces. The outer layers of woody dicots are more complex and are similar to the outer layers of conifers such as pine. These stems are covered with bark. **Bark** consists of cork cells, which protect the underlying tissues and prevent water evaporation. New cork cells are produced by a layer of cork cambium that is located just inward from the cork. As the cork cambium produces new cork cells, it increases the diameter of woody dicots and conifers. As the older cork cracks with the increase in girth, new cork replaces it.

Stems increase in length at their tips from the activity of the apical meristem and cell elongation in a way similar to the growth of roots. The tips, called the **terminal buds,** contain the apical meristem and the surrounding embryonic leaves.

EXPLORING

Growth of stems

Return to the plates of germinating seeds. Observe stem growth under the varying conditions. Draw your observations in Table 25-6.

Compare the response of the stem with the response of the root with respect to light and gravity. In what ways do these two organs differ in their responses to light?

26 ______________________________

In what ways do these two organs differ in their responses to gravity?

27 ______________________________

CONCEPTS/TERMS

Tropisms and stems

In contrast to roots, stems display **positive phototropism** and **negative gravitropism.** The environmental stimuli of light and gravity change levels of a hormone that promotes cell elongation in plants. Although stems respond to the same plant hormones as roots, roots and stems have different sensitivities to the hormone.

APPLYING

Cohesion and adhesion

At the demonstration table, find a set of three glass tubes with different inside diameters. Obtain a millimeter ruler and a small beaker half filled with water.

Table 25-6 Directional growth of stems

	Light conditions		
Orientation	**Dark**	**Light from above**	**Light from below**
No rotation	(Plate A)	(B)	(C)
90-degree rotation every 24 hours	(D)	(E)	(F)

Measure the inside diameters of each tube and record the information in Table 25-7. Stand the glass tubes in the beaker. Measure the difference in water levels inside each tube compared to the water level in the beaker and record it in the table.

What is the relationship, if any, between the diameters of the tubes and the height of their water columns?

28 ______________________________

Using your knowledge of the interactions among molecules of water and between water and other types of molecules, explain the rise of the water in the tubes.

29 ______________________________

____

In general terms, what is the size of the water-conducting tubes in plants compared to the inside diameters of the glass tubes you have just explored?

30 ______________________________

Even if the plant's water-conducting tubes are very small (for example—only 0.01 millimeter in diameter), do the factors that explain the differences in water levels explain the transport of water to the tops of trees that are about 100 meters (greater than 300 feet) tall?

31 ______________________________

Root pressure

Working in a group or as a class, obtain a well-watered healthy plant and a small piece of floral clay.

Form a cup of floral clay around the stem of the plant 6 to 8 cm above the soil line so that it can catch and hold fluids. Gently but firmly seal the base of the cup to the stem. Remove all leaves attached to the stem between the soil line and the cup. Cut the stem 1 cm above the cup. Set the upper stem and leafy top aside. (Your instructor may ask you to place the stem in water for future use.) Observe the cut surface of the stem that is still attached to the roots. Describe any changes that occur in the next half hour.

32 ______________________________

____

If a fluid seeps from the stem, what is the original source of the fluid?

33 ______________________________

Using the diagrams of the cross sections of a root and a stem in Figure 25-4, show the path of water transport in plant roots and stems. Add labels and notes to the diagram.

Table 25-7 Tube diameters and water level

Tube inside diameter (mm)			
Difference in water levels (mm)			

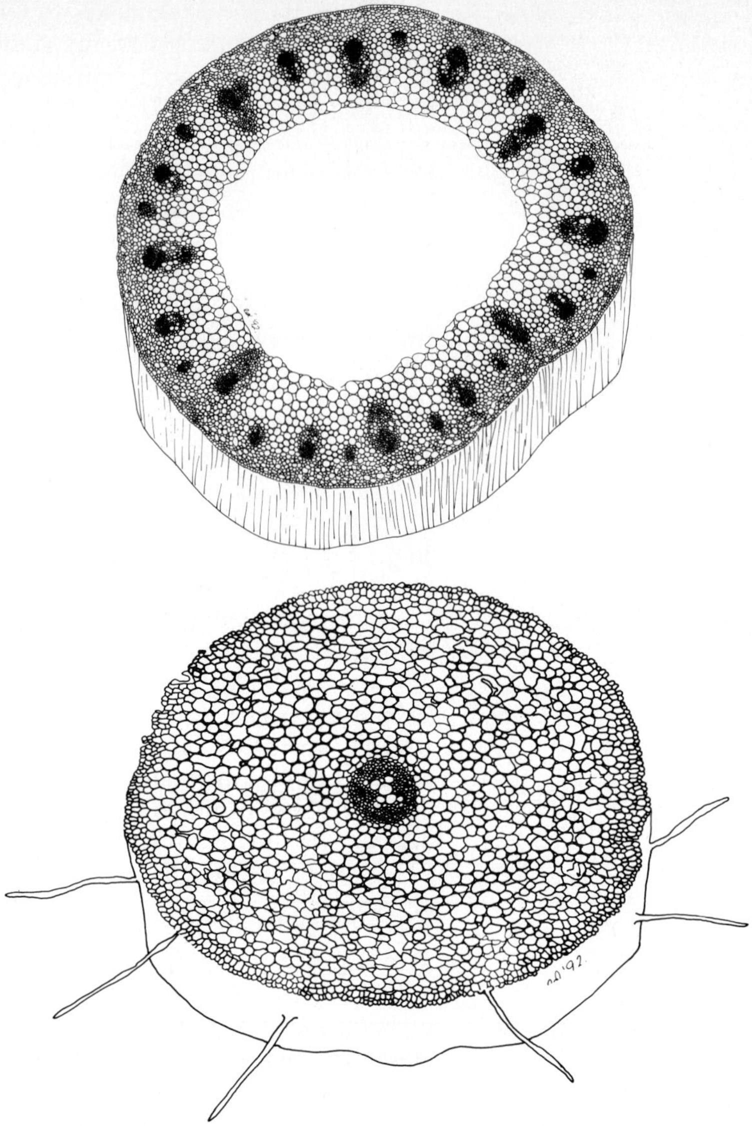

Figure 25-4
Water transport in root and stem.

CONCEPTS/TERMS

Root pressure

As the epidermal cells of the roots actively pump minerals into the cells, a concentration gradient forms between the soil and the cells' cytoplasm. Water then enters the cells by osmosis. The increase in water in the cells increases water pressure, which causes a pressure gradient with other parts of the root. Water with dissolved minerals flows to the region of lowest pressure—the xylem tubes in the vascular tissue of the root and stem. This movement, which is the result of the osmotic uptake of water in the roots, is called **root pressure.**

APPLYING

Experiments: Factors affecting root pressure

Form hypotheses about the effect of the following conditions on root pressure.

34 **Dry soils** ____________________

Partial removal of roots ____________________

High levels of soil fertilizers ____________________

Working in groups or as designated by your teacher, design an experiment to test a prediction from one of the hypotheses. Consider the need for controlled conditions and repeated trials. After designing your experiment, seek suggestions and approval from your teacher. When the experimental design is completed, prepare a materials and equipment list.

When you have all of your needed materials and equipment, conduct your experiment. Write a report of your experiment and attach it to this manual.

LABORATORY

26

Plants

Leaves

Role of leaves

Leaves are the plant's factories for the photosynthetic production of sugars. During photosynthesis, carbon dioxide, water, and radiant energy from the sun are used. Carbon dioxide from the air enters plants through their leaves, and sunlight is absorbed by leaf pigments. Leaves are structured in ways that can provide these functions.

The roots and stems of plants also play a role in photosynthesis; they supply the water. Water is absorbed by the roots and transported to the leaves through the xylem tubes in the roots and stem. In Laboratory 25, you explored some factors that affect water transport: cohesion, adhesion, and root pressure. These factors, however, are not sufficient to explain how water is transported to leaves in some trees that are 100 meters tall or even to leaves in vines that grow to the roof of a two story house. Processes that take place in the leaves complete the story of how water is transported in plants.

Today you will explore leaf structures and functions to gain an understanding of how leaves take up water, carbon dioxide, and light.

EXPLORING

Water uptake

Working in small groups or as a class, set up a **potometer** using the following materials and directions.

A healthy plant or leafy tree branch with its cut end submerged in water
1 5-ml pipet with 0.1 ml markings
1 length of flexible tubing, 3-4 cm
1 ring stand with a ring stand test tube clamp
1 lamp
1 long, water-filled pan
1 pair of pruners or a single-edge razor blade
1 fan or hair dryer
Florist clay

CAUTION
Wear gloves or wrap a towel around your hands to protect them in case the pipet breaks when pushed.

Slip the tubing over the dampened, blunt end of the pipet. In a long pan of water, fill the pipet and the tubing completely with water. (Remove all air bubbles from the tubing and pipet.) While keeping the leaves dry, submerge the stem of the plant or the tree branch in the water. Working under water, make a slanting cut across the stem. If you are using a previously cut tree branch, cut at least 2 inches from the stem's end. Push the free end of the tubing over the end of the freshly cut stem. If the tubing fails to fit tightly around the stem, seal the connection with florist clay. As you remove the plant, tubing, and pipet from the water, keep the pipet horizontal. Anchor the plant in an upright position with the ring stand clamp as shown in Figure 26-1. Raise the tip of the pipet slightly from the horizontal. If the leaves of the plant are wet, dry them off by gently blotting with a paper towel. Shine a bright light

Figure 26-1
Potometer.

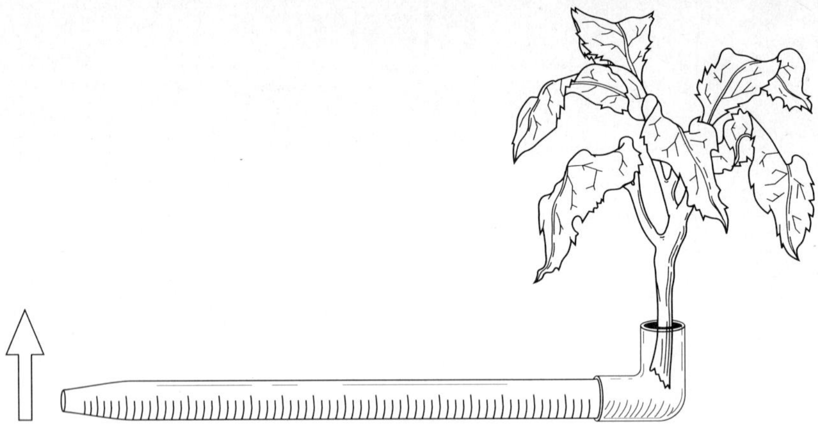

on the plant from a distance of about 2 feet. (Do not let the heat from the lamp damage the leaves.) After 10 minutes, note the water level in the pipet. Also note the temperature at leaf level. What is the temperature?

1 ______________________________

Based on directions from your teacher, read the water level in the pipet at 5- or 10-minute intervals. Record the elapsed time and water readings in Table 26-1. Continue with other parts of the laboratory between observations.

After you record each pipet reading, subtract the current reading from the previous reading and divide the difference by the number of minutes in the interval. These calculations give you the number of milliliters of water uptake per minute.

$$\text{ml water uptake/min} = \frac{\text{Previous reading} - \text{Current reading}}{\text{Number of minutes in the interval}}$$

Record the ml water uptake/min in the third column. After water uptake has been measured for four or five intervals, determine an average (mean). Your teacher will assist you if you need help with your calculations.

Over the time you took your readings, did the water uptake remain constant?

2 ______________________________

If not, what factors could explain changes in water uptake?

3 ______________________________

____

____

Now measure the uptake of water as a fan blows air through the plant. With an increase in air movement, do you expect the plant's water uptake to remain the same, increase, or decrease?

4 ______________________________

Explain the reasoning for your prediction.

5 ______________________________

Table 26-1 Water uptake by leaves

Time (min)	Water uptake (ml)			
	With light only		With light and air flow	
	Readings	ml/min	Readings	ml/min
0		X		X
Mean water uptake				

Refill the pipet and tubing while working under water. Record your observations in Table 26-1 and calculate the ml water uptake/min for each interval. Calculate the mean water uptake under these "windy" conditions.

Compare the means of the water uptake during the two types of environmental conditions. Under which condition was the rate of water uptake faster?

6 ______________________________

What explanations do you have for this difference?

7 ______________________________

____

A leaf sandwich

At the demonstration table, examine the sandwich of a leaf and two curved glass dishes. Describe your observations.

8 ______________________________

____

What is the probable source of the material you observed?

9 ____

Leaf structure

Obtain a prepared slide of a dicot leaf cross section and a light microscope. Search for structures that may be important to photosynthesis and the acquisition of materials. Also look for structures through which leaves may lose water and those structures that may help prevent its loss. In Figure 26-2, sketch the cross section of the leaf and draw an example of each type of cell you see in the

Figure 26-2
Leaf structure.

Table 26-2 Relationship between leaf function and structure

Function	Description of structure and reasoning
Photosynthesis	
Light capture	
Carbon dioxide uptake	
Water transport	
Increases water loss	
Decreases water loss	

leaf. As you draw the different cells, indicate their locations on the leaf cross section and propose their likely roles in leaf functions.

In Table 26-2, describe the leaf structures that you think are important to the listed functions. Explain your reasons for associating the structure with the function.

With a dissecting scope, examine both surfaces of a leaf from each of several plants on the demonstration table. Add this information to Table 26-2.

Using information from both Laboratory 25 and today's observations, answer the following questions: Does the arrangement of leaves on the stem affect leaf functions? If so, which functions does leaf arrangement affect?

10 ______________________________

Explain your reasoning and give examples of the observations and information on which you based your answer.

11 ______________________________

____

____

CONCEPTS/TERMS

Relationship between leaf structure and function

A leaf is composed of four tissue layers: two interior parenchymal tissue layers bounded by upper and lower epidermal tissues. The two epidermal layers consist of closely packed, thick-walled cells that usually lack chloroplasts. The upper epidermis (and sometimes the lower epidermis) usually has a waxy covering, the cuticle, on its outer surface. The cuticle may have been dissolved in the slide preparation. The lower epidermis (and the upper one in some plants) has a number of structures consisting of two kidney-shaped cells surrounding a slit-like opening. The opening is called a **stoma** (plural, stomata), and the surrounding kidney-shaped cells are called **guard cells.** In grasses, the guard cells are shaped like dumbbells. Plants can change the sizes of the stomata. When the stomata are open (enlarged), carbon dioxide gas enters and water (vapor) exits the leaf with their concentration gradients. When the stomata are closed, carbon dioxide cannot enter and water is kept within the plant. What are the consequences if a plant cannot take in carbon dioxide?

12 ______________________________

What are the consequences if water is lost from a plant more rapidly than it can be taken in by the roots and transported to the leaves?

13 ______________________________

The internal parenchymal cells, called the **mesophyll,** contain many chloroplasts. The mesophyll is organized into two layers: the upper layer of column-like cells—the **palisade mesophyll**—and the lower, loosely arranged layer of cells—the **spongy mesophyll.** The sponge-like arrangement of the mesophyll with its air passageways provides a vast surface for gas exchange between the cells and air spaces. The air spaces connect with the stomata. This structural arrangement speeds the movement of gases within the leaf and between the internal parts of the leaf and the external environment.

The structure of the leaf maximizes access to light. The broad (or long), flat leaf blade increases the surface area that receives light. In addition, the arrangement of leaves on the stem and the internodes separate leaves from one another; this increases the amount of light reaching each leaf. The location of the mesophyll layers close to the leaf surface increases exposure of the pigments within the chloroplasts of the cells to light energy. All of these increase the acquisition of light energy.

Water is transported from the stem into the leaf through the xylem in the veins. From the xylem tubes, the water diffuses into the mesophyll cells and air spaces of the leaf. Water is lost from the plant as water diffuses through the stomata from the air spaces and mesophyll cells. The loss of water from the leaves of a plant is called **transpiration.** Just as clothes on a line dry more rapidly on a windy day than on a calm day, transpiration rates increase when air flows over the plants.

EXPLORING

Structure and function of stomata and guard cells

Obtain the plant leaf designated by your teacher. Peel back a piece of its lower epidermis. Prepare a wet mount of the epidermis using distilled water. To prevent the trapping of air bubbles beneath the epidermis, try slipping the tissue into the water puddle on the slide. If needed, add one more drop of water to cover the surface of the specimen. Lower the cover slip from an angle as usual. Examine the mount with the scanning and low power objectives of your microscope. How many stomata are in your field of view at 100 times magnification?

14 ______________________________

Note the sizes and shapes of the guard cells and the stomata. Are all stomata opened or closed to the same degree?

15 ______________________________

If not, describe the extremes of the stomata widths.

16 ______________________________

In Figure 26-3, draw an average stoma with its guard cells, showing their size and shape relationships.

Figure 26-3
A stoma with its guard cells.

Place a drop of 5% salt solution on one edge of the cover slip and draw the salt beneath the slip by touching the opposite edge with a paper towel. After a few minutes, describe the effect of the salt solution on the size and shape of the stomata and their guard cells.

17 ______________________________

____

____

Next replace the salt solution with distilled water; you *may* be successful in reversing the effects of the salt.

At the demonstration table, observe two pairs of cylindrical balloons that have been tied together at their ends. One pair is filled to capacity with air. The other set is only partially filled with air. In addition, a lengthwise strip of each balloon has been coated to thicken it more than the rest of the balloon. Describe the differences in the "stomata" of these balloon "guard cells."

18 ______________________________

____

What effects do you think that open stomata have on carbon dioxide and water levels in the leaf?

19 ______________________________

____

CONCEPTS/TERMS

Stomata regulation and transpiration

The size of the stomata depends on the water pressure in their guard cells. When carbon dioxide is low, water is available, and the sun is shining, the guard cells actively take up potassium ions from the surrounding tissue fluid. When the concentration of ions in these cells increases, water follows the potassium by osmosis. The increased water pressure within the cells stretches them; the thinner, outer side of each guard cell stretches more than the thicker inner side. This difference in stretching causes the cells to curve. On curving, the middle portions of the guard cells do not touch, forming a gap. The gap is the stoma. When carbon dioxide is high and water is low, the guard cells lose potassium ions, water moves out, the cells become flaccid (limp), and their sides touch, closing the stomata.

Now let's consider the last factor in the movement of water to the top of the tallest trees. From Laboratory 25 you know the following: Water enters the root cells by osmosis as minerals are actively pumped into the roots by their epidermal cells. The water flows across cells or between cells to areas of lower pressure in the xylem of the root cylinder. Pressure from the accumulation of water in these tubes (root pressure) pushes the water up the tubes of the xylem against the force of gravity. The water molecules adhere to each other (cohesion) and to the sides of the tubes (adhesion), forming a column of water in the xylem tubes of the root and stem. From today's activities and your understanding of photosynthesis, you know that leaf cells use water and lose water to the leaf's air spaces and the open stomata (transpiration). This loss creates low water concentrations in the mesophyll cells, which "pull" on the continuous column of water in the xylem of the leaf veins, stem, and roots. Botanists consider transpiration of the water at the end of its column as the primary force in the movement of water to the tops of the tallest trees.

APPLYING

Experiment: effects of environmental conditions on stomatal opening

Currently, biologists accept the following as accurate statements about plants:

1. Photosynthesis is required for the survival of plants.
2. The light-dependent reaction of photosynthesis requires light and water to make ATP and NADPH. These products are used in the light-independent (dark) reaction to convert carbon dioxide to sugar.
3. Carbon dioxide enters the plant through open stomata at the same time that water is lost through the stomata.
4. Stomata are open when water pressure inside the guard cells is high. Stomata are closed when guard cell water pressure is low.
5. Plants wilt when transpiration is greater than water movement to leaves.
6. The contents of plant cells shrink and die if water loss is too great.

Plants have a problem. They need opened stomata to acquire carbon dioxide and closed stomata to prevent water loss. Considering this problem, what is your prediction of the opened/closed condition of the stomata under the following environmental conditions?

20 High humidity ______________________

Low humidity ______________________

Hot temperatures ______________________

Low temperatures ______________________

Windy ______________________

Calm ______________________

Light ______________________

Dark ______________________

Propose two alternative hypotheses concerning the effects of soil moisture conditions on the opened/closed state of stomata. State the hypotheses and give the reasoning on which each hypothesis is based.

21 Hypothesis I ______________________

Reasoning ______________________

____

Hypothesis II ______________________

Reasoning ______________________

____

Design a test of the hypotheses using the watered and unwatered plants available in the laboratory. A technique for observing stomata is described below. You need to consider how you will record the opened/closed state of the stomata. After consulting with your teacher, conduct the test of the hypotheses. Attach the report of your experiment to this manual.

You can make an imprint of the leaf surface and its stomata using clear fingernail polish (Figure 26-4). Paint a layer of clear fingernail polish on the bot-

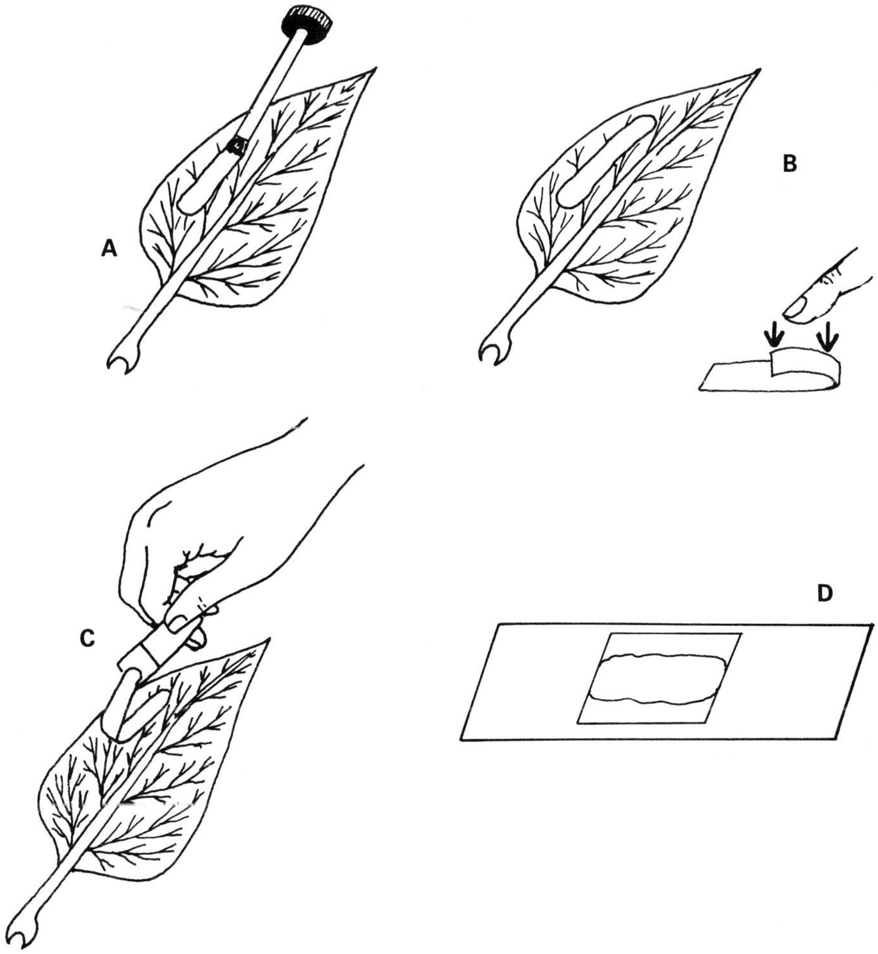

Figure 26-4
Preparing a leaf imprint.

tom surface or both top and bottom surfaces of a leaf. While the polish is drying, prepare a scotch tape tool for lifting the polish from the leaf surface. As shown in Figure 26-4, *B,* fold one third of a 2-cm piece of tape over, sticking it to the middle third of the tape. The other one third of the tape remains with its sticky surface exposed. When the polish is dry, touch the free end of the tape to the edge of the polish layer. Pull back over the polish, peeling the polish imprint from the leaf. Place the imprint, with the leaf contact side up, on a microscope slide. Either tape the edge of the imprint to the slide or lay a coverslip on top to flatten it. Using your microscope, observe the imprint at 40 and 100 magnifications.

Adaptation of leaves

Structural modifications and adaptations

In your exploration of plant leaves, did you find modifications in the surfaces that may adapt the plants to different environments? The leaves of some plants have their stomata only on the lower epidermis. Other plants have stomata at the bottom of epidermal pits. Plants with stomatal pits grow well in sunny, moderately dry environments. How do you think the pits help survival in dry environments?

22 ____

____

The leaves of some plants are covered with hair-like structures. What effects might this modification have on transpiration?

23 ______________________________

If all other factors are equal, how well do you think a hairy-leaf plant competes with similar plants that lack leaf hairs in windy, prairie-like environments?

24 ______________________________

Explain your reasoning.

25 ______________________________

____

Leaf structure and environments

Compare the leaves of three plants. At the demonstration table, examine a cactus. The thorns are the leaves of the cactus. The large fleshy part is the stem. Also observe slides of the cross sections of a pine leaf and the leaf-like structure of a moss. Write a paragraph about the adaptation of the cactus to the desert, the pine to the frozen north, and the moss to a damp environment. Attach the paragraph to this manual.

LABORATORY
27

Animals

Porifera to nematodes

Adaptive radiation

If you look around your classroom, you see unique individuals. People are similar to one another, but no person is exactly like any other. Some of these individuals will have two or three children, a few will have six to eight children, and some will have no children. The individuals without children will not contribute genes to the next generation or to the generations thereafter. The special inherited traits they possess will be lost. The individuals with six to eight children will contribute more genes to the next and following generations than those with two or three children. Their special inherited traits will increase in the human population. With this difference in reproductive rates, the population as a whole slowly changes. Over long time periods, populations of all organisms change, whether they are humans, plants, bacteria, protists, or animals.

The raw material for this long-term change is the inherited uniqueness in individual members in the populations. This uniqueness originates from **mutations** in individuals. Mutations are changes in the genetic material, which holds the instructions for directing cellular activity.

The environment greatly influences which members of the population survive long enough to reproduce. In its particular environment, each successful organism must have structures that perform all activities of life even in competition with other organisms. Otherwise, the organism dies and none of its traits are passed to future generations. The accumulation of many generations of the "fit" organisms surviving to reproduce slowly changes succeeding generations. Eventually, this **differential reproduction** leads to more individuals with the structures that function well in their particular environments. The structures, functions, and behaviors that promote survival of an organism in a specific environment are called **adaptations.**

In the different and differing environments on Earth, organisms have changed over millions of years from an ancient, common ancestor to the kinds of organisms now present. This is called **adaptive radiation.** Today, you will examine some of the results of this adaptive radiation as you explore representatives of the animal kingdom.

Life activities of animals

Animals are a diverse array of eukaryotic, multicellular organisms. They differ from plants in that animals eat their complex food rather than producing it from simple inorganic substances and light energy. Animals differ from fungi (also heterotrophs) in the way they take food into their bodies. Animals **ingest,** or take in, whole or pieces of organisms and then digest the pieces. Fungi digest food first and then **absorb** the digested products into their bodies.

To survive and reproduce, each animal has to carry out certain activities. All animals must locate and ingest other organisms to obtain nutrients and energy. At the same time, animals must avoid becoming food and energy resources for other animals. Animals require oxygen to help release energy from the food they ingest. In addition, they must eliminate toxic waste products of carbon dioxide and nitrogen compounds. At the same time that they are eliminating wastes, animals living in sea water or on land must conserve water. In contrast, animals living in fresh water must eliminate excess water that enters from the environment.

Large animals and terrestrial animals have additional problems. Large animals have to move digested nutrients from the parts of the body where they are digested to all cells of the body where the nutrients are used. Large animals also need to transport wastes produced by cellular metabolism to structures that move the materials out of the body. Land animals must also have structures to support their bodies against gravity and methods to cope with the wide swings in temperature in many terrestrial habitats.

Organisms in the animal kingdom are placed into phyla based on their symmetry, patterns of development, and the complexities of their forms. Let's begin the study of animals by exploring some of the animal phyla that biologists think are more closely related to their simple protist ancestors. As you proceed, look for the similarities and differences in the functional structures of animals that adapt them to their environments.

PART ONE Porifera, the sponges

EXPLORING Sponges

At the demonstration table, examine several sponges, some preserved and some dried. Sponges are aquatic organisms; most of them live in the ocean. In how many ways can you cut across or down these organisms to produce halves that are mirror images of each other (such as your right and left hands)?

1 ____________________

Feel the texture of the sponges' surfaces. Do the surfaces of all preserved and fresh sponges feel the same?

2 ____________________

Observe the surfaces with a hand lens or dissecting scope. Describe the surfaces.

3 ____________________

____

Are all of the surface openings similar?

4 ____________________

If not, describe their differences and general locations of the different types of openings.

5 ____________________

____

If a living sponge is available, add a small drop of india ink near the base of the sponge. Observe for a few minutes. Describe the response of the sponge.

6 ______

What kind of cell(s) could be responsible for the sponge's response?

7 ______

With your microscope, examine a prepared slide of a cross section and a longitudinal section of the sponge *Grantia*. Describe the general structural plan of this organism.

8 ______

______

With high power, examine the cells on the surfaces of the sponge. Find a type of cell that can move water (and ink). Where is this type of cell located?

9 ______

Using your observations of sponge structure, give a possible explanation of how water (and ink) is moved through the body of the sponge. Also describe one possible route of water through the sponge.

10 ______

______

______

Examine other sponge cells. How many different types of cells can you distinguish?

11 ______

CONCEPTS/TERMS

The structure and function of sponges

For most of their lives, sponges are **sessile** (attached) like plants. Essentially, sponges are clusters of a few types of cells around a central cavity called a **spongocoel** (Figure 27-1 and Plate 15). One cell type that lines the cavity has a collar and a flagellum. As the flagellum whips, water is swept from the outside, through channels in ring-shaped cells (pore cells), and into the cavity. The collar cells filter food particles from the water and engulf them by phagocytosis. Water exits from the cavity through a large opening. A jelly-like matrix separates the cells lining the spongocoel and the cells covering the external surfaces. Some wandering amoeboid cells secrete sharp **spicules** that serve as a skeleton for the sponge and make it undesirable to eat. The walls of some sponges are indented from both sides, forming a double set of finger-like canals in the wall.

Sponges have minimal organization. The cells of a sponge do not work together, so sponges lack tissues and more complex levels of organization. Most sponges are not constructed with any one part in balance with another part. This lack of balance is called **asymmetry,** meaning "without symmetry."

APPLYING

Other sponges

Examine a second set of sponges that are on the demonstration table. Describe each sponge and its characteristics that place it within the phylum porifera.

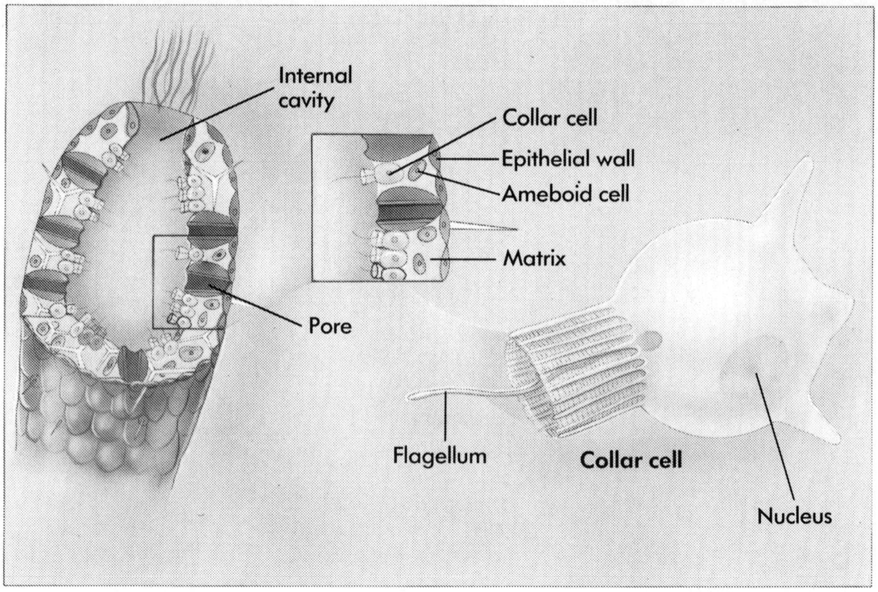

Figure 27-1
Diagram of a sponge.

12 ______________________________

____

____

PART TWO The cnidarians

EXPLORING

A polyp, the *Hydra*

Without disturbing the organisms, visually examine a culture of *Hydra*. Are more of these cnidarians (nigh-DARE-ee-ans, the "c" is silent) attached to a surface than moving freely in the water?

13 ______________________________

Describe the shape of one hydra.

14 ______________________________

What is the approximate length of the longest hydra?

15 ______________________________

Describe their body movements.

16 ______________________________

____

Do any hydras in the culture have a smaller attached hydra?

17 ______________________________

If so, describe the shape of a smaller hydra, its location, and how it is attached.

18 ____________________

____

CAUTION
Keep the hydra covered with water at all times.

Using a wide-mouth dropper, gently squirt a jet of water at the base of an attached hydra. When the hydra is dislodged, gently suck it into the lower portion of the dropper and transfer it to a depression slide. Observe the hydra with a dissecting microscope or the lowest power of your compound light microscope. When you are not observing the slide, turn off your microscope light to prevent overheating the hydra. What is the shape of the hydra immediately after the transferring process?

19 ____________________

Based on your observations of the hydra's response, speculate for a moment about the types of cells, tissues, organs, and systems of this animal. Describe some of the possible structures.

20 ____________________

____

In how many ways can you cut this organism to produce halves that are mirror images?

21 ____________________

Is there evidence of a digestive cavity?

22 ____________________

If so, what is the evidence?

23 ____________________

How many openings are there into or out of the hydra?

24 ____________________

After the hydra has attached to the slide and elongated again, add a few small organisms such as fresh water fleas, *Daphnia,* or rinsed brine shrimp, *Artemia.* What is the size of these organisms compared to the hydra?

25 ____________________

What is their rate of movement compared to the hydra's?

26 ____________________

Observe the interactions between the hydra and the organisms. Describe the structures that the hydra uses in these interactions and how they function.

27 ____________________

____

What is the maximum number of water fleas or brine shrimp ingested by any one of the hydras observed in the laboratory?

28 ______________________________

When you are through with the hydra, gently return the hydra to a culture dish marked "used *Hydra*." Do not place any excess water fleas or brine shrimp into the dish.

At a microscope on the demonstration table, observe a hydra that is in water containing a small amount of glutathione. Describe the hydra's response to this chemical.

29 ______________________________

Sections of *Hydra*

Obtain a prepared slide with a cross section and a longitudinal section of a hydra. Examine the longitudinal section. What is the relationship between the central cavity and the finger-like tentacles?

30 ______________________________

On the cross section, explore the cellular structure of the body wall. How many layers of cells make up the body wall?

31 ______________________________

How many different types of cells can you distinguish in each layer?

32 ______________________________

Considering the distance of each cell to the watery environment, does a hydra require specialized structures for transporting food, gases, or wastes?

33 ______________________________

Explain your answer.

34 ______________________________

Relating *Hydra* to other animals and its environment

In comparison to the complexity of a sponge and a pig, how would you rank this organism?

35 ______________________________

Explain your answer.

36 ______________________________

____

Based on your explorations, how well adapted is hydra to its habitat in fresh water streams?

37 ______________________________

Explain your answer.

38 ______________________________

____

____

A medusa, the jellyfish

Obtain a dish or jar with a jellyfish. In what ways is this cnidarian similar to the polyp? (Hint: Think of a flattened hydra.)

39 ______________________________

____

In what ways does the jellyfish differ from the hydra?

40 ______________________________

____

CONCEPTS/TERMS

The structure and function of cnidarians

The polyp

A hydra is essentially a cylindrical sac with its opening, the mouth, surrounded by tentacles. At the base of the sac is a disk that attaches to surfaces. Like all other cnidarians, hydras have radial symmetry. You can cut organisms with radial symmetry through many different planes that include the central axis (just as you can cut a round layer cake many different ways) to produce mirror-image halves. This sessile cnidarian form is called a **polyp.**

The body wall of the hydra and other cnidarians are composed of two tissue layers separated by a jelly-like material, the **mesoglea** (Figure 27-2 and Plate 15D). Among cnidarians, the thickness of the mesoglea varies. The outer layer,

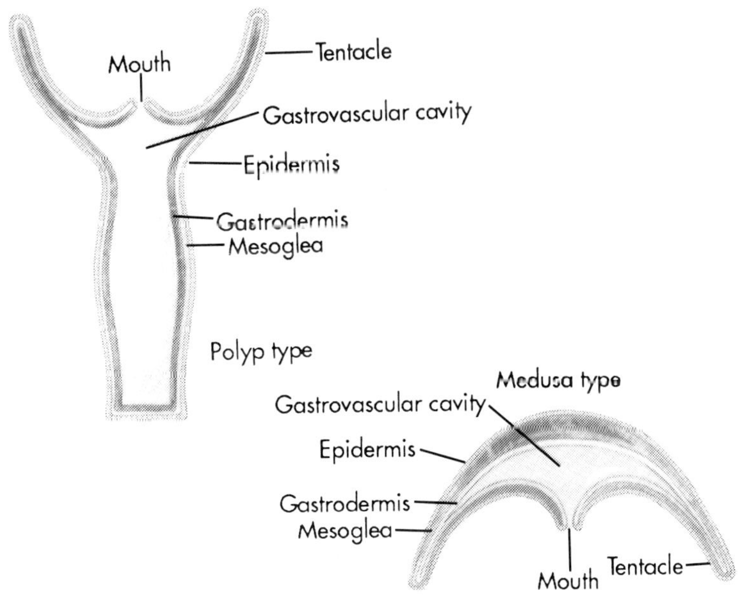

Figure 27-2
Two kinds of cnidarian forms.

the **epidermis,** is composed of several types of cells. Among these cells are those with contractile fibers, small cells that give rise to other cells and the gametes, mucus-secreting cells, receptor-neuron cells, and stinging cells called **cnidocytes.** Particularly abundant on the tentacles, the cnidocytes contain a thread-like tube that injects a paralyzing toxin when discharged into prey. The name cnidarians comes from these unique stinging cells. The inner tissue, the **gastrodermis,** lines the cavity of the hydra's sac and consists of cells similar to the epidermal cells. Some cells have flagella that stir the sac's watery contents. Other cells release digestive enzymes into the sac's cavity, the **gastrovascular cavity.** After partial digestion in the cavity, the gastrodermal cells take in the food and complete digestion intracellularly. The undigested material exits through the same "mouth" opening that it entered.

The neurons of the hydra are linked in a net-like pattern called a **nerve net.** They interact with contractile cells, which lets the hydra respond simply to environmental stimuli, move captured prey to its mouth, and occasionally move by somersaulting across a surface.

The two-layer construction of the body wall puts all cells in direct contact with either the oxygen-holding water in the gastrovascular cavity or the surrounding water. Therefore specialized structures for acquiring oxygen and eliminating waste products are unnecessary. Because all cells are close to the nutrient-acquiring gastrodermis, the hydra survives without specialized food-transporting structures.

The hydra can reproduce both sexually and asexually. During sexual reproduction, cells of the epidermis produce gametes that unite, forming the zygote. The zygote develops into a solid larva **(planula)** that consists of a layer of ciliated epidermal cells surrounding a solid mass of cells. The larva becomes sessile and forms a polyp. In asexual reproduction, outgrowths of the polyp's body wall form new polyps. In some cnidarians the new polyps remain attached to the "mother" polyp, forming a colony of polyps.

The tissues of cnidarians do not interact; they lack organs and their complexity reaches only to the tissue level.

The medusa

The umbrella-shaped jellyfish is a free-swimming cnidarian form called a **medusa.** The medusa swims with a pulsating motion. It has the same basic sac-like structure as a polyp, with a body wall of two tissue layers sandwiching a thick mesoglea. The tentacles have many cnidocysts, which are the reasons for common names such as sea nettle and sea wasp of some medusa cnidarians.

Life cycles of cnidarians

The life cycles of cnidarians vary. Some cnidarians alternate between polyp and medusa forms. Other cnidarians exist only as one of these two forms. In those cnidarians that alternate between both forms, the medusa is the one that reproduces sexually. It produces gametes that unite, forming the zygote, which develops into the plannula larva and polyp. Parts of the polyp produce new medusae by mitosis and differentiation.

APPLYING

Other cnidarians

Examine a preserved or living sea anemone. Is the form of the sea anemone a polyp or a medusa?

41 ______________________________

In what ways is it similar to the hydra and the jellyfish?

42 ______________________________

____

In what ways is it different?

43 ____________________

Obtain two prepared slides of *Obelia*. Using your microscope, explore both slides. Describe the forms of this cnidarian.

44 ____________________

____

Examine a piece of coral. Where did the animals live in this structure?

45 ____________________

Were the cnidarians that constructed this coral the polyp or medusa form?

46 ____________________

Explain your reasoning.

47 ____________________

____

PART THREE Platyhelminthes, the flatworms

EXPLORING

A living planarian

Partially fill a small, flat dish with aged or conditioned water (chlorine removed). With a wide-mouth dropper, transfer a slow-moving planarian from a culture dish of flatworms to your dish. Examine this animal with a dissecting microscope. Gently touch a part of the organism; describe its behavioral response to the touch stimulus.

48 ____________________

Does it appear to sense light?

49 ____________________

What is the evidence for your answer?

50 ____________________

How does the symmetry of the planarian compare to the cnidarians?

51 ____________________

Does the same end of the animal usually, if not always, go forward?

52 ____________________

If so, describe that end.

53 ____________________

By looking through the bottom of your dish, determine if there are distinct upper and lower surfaces. If so, describe the differences between the surfaces.

54 ______________________________

Describe any openings on the surfaces of the planarian.

55 ______________________________

In Figure 27-3, draw the planarian and add your notes.

Place a small piece of liver or cooked egg yolk in the dish. Does the planarian behave in ways that suggest it senses the presence of this material?

56 ______________________________

If so, describe these behaviors.

57 ______________________________

If the planarian moves to the liver or egg and stays for a few minutes, gently roll it over using the side of a small paint brush. Describe the structure protruding from its lower surface.

58 ______________________________

Compare the cnidarian's sessile polyp to the slowly moving planarian. Are both well adapted for obtaining food from their fresh water stream habitats?

59 ______________________________

Figure 27-3
Planarian, a flatworm.

What evidence supports your answer?

60 ______________________________

____

Return the planarian to a "used" planarian dish without extra food particles.

Sections of a planarian

Obtain two prepared slides: a whole mount of a planarian with a dye-stained interior and a cross section taken from the anterior region of a planarian.

Examine the whole mount with the microscope's lowest magnification. Explore the dye-filled structure in the interior of the planarian. Identify this structure if you can.

61 ______________________________

Working between the two slides (the whole mount and the cross section) and the labeled cross-sectional diagram in Figure 27-4, try to identify all structures visible in the whole mount. In Figure 27-5, outline the shape of the whole planarian from the dorsal view and add the internal structures that you identified.

Based on your observations of the planarian cross section, how many embryonic germ layers do you think are in the gastrula of a planarian?

62 ______________________________

On what evidence did you base this answer?

63 ______________________________

____

What organs did you observe in the planarian?

61 ______________________________

____

Figure 27-4
A planarian in cross section.

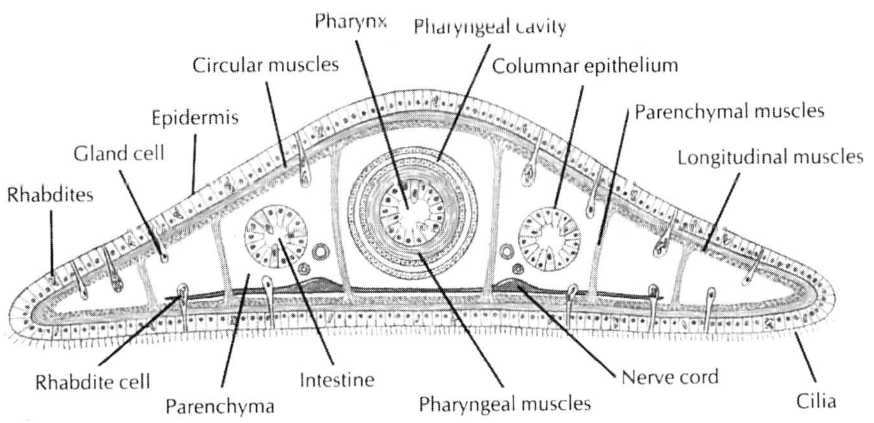

Figure 27-5
Internal structures of a planarian.

Which organism, the *Hydra* or the planarian, seems to have the most effective way of avoiding predators?

65 ____________________

Explain your answer.

66 ____________________

____

CONCEPTS/TERMS

The structure and function of flatworms

Planarians are free-living flatworms and relatives of the flat parasites, tapeworms and flukes. Flatworms exhibit **bilateral symmetry** and not radial symmetry as cnidarians. (Organisms with bilateral symmetry can only be cut in one longitudinal plane to form mirror-image halves—the right and left.) They develop three germ layers: (1) the ectoderm that forms the epidermis of the flatworm, (2) endodermis that forms the lining of the digestive cavity, and (3) mesoderm that forms the solid mass of organs such as muscles, reproductive organs, and water-regulating organs in the area between the digestive structure and the epidermis. Flatworms are the first organisms in the descent from the protist ancestors to have bilateral symmetry, three germ layers, and an organ system level of complexity.

A planarian has many organs and organ systems. Digestion in a planarian occurs both within the branched gastrovascular cavity (dye filled in the whole mount) and in food vacuoles within the gastrodermal cells. The tubular **pharynx** sucks in food pieces when it is extended from the mouth on the midventral surface and egests undigested materials. A planarian moves to the food when the chemoreceptors at the anterior end, nerve cords, and small brain detect food diffusing through the water. The brain consists of two clumps of nerve cells in the anterior end. The two nerve cords with cross connections (similar to a ladder) extend laterally, ventrally, and posteriorly from the brain. A specialized tube system toward the sides of the planarian regulates water volume.

A planarian moves slowly. After it secretes a layer of mucus on its traveling surface, cilia on the lower epidermis propel the planarian over the mucus layer. Internal muscles move body parts. Other free-living flatworms use their muscles to swim with an undulating motion.

With the development of bilateral symmetry, an accumulation of sense organs and neurons is associated with the anterior end. This distinguishes the anterior (forward) from the posterior end. What advantage do you think this **cephalization** gives to the organism?

67 ______________________________

In planarians, the eyespots are sensitive to light and the organisms usually respond by turning away from the light. A body orientation in response to an environmental stimulus is called a **taxis.** Therefore this response of planarians to light is a negative (away from) phototaxis.

APPLYING

Other flatworms

At the demonstration table, examine a preserved fluke and the anterior portion of a tapeworm. What characteristics do each of these organisms possess that place them in the phylum with planarians?

68 Fluke ______________________________

Tapeworm ______________________________

What modifications does each organism have that adapt it to parasitic life on or inside other animals?

69 Fluke ______________________________

Tapeworm ______________________________

PART FOUR Nematodes, the roundworms

EXPLORING

A living nematode

Place one or two worms that live in vinegar sediments on a depression slide; add a cover slip. Explore these worms with the lowest power of your microscope. In general terms, what is the shape of these organisms?

70 ______________________________

Is the body's length separated into sections?

71 ______________________________

If so, describe the arrangement of the sections.

72 ______________________________

Flatworms, which you explored earlier, have both longitudinal and circular muscles. Based on your observations of the movement of nematodes, how are the nematode's muscle groups arranged?

73 ______________________________

A preserved nematode

At the demonstration table, examine the dorsally opened roundworm *Ascaris* with a hand lens or dissecting microscope. (Many species of parasitic roundworms inhabit plants and animals, and produce many offspring [see Laboratory 6].) What is the general form of its digestive system?

74 ______________________________

Using touch and vision, explore the digestive system and note any changes from one end of the organism to the other. Beginning at the anterior end, describe all observable differences, if any, along the digestive system.

75 ______________________________

____

Examine the space between the digestive system and the body wall. Particularly note the positions of tissue layers that may line this cavity.

Continue your observation of the body cavity by examining a prepared slide of a cross section through the anterior portion of *Ascaris*. What type of symmetry does this organism exhibit?

76 ______________________________

On what evidence did you base your answer?

77 ______________________________

In Figure 27-6, draw the cross section, emphasizing the body space and its relationship to the wall of the digestive cavity and the outer body wall.

Figure 27-6
Internal relationships in a roundworm.

CONCEPTS/TERMS

The structure and function of roundworms

The roundworms are the first phylum of animals in the line of descent from the protist ancestors that have a tubular digestive system. Zoologists refer to animals with a two-opening digestive tube as having a **tube-within-a-tube** body plan. This plan contrasts with the sac-like body plan of animals that have a gastrovascular cavity. With food ingested through one opening (the **mouth**), passed along a specialized tube, and egested through a second opening (the **anus**), food can be efficiently processed in an assembly-line manner. In the gastrovascular cavity where freshly ingested food is mixed with food in various stages of digestion, sequential specialization of the digestive structure is impossible. During development, the roundworm exhibits developmental patterns of protostomes such as the mouth forming from the blastopore of the gastrula (see Laboratory 22).

Another new feature in the line of descent is the body cavity of the roundworms. The cavity between the digestive tube and the body wall is in contrast to the solid mesodermal layer between the gastrovascular cavity and body wall in flatworms. This cavity provides room for greater organ development. In organisms that have a body cavity completely lined by tissues from the middle germ layer (the mesoderm), the cavity is called a **coelom** (SEE-lome). In the roundworms, however, the body cavity is not completely lined by tissues of mesodermal origin. Therefore biologists call the body cavity of roundworms a **pseudocoel** (*pseudo* meaning "false"; *coel* meaning "cavity".)

APPLYING

Other nematodes

At the demonstration table, explore the slide of the larva of *Trichinella,* a roundworm that parasitizes vertebrates such as pigs. The tissue around the larva is muscle tissue of the vertebrate host. Describe the larva's characteristics that place this organism in the roundworm phylum.

78 ____________________

Also explore the slide of the dog hookworm *(Ancylostoma canium)*. What modifications of this roundworm adapt it to its usual habitat in the dog's digestive tract?

79 ____________________

Table 27-1 Comparison of the simpler animal phyla

Phylum	Symmetry	Number of germ layers	Level of organization	Body plan	Digestive structure
Sponges					
Cnidarians					
Flatworms					
Roundworms					

Figure 27-7
Phylogenetic tree of simpler animals.

Animal phyla

Over long time periods, animals have changed as they descended from their protist ancestors. Review the characteristics of the more ancient animal phyla that you explored today. As you review, compare the structural organization and developmental patterns of these groups and complete Table 27-1.

Phylogenetic relationships

In Figure 27-7, draw the probable descendant relationships of the animal phyla in Table 27-1 as a phylogenetic tree. At the base of the tree, place the likely ancestor of the animals. On the tree, add the characteristics that biologists think are important in showing close relationships, those that separate each phylum from the others, and those that are new.

LABORATORY 28

Animals

Annelids, mollusks, and arthropods

Importance of some complex invertebrates

Although we have a tendency to squash bugs, spray mosquitoes, and ignore earthworms, all these organisms and their relatives affect our lives. Earthworms are excellent soil conditioners. They form tunnels and move soil as they eat their way through the soil. Water and air reach plant roots through these tunnels, increasing plant growth. Many biologists consider insects to be the most successful of all land animals. Insects have the greatest number of species (estimated to be in excess of five million) and the greatest number of organisms compared to all other groups. About 75% of all flowering plants depend on insects for pollination. Other insects cause misery by biting and transmitting diseases such as malaria and some forms of encephalitis. Many people consider other complex invertebrates such as clams, oysters, shrimp, and lobsters as gastronomic delicacies. Many complex invertebrates are primary food sources for other animals in both aquatic and terrestrial environments.

Three phyla of complex invertebrates have a **coelom** (body cavity lined with mesoderm) and the same pattern of early development. Their developmental similarities include their mouth forming at the blastopore, spiral cleavage, and the mesoderm forming from cells near the blastopore. (You may have explored these and other features of this developmental pattern in Laboratory 22.) Biologists use the developmental similarities to place annelids, mollusks, and arthropods on the same branch of the animal family tree and to separate them from the echinoderms and chordates. Some biologists refer collectively to the annelids, mollusks, and arthropods as **protostomes** (first opening), although other animal phyla (such as the roundworms and flatworms) also have protostomic developmental patterns but lack coeloms. In Laboratory 27, you explored the simpler animals that zoologists think descended from flagellated, heterotrophic protists. Today, you will examine the forms and functions of the more complex protostomic animals and the adaptations that help them succeed in their particular environments.

PART ONE Annelids, the segmented worms

EXPLORING

A living earthworm

Obtain a living earthworm in an observation dish lined with a damp paper towel. Examine the external features of the worm. Describe the differences between the anterior and posterior ends.

1 ______________________________

In its normal habitat, what is the food source of the earthworm?

2 ______________________________

Does an earthworm look as though it is separated into sections?

3 ______________________________

If so, describe the sections and their arrangement.

4 ______________________________

Watch the earthworm move. Do you think it has both longitudinal and circular muscles?

5 ______________________________

What is the evidence for your answer?

6 ______________________________

____

Run your dampened fingers up and down the earthworm's surfaces. In addition to muscles, can you feel other structures that may aid the earthworm in its locomotion?

7 ______________________________

If so, describe these structures and how they could be used.

8 ______________________________

____

Examine the earthworm with a dissecting microscope. Describe its surface and include a description of any openings.

9 ______________________________

____

From your observations, have you gathered any evidence to suggest that an earthworm has a circulatory system?

10 ______________________________

If so, what is your evidence?

11 ______________________________

Do you have evidence that suggests an earthworm has a gas-exchange system?

12 ______________________________

If so, describe your evidence.

13 ______________________________

Did you observe any structures that may be sense organs?

14 ______

If so, describe these structures.

15 ______

Explore the behavior of the earthworm. If it is sensitive to light and/or touch, describe its responses.

16 ______

______

______

Do earthworms exhibit positive or negative phototaxis (orientation towards or away from light)?

17 ______

Do they exhibit positive or negative thigmotaxis (touch)?

18 ______

Place a drop of alcohol directly in front of but not on the worm. Describe its response to alcohol.

19 ______

If any alcohol gets on the earthworm, rinse the worm with water. When you are finished with your observations, return the worm to its original container.

A preserved clam worm

Examine a preserved *Neris* ("clam worm"). Describe how its external features differ from the earthworm's features.

20 ______

______

What are the differences, if any, in the appendages (organs or limbs that extend from the body) of this worm?

21 ______

Explore a clam worm that has been opened on its dorsal side. Do the surface sections of the worm continued into its interior?

22 ______

In Figure 28-1, draw the internal contents of one section in three dimensions.

A section of an earthworm

Obtain a prepared slide of a cross section through the intestine of an earthworm. Carefully examine the body cavity, the organs within the cavity, and the body wall.

Figure 28-1
Segment of a segmented worm.

Describe the layers of the body wall beginning with the outermost layer.

23 ______________________________

____

On the basis of the earthworm's structure, describe the characteristics of the environment in which the earthworm lives. Also explain why you linked the environment to the structures.

24 ______________________________

____

____

Was your prediction of the muscle arrangement of the living earthworm supported by your observations of muscle tissues in the cross section?

25 ______________________________

If not, describe the arrangement of muscles in the cross section and how they differ from your prediction.

26 ______________________________

____

List the internal organs that you can identify and give their positions in one segment of the earthworm. If these structures differ from those in the clam worm, note it. Some structures you can look for are the digestive tube, blood vessels, nerve cords, and water-regulating tubes.

27 ______________________________

____

____

In contrast to people, an earthworm has no bones providing stable muscle attachment sites. What structures may provide the attachment sites for the earthworm's muscles?

28 ______________________________

____

Did all parts of the worm seem soft and flexible?

29 ______________________________

If so, what do you think the earthworm's muscles pull against? (Hint: Think of squeezing an air- or water-filled balloon.)

30 ______________________________

CONCEPTS/TERMS

The structure and function of segmented worms

The segmented worms have a **tube-within-a-tube** body plan, with their bodies divided into similar repeating segments by partitions called **septa.** This division of the body is called **segmentation.** If the worms have appendages extending from the segments, the appendages are very similar to each other. Annelids are protostomes with coeloms.

In a segmented worm, fluid confined within the segments of the coelom functions as a **water (hydraulic) skeleton.** Contractions of longitudinal muscles cause the fluid in the segment to push laterally, widening the body. Contractions of circular muscles cause the fluid to push anteriorly and posteriorly, elongating segments. If the bristles (the **setae**) and the laterally expanded segments anchor the worm in the soil, the more anterior segments move forward with the elongation.

Segmented worms have several organ systems. The digestive system is a muscular, specialized tube. The circulatory system has five pairs of muscular vessels that circle the esophagus and connect a large dorsal vessel to a ventral vessel. The vessel pairs pump the blood posteriorly in the ventral vessel and anteriorly in the dorsal vessel. A network of connecting vessels distributes and returns the blood in each segment in this closed system. The excretory system in each segment consists of paired tubes with ciliated funnel-shaped openings in the coelom of the adjacent segment. From the funnels, the wastes are swept through the tubes to minute openings on the surface of the earthworm. The nervous system has two solid cords that are fused together in the ventral midline and two groups of neurons (brain) dorsal to the pharynx. Some annelids have both male and female reproductive organs in the same organism (earthworm). Other annelids have only organs of one sex *(Neris)*. The thickened area or band around the earthworm is a mucus-secreting reproductive organ called the **clitellum.** Its mucus holds two mating worms together and forms a cocoon around the fertilized eggs. Although there is no special gas-exchange system, gas exchange does takes place between the capillary beds and the moist, cuticle-covered surface of the body wall or the segmental appendages.

APPLYING

Other segmented worms

Obtain a prepared slide of a leech or, if available, a living leech. What characteristics does it have that places it in the annelid phylum?

31 ______________________________

What adaptations does it have for parasitic life?

32 ______________________________

PART TWO Mollusks

EXPLORING

Clams

Obtain the following materials:
- 2 blunt probes
- 1 dissecting pan
- 1 opened clam
- 1 unopened clam

Explore the external features of the unopened clam. What type of symmetry does the clam have?

33 ______________________________

Describe its covering.

34 ______________________________

What do you think are the advantages of this type of covering?

35 ______________________________

Try to separate the two parts of the shell by pulling them apart. Describe what happens.

36 ______________________________

Position the clam with the hinge of its shell upward, the wider portion of the clam away from you, and its pointed region facing toward you. Between the two edges of the shell, find two spaces defined by darkly pigmented soft tissue **(siphons).** Think about the clam's normal habitat and try to predict what passes in the lower and out the upper siphon.

Explore the opened clam.

What structures were cut before the clam shell could be opened?

37 ______________________________

Describe the soft structure that is most closely associated with the inner surface of the shell.

38 ______________________________

Search for an organ (or organs) that has some characteristics that can be associated with gas exchange. Where is this structure located?

39 ______________________________

What are the characteristics of the gas-exchange surface?

40 ____

Based on this structure, do you think a clam could live on land?

41 ____

Explain your answer.

42 ____

____

On the side opposite the siphons, try finding a small mouth between two flaps of soft tissue. Does this mollusk appear to have a head?

43 ____

If so, describe its features.

44 ____

How do you think the clam ingests its food?

45 ____

Locate a space near the hinge of the two shell halves that contains an elongated structure with a tube running through it. Describe the structure.

46 ____

What function could this structure have?

47 ____

Explain your answer.

48 ____

Obtain a snail (or slug) in a small, flat dish and examine it. Observe the movement of the snail. On the opened clam, locate a similar locomotor structure. Describe this organ and its location in the clam.

49 ____

What methods do you think a clam and a snail use to avoid predators?

50 ____

Explain your answer.

51 ____

Snails

Observe the snail (or slug). Describe the ways in which the clam and snail are structurally similar and different.

52 Similarities

Differences

In what kind of habitats would you expect to find snails?

53

Explain your answer.

54

CONCEPTS/TERMS

The structure and function of mollusks

Mollusks are soft-bodied animals that may have protective shells. They include organisms that are very diverse, but each has a **foot** that is used for locomotion, a **visceral mass** that contains most of the internal organs, and a **mantle** that covers the visceral mass and may secrete a calcium-containing shell (Figure 28-2 and Plate 16). In many mollusks a toothed, tongue-like feeding structure, the **radula,** scrapes algae from surfaces.

Mollusks are protostomes with coeloms and a modified tube-within-a-tube body plan. The clam shell has two parts or valves. The two valves are held together by large anterior and posterior muscles. Inside the shell, the thin mantle lies over the body of the clam. The mantle secretes the shell, moves water with its surface cilia, and is one site for gas exchange. Inwardly from the mantle are large gills, which are sites for gas exchange and which trap food particles in its mucus. Surface cilia move the trapped food toward the mouth located between two flaps of tissue (palps) near the large, anterior, shell-holding muscle. Clams lack heads. A clam's small mouth leads into a specialized digestive tract that winds around in the visceral mass just dorsal to the foot, extends through the heart in the dorsal space near the shell's hinge, and terminates in an anus close to the posterior, shell-holding muscle and the upper siphon.

Unlike the earthworm and human, the mollusk circulation is an **open system;** it lacks vessels that return blood to the heart. The heart is within the re-

Figure 28-2
Mollusk body plan.

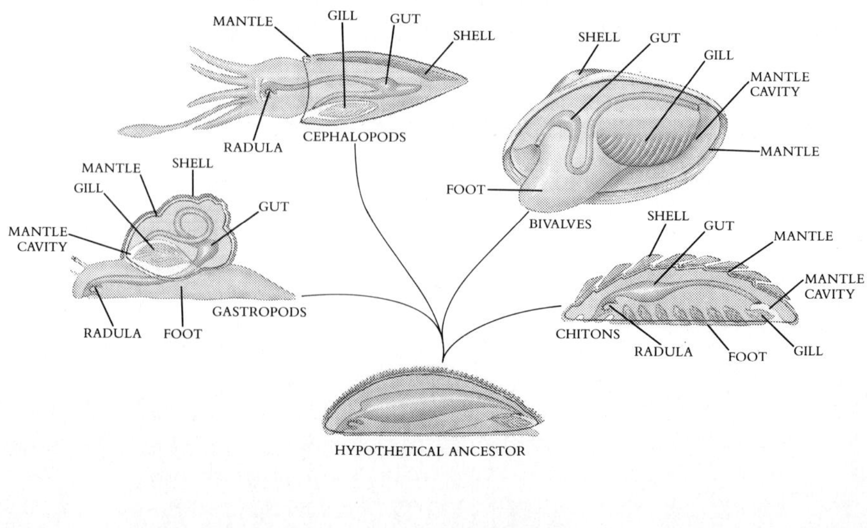

maining remnant of the coelom (space near the hinge). From the heart, blood flows through arteries to the tissue spaces. It slowly returns to the heart.

The clam has additional organs and structures. A primitive kidney removes nitrogen waste products from the blood and releases them near the upper siphon. Cilia on the gill, mantle, and other body surfaces circulate water from the environment. Water comes in through the lower **incurrent siphon** and passes through gills (which remove food and oxygen) and out the upper **excurrent siphon.**

Other mollusks have variations on the same body structure of mantle, visceral mass, and foot. Snails and slugs (gastropods) possess heads with sense organs, and during development, their visceral mass twists. Many gastropods inhabit aquatic environments; others inhabit land. Land snails use the mantle for gas exchange.

APPLYING

Other mollusks

Examine a chiton. Describe the visible similarities and differences of the chiton to the clam and snail.

55 Similarities ______________________________

____

____

Differences ______________________________

____

____

What other structures would you expect to find inside the chiton if it were dissected?

56 ______________________________

____

Many zoologists think that the chiton is more like the common ancestor of both mollusks and annelids than most other living mollusks. Are any characteristics of the chiton similar to some of the characteristics of segmented worms?

57 ______________________________

If so, what are these characteristics?

58 ______________________________

Explore a squid or octopus. Try to identify the highly modified molluscan structures on these more complex mollusks. Describe their versions of the foot, mantle, and visceral mass.

59 ______________________________

____

____

EXPLORING

Millipedes

Obtain a millipede and examine it with a hand lens or dissecting microscope. In what ways is its external anatomy similar to an earthworm's?

60 ______________________________

____

In what ways are its external structures different from the earthworm's?

61 ______________________________

____

Describe its appendages.

62 ______________________________

____

In what ways are the millipede's appendages different from appendages on the segmented worm *Neris?*

63 ______________________________

Examine its mouth. Based on the structure of the parts around the mouth, predict the type of food it ingests.

64 ______________________________

On what evidence did you base your prediction?

65 ______________________________

CONCEPTS/TERMS

The structure and function of millipedes

The slightly flattened body of the millipede is distinctly divided into doubled segments, most with two pairs of appendages. In contrast to segmented worms such as *Neris,* the millipede's appendages are jointed. Jointed appendages are characteristic of all **arthropods** (meaning "jointed feet") and give this phylum its name. Some segments are without appendages. A few segments may have slight differences in their appendages, particularly with regard to the pair of antennae on the head. Most other arthropods have more specialized appendages. In addition, millipedes are covered with a cuticle of chitin, the **exoskeleton,** which is also a characteristic of all arthropods.

At the mouth, millipedes have two big, laterally moving jaws (the **mandibles**) with teeth on the biting edge and a raspy surface. Millipedes eat decaying leaves and other plant material.

EXPLORING

Insects

Using a hand lens or dissecting microscope, examine the external features of a preserved grasshopper. Is there fusion of some body segments into larger, more complex structures?

66 ______________________________

If so, describe these fused parts.

67 ______

How many regions can you distinguish on the body of the grasshopper?

68 ______

Explore the most anterior body region, the **head.** The appendages of the head include antennae and mouth parts. Describe at least three of these appendages.

69 ______

______

______

Based on its mouth parts, what type of food does the grasshopper ingest?

70 ______

How do their eyes differ from humans'?

71 ______

Explore the appendages attached to the second body part, the **thorax.** What are the functions of the thoracic appendages?

72 ______

In Figure 28-3, draw two distinctly different legs. On your diagrams, identify the joints with small arrows.

For the grasshopper to jump several feet, muscles need to be anchored to body parts. In humans, muscles are anchored to bones. In general, what are the possible attachment sites of the grasshopper's muscles?

73 ______

Figure 28-3
Thoracic appendages of the grasshopper.

Explain your reasoning.

74 ______________________________

____

Explore the wings of the grasshopper. How many wings are present?

75 ______________________________

Based on the wing structures, do you think that all of the wings are used in flight or are some modified for other functions?

76 ______________________________

On what evidence did you base your answer?

77 ______________________________

Examine a series of small openings along the side of the third body section, the **abdomen.** Considering the impermeable covering of the grasshopper and the dry terrestrial environment in which it lives, what is a possible function of these openings?

78 ______________________________

Explore the tip of the abdomen. What evidence, if any, indicates that the grasshopper has a tube-within-a-tube body plan?

79 ______________________________

The structure of the tip varies with the sex of the grasshopper. You might want to look at other grasshoppers in the class to observe the differences in them.

With a single-edged razor blade or scalpel, cut the grasshopper into right and left halves (saggital section). Utilizing your knowledge of internal structures of other animals, try to identify the digestive tube, circulatory vessels, a ventral nerve cord, and other internal organs.

CONCEPTS/TERMS

The structure and function of insects

Insects have three body regions: the head, thorax, and abdomen. The head consists of more tightly fused segments than the thorax or abdomen. Unlike millipedes, the insect's appendages are highly modified in structure and function. Three pairs of legs attached to the thorax are important in locomotion. Their muscles are attached to the inside of the exoskeleton on opposite sides of the joints; as the muscles contract, the jointed parts move. Some other appendages form mouthparts and antennae (feelers).

The large **compound eye** of the insects is composed of many separate visual units, each with a lens (see Plate 9C). Most insects also have one or more simple eyes.

Insects have a unique gas-exchange system, the **tracheal system.** Trachea are branching tubes that extend internally from the openings **(spiracles)** along the sides of the abdomen to the vicinity of all body cells. The spiracles can be closed, reducing water loss when the need for gas exchange is satisfied.

Other systems include a sluggish, open circulatory system with a dorsal vessel (the heart) that alternately contracts and relaxes, a unique excretory system that empties into the digestive tube from which some materials are reabsorbed, and a nervous system consisting of an anterior brain, two ventral nerve cords fused into one, and several well-developed sense organs. Insects also have a fairly straight, specialized digestive tract.

APPLYING

Other arthropods

Explore an intact, preserved crayfish. What external structures group this animal with other arthropods?

80 ______

Compare the main body region of the crayfish to the regions of the insect. What are the similarities and differences?

81 ______

______

Describe at least three different appendages of the crayfish and include the location and a possible function of each.

82 ______

______

______

At the demonstration table, explore a crayfish from which the exoskeleton on the sides of the head-chest (the **cephalothorax**) has been removed. Describe the structure of the gas-exchange system and explain how it is adapted to the crayfish's watery environment and how it differs from the gas-exchange system of the land-dwelling grasshopper.

83 ______

______

______

On another crayfish that lacks the dorsal part of the exoskeleton, observe the exposed digestive system. Note modifications along the tract and the mass of the digestive gland. Describe some of the specialized areas and predict their functions in the digestive process.

84 ______

______

Examine a third crayfish with the dorsal exoskeleton and most of the internal contents removed. Is the crayfish's nervous system more similar to that of the earthworm or that of the fetal pig?

85 ______

Explain your answer.

86 ______

______

Observe a preserved spider with a hand lens or dissecting microscope. What characteristics does it have that places it within the arthropods?

87 ______

How many body regions does a spider have?

88 ______________________________

Are these body regions more similar to the body regions of the crayfish or the grasshopper?

89 ______________________________

Explain your answer.

90 ______________________________

____

How many pairs of appendages are used for locomotion?

91 ______________________________

Are antennae present?

92 ______________________________

How many eyes are there?

93 ______________________________

Based on the structure of the more anterior appendages, do you think that spiders are herbivores or carnivores?

94 ______________________________

Explain your answer.

95 ______________________________

____

A unique activity of some spiders is the construction of sticky webs. Observe the tip of the abdomen, searching for its nozzle-like **spinnerettes** from which the webbing fluid is ejected.

Table 28-1 Comparison of annelids, mollusks, and arthropods

Phylum	Symmetry	Number of germ layers	Body plan	Segmentation	Types of systems	Unique characteristics
Annelids						
Mollusks						
Arthropods						

Figure 28-4
Phylogenetic tree of complex protostomes.

Spiders and some other arthropods such as ticks and chiggers have the anterior appendages modified as fangs or pincers. On attacking their prey, they frequently inject a paralyzing poison or viruses and bacteria. The black widow and brown recluse spiders can inject poisons that affect humans (see Plate 16). Some ticks can inject organisms that cause Rocky Mountain spotted fever and Lyme disease.

Animal phyla

Compare the characteristics of the segmented worms, mollusks, and arthropods. List these characteristics in Table 28-1.

Organisms

Obtain a tray of animals from the demonstration table. Examine each one of these numbered organisms. Using the information you have about the annelid, mollusk, and arthropod phyla, sort these animals into three groups. Name each group of animals and list the number of all organisms in each group.

96 ______________________________

____

____

Phylogenetic relationships

In Figure 28-4, draw the probable descendant relationships of these complex animal phyla as part of a phylogenetic tree. At the base of this part of the tree, place the simpler animal type that seems to be the closest to the ancestor of these phyla. On the tree, add the characteristics that biologists think are important in showing close relationships among these phyla and those characteristics that separate the phyla from one another.

LABORATORY 29

Animals

Echinoderms and chordates

The relationship between echinoderms and humans

Echinoderms such as a sea stars and sea urchins certainly do not look like us, but they do have something in common with us. In development, the first opening in the echinoderms and chordates (which include humans) becomes the anus and a second opening becomes the mouth. In contrast, all the animals that you explored in Laboratories 27 and 28 have the protostomic type of development in which the first opening (the blastopore) develops into the mouth of the organism. Other developmental differences in cleavage, the timing of cell specialization, the formation of the coelom, and the type of larvae distinguish echinoderms and chordates from the annelids, mollusks, and arthropods. Biologists recognize these developmental differences as sufficiently important to put the echinoderms and chordates on their own branch of the animal family tree. This separates them from the protostomes. The organisms that develop the first opening into the anus and the second opening into the mouth are called **deuterostomes.** (You may want to review Laboratory 22, where some of these developmental differences were examined.)

Echinoderms and chordates represent only a small percentage of all animals, but there is a great diversity within these phyla. Let's begin by examining the spiny-skinned echinoderms.

PART ONE Echinoderms

EXPLORING

External structures of sea stars

Explore the external structures of a preserved sea star (starfish). Describe its general shape.

1 ______________________________

What type of symmetry does it exhibit?

2 ______________________________

Do you see any evidence of missing parts or replacement of parts?

3 ______________________________

If so, what is your evidence?

4 ______________________________

Does this animal have repeating structural units?

5 ______________________________

If so, describe the number of repeating units and how they are arranged with respect to one another.

6 ______________________________

In general, what are the differences between the upper and lower surfaces?

7 ______________________________

____

____

Does the sea star have a tube-within-a-tube body plan?

8 ______________________________

If so, describe your evidence.

9 ______________________________

Examine all surfaces with a hand lens or dissecting microscope. (Placing the sea star in water will float surface structures and make them easier to observe.) Scrape a few of the surface structures from the sea star and prepare a wet mount for closer examination with a microscope. Describe the sea star's surface structures and give a possible function of each.

10 ______________________________

____

____

____

If a living sea star is available, observe its locomotion. Describe the structures it uses for movement.

11 ______________________________

What structures does it use to attach to vertical surfaces?

12 ______________________________

Where are the locomotor and attachment structures located?

13 ______________________________

Describe how these structures work.

14 ______________________________

____

Internal structures of sea stars

If previously opened sea stars are not available, obtain dissecting tools and a sea star. Remove and set aside a 1- to 2-cm tip from one of the arms of the sea star. With the mouth surface of the sea star downward, insert one blade of the scissors just inside the lateral wall and cut along each side of the arm to the central disk. Avoid cutting into underlying tissues. Cut from one side of the arm to the other by curving your cut into the central disk while avoiding a hard, small disk. Carefully lift the cut wall from the arm.

Describe the composition of the wall.

15 ______________________________

____

Describe the different layers of structures and spaces within the arm.

16 ______________________________

____

____

Carefully remove the wall on the upper surface of the central disk leaving the small disk in place. What other structures are anchored to this wall and/or open onto the surface of the wall?

17 ______________________________

____

Follow the structures first located in the arms into the central disk. The first layer, which is soft and bumpy, consists of digestive glands. Describe the structures of the central disk to which these digestive glands connect.

18 ______________________________

____

The next layer consists of reproductive structures. The size of the reproductive organs depends on the fertility state of the sea star. Move this layer aside and inspect the lowest layer within the arm that also extends through the wall of the mouth surface. Trace its connections into the central disk. What is the relationship between these central disk structures and the small disk on the surface opposite the mouth (aboral)? In Figure 29-1, draw the structures of this lowest layer in one arm and the connecting central disk structures.

Echinoderm larvae

At the demonstration table, examine a prepared whole mount of the sea star or sea urchin larva. Describe the body plan of the larva.

19 ______________________________

____

What is the symmetry of the larva?

20 ______________________________

Figure 29-1
Water-vascular system in the sea star.

CONCEPTS/TERMS

The structure and function of sea stars

Adult sea stars and most other echinoderms have five repeating units that are arranged radially around a central axis. Zoologists, however, do not place the echinoderms with the radially symmetrical cnidarians because the echinoderms are deuterostomes, have much greater complexity, and have bilaterally symmetrical larvae. In sea stars, the repeating units extend from the center of the central disk and are called **arms** or rays.

If one or more of the sea star's arms is removed, the sea star grows a replacement arm. People protecting valuable coral reef habitats from destruction by an overabundance of sea stars caught sea stars and killed them by cutting them into pieces. After they threw the pieces back into the ocean, the pieces regenerated missing parts, producing many new sea stars.

The sea star has many organ systems. The mouth, which forms as the second opening during early development, is in the center of the lower (oral) surface. The mouth leads into an esophagus, a two-part stomach, and an intestine with an inconspicuous anus on the upper surface of the central disk. (The small disk is not the anus.) Hence sea stars as well as other echinoderms have a **tube-within-a-tube** body plan. The body cavity containing the internal organs is a **coelom.** In early development the coelom forms from pouches of the digestive tube; in contrast, the development of the coelom in protostomes arises from cells near the blastopore.

A system unique to the echinoderms is the **water-vascular system,** which they use for locomotion and grasping. This system consists of a tubular ring in the central disk that connects to tubes running the length of each arm. The tubular ring also connects through a vertical canal to the small, hard disk (the madreporite) on the upper surface of the sea star. Many connecting sucker-like **tube feet** are lateral to each radial arm tube. Water enters the madreporite and passes through the tubes to the tube feet. A muscular bulb (the ampulla) at the top of each tube foot pumps water into the tube foot, extending it. When the bulb relaxes, the tube foot retracts and can form a suction cup-like attachment to a surface. The extension, attachment, and retraction of many tube feet move the sea star. Sea stars also use them to open clam shells. The sea star clamps its tube feet on the clam shell and retracts the feet, which begins a tug-of-war with the clam. The tube feet working together are stronger than the clam's muscles and the clam loses the battle to keep its shell closed. The sea star everts its esophagus and stomach inside the shell and secretes digestive enzymes. The sea star takes the partially digested particles into the digestive tract where digestion is completed.

The walls of echinoderms are complex. They have calcium plates or rods embedded within them. Many jointed spines protrude from their surfaces, giving this phyla the name **echinoderm,** which means "spiny skinned." In addition, small pincer-like structures and gill-like projections protrude from the surface. The small pincers discourage other organisms from living on the echinoderm's surface. The gills help in gas exchange.

APPLYING

Other echinoderms

At the demonstration desk, examine a sea urchin. What are the characteristics of the sea urchin that place it within the echinoderm phylum?

21 ______________________________

____

Compare the sea urchin to the sea star. Describe the ways in which you think these two echinoderms are similar and different.

22 Similarities ______________________________

____

Differences ______________________________

____

Examine a sea cucumber. What characteristics of the sea cucumber place it within the echinoderm phylum?

23 ______________________________

____

Compare the sea cucumber to the sea urchin and to the sea star. Describe the ways in which these three echinoderms are similar and the ways in which the sea cucumber differs from the other two.

24 Similarities ______________________________

____

Differences ______________________________

____

EXPLORING

Lancelets

Obtain a light microscope and the following prepared slides of lancelets:

1 pharynx region, x.s.
1 digestive tube, x.s.
1 lancelet, w.m.

Using the scanning objective of the compound light microscope, explore the whole lancelet. (In many whole mounts of lancelets, the muscular tissues are treated to make them transparent, thus making internal structures more visible.) Begin at the end of the organism that has a feather-like fringe. The fringe is close to an area of alternating diagonal bars and slits. The fringe is on the anterior-ventral side of the lancelet. (Turn your slide around if the fringe is pointing upward in your field of view.) Beginning with the fringe, follow the structures through the diagonal bar area into the digestive tract that extends posteriorly to the anus. Does the tail extend past the anus?

25 ____

Work between the whole mount and the two cross sections. Examine the cross section through the pharynx region. If you have examined the pharynx and gill region of a fish, use this information to identify the pharynx of the lancelet, or use your explorations of the pharynx in the fetal pig. (Your teacher can help you convert your thoughts from a three-dimensional pig or fish to the two-dimensional cross section.) Using your identification of the pharynx on the two-dimensional cross section, identify the pharynx region on the whole mount (a mental exercise in the reverse order). Your teacher can verify your identification of the pharynx. In Figure 29-2, draw the lateral view of the lancelet and label the pharynx.

Examine the more posterior cross section though the digestive tube. Based on the position of the digestive tube in this cross section, confirm your previous identification of the digestive tube on the whole mount and label this structure in Figure 29-2.

Return to the pharynx cross section and find an oval structure that is dorsal to the pharynx. Check the digestive tube cross section and identify the same

Figure 29-2
Lancelet.

oval structure (the size may be somewhat different). These two ovals are sections of the same rod that runs from near the anterior end of the lancelet to near its posterior tip. Where is this rod located with respect to the digestive tract?

26 ____________________

Identify this structure as the notochord in Figure 29-2.

Examine the structure just dorsal to the notochord in both cross sections and in the whole lancelet. In general terms, describe this structure.

27 ____________________

In both cross sections, large masses of a single type of tissue wrapped in connective tissue are lateral to the notochord. What is the type of tissue that composes these large structures?

28 ____________________

What is the structure in the ventral-lateral portion of the cross section? (It may be stained darker than the dorsal-lateral masses.)

29 ____________________

CONCEPTS/TERMS

The structure and function of lancelets

The adult lancelet is a filter feeder, which is an organism that sifts food from the water. It has bands of cilia in the vestibule (an entry space posterior to the fringe) that sweep water from the environment into the vestibule and then through the mouth into a basket-like pharynx. As water passes through the pharyngeal basket, the water exits through slits between the diagonal gill bars into a surrounding chamber. A ventral pore in that chamber returns the water to the environment. As the water passes the gill bars, food in the water is trapped in a layer of mucus and oxygen diffuses into the blood vessels of the bars. The mucus with the food streams posteriorly into the digestive system. You may have seen parts of small organisms such as diatom shells in the digestive tube.

In addition to having **pharyngeal gill bars** and **gill slits,** the lancelet has a flexible cartilaginous rod, the **notochord.** This support rod, just dorsal to the pharynx and digestive tract, runs the length of the lancelet from the anterior projection (rostrum) to near the tip of the posterior fin. The notochord prevents shortening of the lancelet as the large lateral muscles contract.

Another new structure is the **dorsal nerve tube** located just above the notochord. In contrast to the solid and ventral nerve cords of protostomes, this nerve tube forms in the dorsal midline of the lancelet from a single tube of epidermal tissue.

Reproductive organs are located in the ventral, midregion of the lancelet.

PART THREE Chordates, subphylum vertebrates

EXPLORING

Frog larvae

At the demonstration table, observe a cross section of the belly region of a young frog larva and a mid frontal section. (A frontal section separates dorsal and ventral halves.) On the cross section, search for structures that are also present in the lancelet. On the frontal section, concentrate on the internal space that expands laterally toward the gills. If you had gills as the frog larva does, where would your gills be located?

30 ______________________________

What structures are common to both the lancelet and the frog larva?

31 ______________________________

Using the frontal section for Figure 29-3, draw the spatial relationships among the pharynx passageways, the digestive tube, and the gills.

CONCEPTS/TERMS

Common structures of the chordates

The frog larva and lamprey share the common structures of the pharyngeal gill slits, the notochord, a dorsal nerve tube, and a postanal tail. In fact, all chordates have these same four structures at some time in their lives.

The lancelet is a member of one of the chordate subphyla, the **cephalochordates.** The cephalochordates retain these four structures through their entire lives.

The frog is a member of a different chordate subphylum, the **vertebrates.** All vertebrates replace their notochords with bones that form the vertebral column. This characteristic vertebral column gives the vertebrates their name.

Some vertebrates maintain the pharyngeal gill slits only during the embryonic or larval stages. During development of these organisms, the gill slits close and the gill bars differentiate into new structures in the head and neck. The frog has gill slits (and gills) only in its larva form (the tadpole). A few vertebrates keep the gill slits through adult life.

In all vertebrates the dorsal hollow nerve tube expands at the anterior end, forming the brain. The remainder of the tube becomes the spinal cord, which is encased by the bones of the vertebral column.

APPLYING

Chordate structures of a chick embryo

At the demonstration table, examine a chick embryo (w.m.) that was incubated for 48 hours. At this point in the chick's development, the anterior half of the embryo has twisted so that it lays on its left side. The "hook" of the chick's question mark–shape (in reverse) houses the brain. You can identify

Figure 29-3
Structural relationships of the frog's pharynx, gills, and digestive system (frontal section).

the first part of the brain by the large eye near the tip of the hook. Just posterior to the tip of the hook is a looped tube that will eventually form the four-chambered heart of the chicken. At the tip of the microscope's pointer is a space that the light shines through. Using information from the frontal section of the frog and your exploration of the lancelet, identify this space.

32 ____________________

Using your knowledge of the location of the notochord in lancelets and frog larvae, locate the chick's notochord. Describe the notochord's position with respect to the curved brain and the pharynx of the chick.

33 ____________________

What evidence places chickens in the chordate phylum?

34 ____________________

Bony fish

CAUTION

The spines on the dorsal fins of the perch can pierce human skin. Your teacher may ask you to remove the fin tips after examining them.

Obtain dissecting tools and a preserved perch. Wash off the excess preservative.

Examine the external surfaces of the perch. Describe the body regions.

35 ____________________

What is one possible relationship between the perch's body shape and movement in its watery habitat?

36 ____________________

List all the fins of the perch, identifying them by location. Also describe the structure of each kind of fin and give possible functions of each.

37 ____________________

____

____

____

____

____

Examine one of the perch's scales under the microscope. Its rings represent growth rings. What is one possible function of the scales?

38 ____________________

Examine the skeleton of a fish. What structure places fishes in the same subphylum as frogs and humans?

39 ____________________

Do fishes have appendages?

40 ____________________

If so, describe them.

41 ______________________________

Are the fins of fishes homologous (originating from a common ancestral structure) or analogous (from different origins) to the fins of whales, which are aquatic mammals?

42 ______________________________

List three ways in which fish skeletons differ from human skeletons.

43 ______________________________

____

Examine the mouth, pharynx, and gill region of the fish. The gills on each side of the head are covered externally by a bony flap called the *operculum*. You may want to make a cut at the angle of the jaws and also remove one operculum for better access to the gills. Do lancelets possess jaws?

44 ______________________________

Compare the pharynx of the fish to the pharynx of the lancelet. Describe their similarities and differences.

45 Similarities ______________________________

____

Differences ______________________________

____

Explore the internal structures of the fish. To expose these organs, insert the point of one scissor blade inside the body wall just anterior to the anus. Cut the ventral wall from the anus to the gill region. Cut dorsally from the ventral midline around the bones at the posterior edge of the gill chamber. Cut dorsally from the anus. Raise the flap of the lateral wall of the fish, revealing the internal organs. Dispose of the excess preservative as directed by your teacher and thoroughly rinse the internal compartment.

Using your knowledge of internal organs and their positions in the fetal pig, find and identify the internal organs of the perch. The digestive system is probably the easiest organ system with which to begin. The following information will assist you with your explorations: (1) the heart is located in a cavity positioned more anteriorly than the pig's, (2) the fish has an air sac (swim bladder) in the dorsal part of the cavity that is used in regulating buoyancy in the water, and (3) the excretory system is less compact than the pig's—it covers a longer length of the fish's dorsal cavity. Think of some other differences you would expect to find between the pig and the perch. Do you expect the perch to have lungs and a diaphragm?

46 ______________________________

Explain your answer.

47 ______________________________

Table 29-1 Comparison of the internal organs of the fish and pig

Organ system/organ	Comparison of fish and pig organs

In Table 29-1, list the organs of each organ system in the fish and describe how they differ from the same organs in the fetal pig.

Phylogenetic relationships

Now that you have observed some of the diversity within the living world, organize your information into one diagram. On a large piece of paper, diagram a tree of the phylogenetic relationships among the five kingdoms and all their major phyla or divisions that you examined. (You may have already constructed specific sections of this phylogenetic tree in Laboratories 23 through 28.) On the tree, place the characteristics that biologists think are important in establishing these relationships. A few of the terms that you may want to use in addition to the names of kingdoms and phyla/divisions are the following:

Absorption
Ascus
Asymmetry
Autotrophic
Basidia
Bilateral symmetry
Colonial
Cone
Cnidocytes
Coelom
Dominant diploid generation
Dominant haploid generation
Hypha
Flower
Gastrovascular cavity
Germ layers (none, two, three)
Heterotrophic
Hypha
Ingestion
Leaf
Multicellular
Nonvascular
One celled
Organizational level (cells, tissues, and organ systems)
Ovule
Photosynthesis
Pseudocoelom
Radial symmetry
Roots
Seed (naked and enclosed)
Seedless
Stems
Tube-within-a-tube
Vascular tissue
Water-vascular system
Zygote forming

LABORATORY

30

A comparison of biodiversity in two ecosystems

Interactions among organisms

You have explored many levels of biological organization from cells with their interacting organelles to the interactions of organ systems that make up an organism. Although an organism is very complex, many biologists explore higher levels of biological complexity. These levels of biological complexity involve interactions of organisms with one another and with the nonliving components of their environment. In these interactions, members of a species and members of many different species compete with and sometimes assist each other as they secure and use the resources necessary for life.

The study of interactions among organisms and between organisms and their nonliving environment is called **ecology.** Biologists who explore these interactions above the organismal level are called **ecologists.** Some ecologists study ecosystems. An **ecosystem** is a unit of nature that includes living organisms (the **biotic factors**) and their nonliving **(abiotic)** environment such as water, soil, light, and oxygen. Ecosystems vary in size and complexity. For example, an ecosystem can consist of the organisms and abiotic factors in a tiny pool of rain water held in a cup-shaped leaf. In contrast, an ecosystem also can be many square miles of an evergreen forest with all its organisms and nonliving components.

Cultivation and biodiversity

Scientists estimate that only 1% of the species that have inhabited the Earth are living today. The effects of losing 99% of species on the remaining biotic and abotic components of the Earth are unknown. In fact, scientists do not know the total effects of the loss of even one species on the remaining organisms and the abiotic environment. Ecologists and many other citizens are concerned about the effects of the elimination of numerous species throughout the world. Today, as the class compares two terrestrial ecosystems, let's focus on a question about the effect of one human activity. What affect does cultivation have on the number of species (the **biodiversity**) in an ecosystem?

EXPLORING

Study sites and predictions of biodiversity

Two study sites have been chosen by your teacher. One is a relatively undisturbed (natural) site typical of your geographical area and the other is a cultivated site. Your teacher will *briefly* describe the two study sites. From the teacher's description, predict which site has the greater number of plant and fungal species.

1 ____

What are some possible reasons for the difference in the plant and fungal diversity?

2 ______________________________

____

Which site has the greater animal diversity?

3 ______________________________

Explain your answer.

4 ______________________________

____

Sampling procedures

For this laboratory, the class will divide into data-collecting teams. If the study sites are close to each other, your teacher may divide the class between the two areas for simultaneous collection of data. Each team will collect the same type of data from the study sites. After data collection is complete, you can share data.

The following is a list of materials you need to take to your study sites: paper and pencils, string, stakes, hand trowel (or soil sampler), pH paper, thermometer, meter stick, hand lens, small paint brushes, plastic spoons, bottle of distilled water, and a collection of jars and plastic bags. If you have a calculator, bring it.

When you arrive at the study site, decide which specific areas, or plots, the teams will explore in the time available. Some questions to consider are the following:

1. What are the outer boundaries of the study site?
2. What should be the size of the sampling plots?
3. How many plots will be sampled?
4. Where will the plots be located?

If your study site is a heterogeneous mix of large organisms, the sampling plots should be large. Explain the reasoning for this statement.

5 ______________________________

If your study site is a homogeneous group of small organisms, the plots can be smaller. For example, you would map out a large plot to study a deciduous forest with small trees and bushes distributed among large oak and hickory trees and map out a much smaller plot to study a densely populated grass lawn. The number of sites depends on the number of teams and the time available. The greater the number of plots, the higher the confidence of investigators that the sample data represent the study site. (You can review some concepts of sampling in Laboratory 13.)

Deciding the locations of sampling plots is a more involved process. From measuring the organisms and abiotic factors in only the sample plots, you want to generalize your conclusions to the entire study site. Therefore the sample plots must be representative of the study site to yield reliable information. Because you have not explored organisms and their distribution in the study site, there is no method to choose specific plots and be certain that they represent the entire study site. This is the same situation that scientific investigators usually experience. To counter this problem, scientists frequently use **random sampling techniques.**

A **random sample** is a subset of a large group in which *each member* (of the large group) *has an equal chance of being in the sample*. In obtaining a random sample to measure, the choice of which members of the large group are measured is left entirely to chance. Letting numbered balls drop through a tube from a shaking container is one random method that some states use to choose winning lottery numbers. In this case, each of the numbered balls in the agitating container has an equal chance of dropping through the tube and becoming a winning number. Random sampling for site selection requires that each unit area of the study site has an equal chance of being included as one of the sampling sites. With your classmates, decide on a random sampling method of plot selection for the study sites. Describe the sampling procedure.

6 ____

____

SCIENCE NOTE *A sample of a large group in which each member of the group has an equal chance of becoming part of the sample is called a* random sample. *The members that become part of the sample are chosen strictly by chance. This method of obtaining samples is a frequent sampling method in ecological studies.*

Figure 30-1
Study sites.

After your team has its designated sampling plots, measure off the plots and mark them with stakes and string. In Figure 30-1, draw a three-dimensional diagram of each study site. On the diagrams, identify each team's sampling plots with a number. With letters, identify the layers of the environment from the topmost plant layer to the soil layer. For example, a site could have a tree canopy from 3 to 20 m above the soil surface, a layer of shrubs up to 2 m above the soil, and a layer of moss that extends to 3 cm above the soil. Include a key describing what your letters represent.

Plant diversity

Explore the plants, fungi, and animals in your sampling area. Begin by exploring all the plants, but because animals are motile and can leave the sampling area, collect them as you encounter them. (You can use paint brushes and spoons to transfer animals to collecting jars and sacks.) Leaving the plants in place, give each plant species an identifying number or name and count the number of each species. If suggested by your teacher, collect one example of each species to identify later in the laboratory. In Table 30-1, list each species of plant by name or identifying number. Record the number of members in each species and the height and vegetation layer, and give a general description. Estimate the height of vegetation that is too tall to measure.

Determine the number of each species in a square meter. Multiply the number of each species in your plot by the number (or fraction) of sample plots necessary to equal one square meter. (Your teacher can help you determine this number.)

$$\frac{\text{Number of the species/}}{\text{square meter}} = \frac{\text{Number of the}}{\text{species/plot}} \times \frac{\text{Number of plots/}}{\text{square meter}}$$

Record the number of each plant species/square meter in the third column. With this conversion of numbers of each plant species to a standard plot size (1 m^2), the numbers of each species can then be compared between study sites.

Animal diversity

In addition to collecting small animals from the vegetation and soil surface, collect animals in the soil. Dig up a 10-cm deep column of soil of known dimensions. Put the soil in a sack. When time permits, spread the soil on a flat surface and examine the soil particles with a hand lens. Multiply the number of each type of animal in the soil by a number that converts the column's volume to a 10-cm deep layer of soil covering the sampling plot. (Your teacher can help you determine the conversion number.) Record the number/plot of each animal species in the second column of Table 30-2. Also watch for large animals and look for signs of animals such as tracks, food remains, burrows, and fecal droppings.

In Table 30-2, list each species of animal by an identifying number or name. Give the number of each species (per plot), the length, and a general description of each species. If the team members have knowledge of the animals' possible food source, add this information to the fifth column. If suggested by your teacher, keep one member of each species for identification in the laboratory. Release all other animals. Multiply the number of each species in your sample plot by the number of sample plots necessary to equal one square meter. Record the number of each species/m^2 in the third column.

Fungal diversity

Examine all fungi in the sample plot and record your observations in Table 30-3. For a measure of size, you can measure the diameter of round fungi and the length and width of other shapes. (Indicate what dimensions are recorded.)

Table 30-1 Plants

Species	Total number	Number/m^2	Height and layer	General description
IN THE UNDISTURBED SITE				
IN THE CULTIVATED SITE				

Table 30-2 Animals

Species	Total number	Number/m^2	Length (cm)	General description
IN THE UNDISTURBED SITE				
IN THE CULTIVATED SITE				

Table 30-3 Fungi

Species	Total number	Number/m^2	Size (cm)	General description
IN THE UNDISTURBED SITE				
IN THE CULTIVATED SITE				

Abiotic components

Measure some of the abiotic factors and record them in Table 30-4. Measure the temperature at 1 m above the soil surface, at the surface, and at 10 cm below the soil surface. Also record the sun and cloud conditions and the time of day. Estimate the percentage of surface that usually receives sunlight. (If it is cloudy, lie down and look up at the leaves to estimate this percentage.)

Table 30-4 Abiotic factors

Factor	Study sites: Undisturbed	Study sites: Cultivated
Temperature 1 m above soil At surface 10 cm below surface		
Sun and cloud conditions		
Time of day		
% of surface with sunlight		
pH of soil At surface 10 cm below surface		

Moisten two 2-cm strips of pH paper with distilled water. Press one side of one strip against a soil sample from the surface and the other to a sample from 10 cm below the surface. Compare the papers' colors to the pH scale on the container.

Data presentation

On a graph for each study site, summarize the data for the plants and fungi. Prepare a bar graph with the x-axis representing each species and the y-axis representing the number of organisms of each species/square meter. Place the name or number of each species on the x-axis in descending order of the number of organisms of the species. For example, the first species on the x-axis should be the species with the highest number of members/square meter at the study site. The last species should be the species with the fewest numbers/square meter. At each species' name or number, draw a vertical bar from the x-axis up to level that corresponds to the number of that species (given on the y-axis). Also summarize the data for animals for each study site in bar graphs similar to the graphs for plants and fungi.

CONCEPTS/TERMS

Examining the data: diversity of organisms

Which study site has the greater number of plant and fungal species?

7 ____

How do these data compare with your original prediction about the diversity of plants and fungi in the two study sites?

8 ____

What factors may be limiting the number of plant and fungal species at each of the two study sites?

9 **Undisturbed site**

Cultivated site

Compare the numbers of the four plant or fungal species that have the largest number of members at each study site. Which study site has species with the largest number of members?

10

Try to explain the difference in the population size (the number of a species in an area) of the four types of plants and fungi.

11

If the study sites have several layers of plants, what effect do the plants of each layer have on the plants of the other layers?

12

Which study site has the greater diversity of animals?

13

How does this data compare with your original prediction about the diversity of animals in the two study sites?

14

What explains the difference in animal diversity between the study sites?

15

What are the effects of the distribution of plants on the distribution of animals?

16

APPLYING

Food production and biodiversity

What are the long-term biological effects of increasing food production by turning all available uncultivated land into farmland?

17 ______________________________

____

Farming is obviously necessary to provide food for people and other animals. As the number of people on Earth increases, production of food will need to increase to prevent starvation. What is your advice to our political leaders about opening uncultivated land for farming to solve both the short-term and long-term food shortages?

18 ______________________________

____

____

What other methods for increasing food production do you recommend?

19 ______________________________

____

____

In what ways can humans decrease the demand for food and the effects of food production on the Earth's biodiversity?

20 ______________________________

____

____

____

____

The need for biodiversity

Considering your knowledge of genetics and interactions among organisms, what are the most important reasons for preserving biodiversity?

21 ______________________________

____

____

____

____

APPENDIX

A

The compound light microscope

PART ONE Microscopes in the laboratory

The instrument and its use

Biologists use microscopes to increase the images of living materials until they are large enough to see. Microscopes were first used to magnify the structure of living material in the 1600s. Since then, newer and more powerful microscopes have been built that let biologists see much of the detailed structure of organisms. Without this technology, much of what we understand about the structure of living material today would be unknown. In these pages, you will explore the structure and function of a compound light microscope, enabling you to use it with ease during the course. Two other appendixes provide additional information associated with microscopes. Appendix B gives information on the preparation of materials for microscopic observation and Appendix C provides tips on the use of the stereomicroscope.

You should gain some expertise with microscopes for two reasons: (1) no instrument functions to its designed capacity if it is improperly used, and (2) the improper use of any instrument can lead to damage that may involve expensive repair. Taking time now to learn how to use your compound light microscope will save you time in the future, increase your observational capabilities, and prevent damage to the instrument. Your teacher will assist you in your learning process. You will find a quick-reference section reminding you of the "do's and don'ts" for microscope use at the end of this exercise.

On finishing this appendix and Appendix B, you should know the functions of the parts of the microscope and be able to (1) determine the magnification of the different lens systems, (2) determine the diameter of the field of view at different magnifications, (3) prepare a wet mount of a suspension of yeasts, and (4) adjust the focusing and lighting mechanisms of the microscope to obtain the sharpest view possible of different parts of a yeast.

If you feel you are already proficient in using microscopes, talk with your teacher about your experience. You will need to take a few moments to become familiar with the microscopes used in this class, measure the diameter of field of view for each lens system on your microscope, and demonstrate your expertise by adjusting the scope for maximum visibility of yeast cells in a wet mount that you prepare.

EXPLORING

The microscope

From the microscope cabinet, obtain the microscope assigned by your laboratory teacher. Place the **base** of the microscope on one hand and hold the upright **arm** of the microscope with the other hand as you carry it to your desk. Set it on the desk with the carrying arm of the microscope toward you. Your microscope may look slightly different from the one in Figure A-1. However,

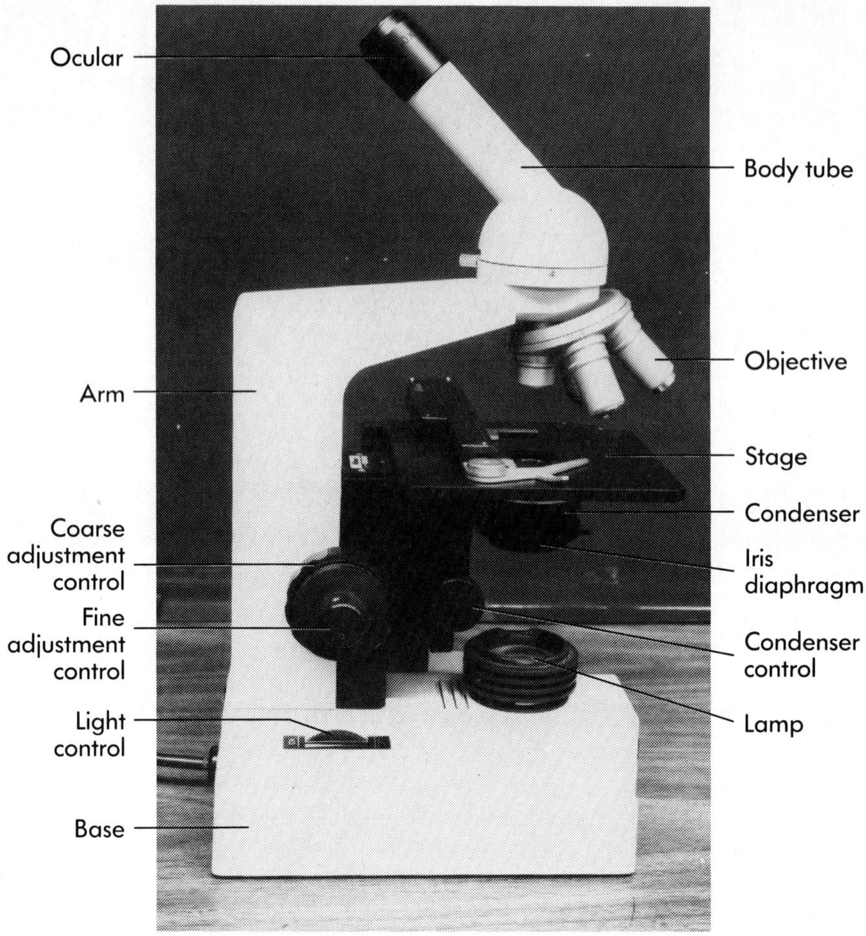

Figure A-1
Parts of the compound light microscope.

you can use the information in Figure A-1 and in the next section to identify and learn about each part of your microscope.

CONCEPTS/TERMS

Parts of the compound light microscope

Arm

The arm is an upright portion of the frame that joins the body tube to the base of the microscope. You hold the arm, along with the base, when transporting the microscope.

Body tube

The body tube is attached to the arm of the microscope. It houses the ocular at the top and the nosepiece at the bottom. Light rays from the objectives pass through this tube to the ocular.

Ocular or eyepiece

The ocular, at the top of the body tube, contains a lens system, which magnifies. The magnification power of the ocular lens is engraved on the outside of the ocular. The ocular can be removed for cleaning but should be removed *only by the teacher*. Clean the outer lens surface by wiping with **lens paper.** Some microscopes, frequently called **binocs,** have two eyepieces, which are adjusted to fit the distance between the observer's eyes (see Appendix C). Other microscopes have only one eyepiece, which is usually used with the eye on the side opposite to your writing hand.

Pointer

You can use this thin wire within the ocular to point to a specific part of the material you are viewing.

Revolving nosepiece

The nosepiece is a circular disk attached to the body tube. The objectives are screwed into the disk. You rotate the nosepiece to position one of the objectives for use.

Objective

The objective is a lens system that magnifies the image of the specimen (material you are viewing). The magnification and a number designating a property of the objective lens (numerical aperture) are engraved on the barrel.

The shortest objective is the **scanning objective.** How many times does it magnify?

1 ______________________________

Of all the objectives, this one gives you the largest **field of view** (the area of the specimen that you see through the microscope). You should always begin viewing the specimen with this objective. Move the portion of the specimen you want to see to the center of the field of view before observing details or moving to higher magnification. Rotate this objective into position when you put the microscope away.

The second shortest objective is the **low power objective.** How many times does it magnify?

2 ______________________________

Use this objective *after* the specimen is in sharp focus with the scanning objective.

The third shortest objective is the **high power objective.** How many times does it magnify?

3 ______________________________

It has the greatest magnification of the objectives that require air between the objective and the specimen. It is used only when you observe minute details. It cannot be used to focus on the lower part of a thick specimen or a specimen covered with a thick cover slip.

If your scope has a fourth objective, it is probably an **oil immersion objective,** which is marked with 97X or 100X. Use it for observing organelles or the smallest of cells, such as bacteria. It requires a special oil in the space between specimen and objective.

Stage

The stage is the horizontal platform attached to the arm. It lies directly beneath the objectives and is the surface on which you place the specimen for observation. Light from below shines through its central opening. When a specimen is placed over this opening, light passes through it. (Specimens should always be placed on a glass slide before placing them on the stage. All liquid materials are topped with a thin glass cover slip [see Appendix B].)

Stage clips

These clips anchor the specimen slide to the stage. In addition, the downward pressure of the clips on the slide makes it easier to move the slide very short distances. In place of the clips, some microscopes have a mechanical slide holder attached to the stage. The slide is moved by rotating two controls knobs just below or above the corner of the stage.

Light source

A lamp fits into the base of the microscope. On some microscopes the lamp is removable, but it should always be kept in a level, untilted position in the microscope base. Other microscopes have lamps that are not removable. These microscopes have a control switch for varying light intensity. Always turn the light off when the microscope is not in use.

Iris diaphragm

The iris diaphragm is a circle of overlapping leaflets located under the stage at the bottom of the condenser. A small lever moves the leaflets in and out, which regulates the size of the diaphragm's opening. The opening determines the diameter of the cylinder of light that passes from the lamp to the condenser and eventually through the specimen on the stage.

Condenser

The condenser holds a series of lenses that focuses a cone of light on the specimen. It is located below the opening in the stage. A knob beneath the stage (on the right) raises and lowers the condenser. For this class, keep the condenser raised so that its upper surface is nearly level with the surface of the stage.

Coarse focusing knob

The coarse focusing knobs (one on each side) are large knobs located between the arm and the base of the microscope. They raise and lower the stage or, in some microscopes, the body tube. As you rotate the focusing knob, the specimen moves in and out of focus. The coarse focusing knob moves the stage (or body tube) rapidly. Use the coarse focusing knob *only* when the scanning and low power objectives are in place. *Do not use it* if the high power or oil objective is in place because you may crash these long objectives into the slide.

Fine adjustment knob

The small knobs located at the center of the coarse focusing knobs are the fine adjustment knobs. They also raise and lower the stage (or body tube) but very slowly. *Always* use these knobs to focus the specimen when the high power and oil objectives are in position. You can use them to sharpen the focus when any objective is in place. When using the microscope, you should continually adjust your fine focusing as you observe.

Measuring with metric units

With a millimeter ruler, measure the width of the period at the end of this sentence to the nearest 0.1 millimeter (mm). What is the width of the period in millimeters?

4 ______________________________

What is this width expressed in micrometers (μm)?

5 ______________________________

(If you need assistance with metric units or converting one unit to another, see Appendix D.) With a very sharp pencil, make the smallest dot you can see in the margin and measure it. What is the width of your dot in millimeters?

6 ______________________________

What is the width of the dot in nanometers (nm)?

7 ______________________________

Using micrometers, what is the smallest dot that you can see with your eye?

8 ______________________________

Visual acuity

The ability to see small things varies with the type of organism. Most humans can see dots as small as 0.1 mm, or 100 μm. Some predatory birds see much more acutely or sharply than humans; many can spot mice in a field from a height of greater than 100 meters, or more than the length of a football field.

The diameters of the basic units of living material generally range in size from 0.5 mm or

9 ______________________________ μm

to 1 μm. Consequently, the invention of the microscope, enabling curious people to first see and then study these small units, provided a tremendous tool for expanding knowledge of the living world.

Field of view, magnification, and image orientation

Obtain a hand lens and a thin glass slide with a rectangle of paper on which an "R" has been typed. Clean the slide with laboratory tissue or a paper towel. Look at the letter through the hand lens. Bring the letter into sharp focus by adjusting the distance between the lens and the letter. In circle A of Figure A-2, draw the letter "R" as you see it through the lens. Make the size of the letter in the circle proportional to the size of the "R" in the field of view. Record the magnification of the "R" below the diagram.

Move the slide to your left. In the field of view, what direction does the letter appear to move?

10 ______________________________

Observe the slide with the compound light microscope using the following procedures:

1. Clean all lens surfaces of the microscope, including the upper surfaces of the condenser lens and lamp, with lens paper. This paper is specially manufactured for this task. Using any other tissue can damage the highly polished lens surfaces. Plug in the lamp and turn on the light.
2. Place the letter slide on the stage and secure it with the stage clips or mechanical slide holder.
3. Move the slide to center the letter "R" over the opening in the stage.
4. While looking from the side and with the shortest and lowest power objective in place, rotate the coarse focus knob until the stage and the objective are close together.
5. While looking through the ocular, slowly roll the stage down (or the body tube up) until the paper and its letters are visible. This adjusts the

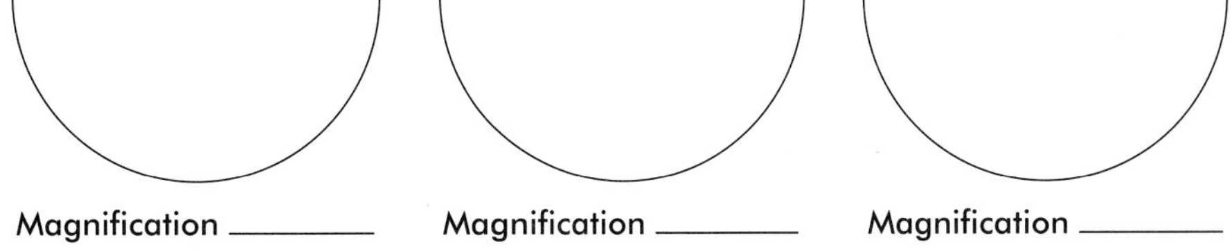

Figure A-2
Letter "R" at differing magnifications.

distance between the microscope's lenses and the slide to bring the letter and paper fibers in focus.

6. If the intensity of the light is very bright, move the light intensity control to a lower setting. If your microscope lacks an intensity control, partially close the iris diaphragm or lay lens paper over the lamp to decrease the light.

Move the slide to your right. In what direction does the letter move in the field of view?

11 ____

Move the slide farther away from you. In what direction does the letter move in the field of view?

12 ____

Practice slowly rotating the coarse focusing knob back and forth, focusing on different levels of fibers in the paper and the ink of the letter. In circle B of Figure A-2, diagram the "R" just as you see it in the field of view. In the box above circle B, draw the letter "R" in the same orientation as you see it on the slide without looking through the ocular.

The image of the specimen is formed by both the objective and ocular lenses. The magnified image that passes from the objective is magnified again by the ocular; the total magnification is the product of these two magnifications. This relationship is expressed in the following equation:

$$\text{Specimen magnification} = \begin{matrix}\text{Magnification by}\\ \text{ocular lens}\end{matrix} \times \begin{matrix}\text{Magnification by}\\ \text{objective lens}\end{matrix}$$

Record the total magnification of the specimen below circle B.

Turn the 10× objective into viewing position by rotating the revolving nosepiece. Turn the coarse focusing knob to sharpen the focus on the letter. How much do you need to turn the knob to obtain sharp focus with the low power objective?

13 ____

In circle C, diagram the portion of the "R" now in view. Record the present magnification of the letter.

You should have found that the image of the "R" when using the compound light microscope is reversed and inverted. Put in simpler terms, the image of the "R" is

14 ____

and

15 ____

This inverted and reversed image also explains why it looks as if the slide is being moved toward you when it is actually moving away from you.

In general terms, how much of the "R" can you see through the microscope when magnifying 40× compared to 100×?

16 ____

What happens to the size of the field of view as magnification increases?

17 ____

The relationship of the size of the field of view to magnification is an inverse relationship. Check the amount of the "R" that you can see at total magnification of 400X. Which objective should be in viewing position?

18 ______________________________

Is the inverse relationship still in effect as the magnification is increased?

19 ______________________________

If not, describe the new relationship.

20 ______________________________

Measurement and estimation of the diameter of the field of view

Determine the diameter of the field of view by doing the following: Obtain a slide containing a small section of a millimeter ruler. Return the scanning objective to viewing position. Center the ruler segment in the field of view and focus on the numbers and marks. What is the distance (in millimeters) between the left edge of one of the ink lines and the left edge of the next line?

21 ______________________________

Move the slide to a position that lines up the edge of one of the lines with the left-hand edge of the field of view. The other lines should extend across the middle of the field of view. Count the number of millimeters across the diameter of the field of view. If there is a fraction of a millimeter left at the right-hand edge, estimate it to the nearest tenth; remember that the millimeter lines have a width that you need to consider in the estimation process. In Table A-1, record the diameter and the appropriate magnification.

Repeat this measurement of the diameter of the field of view using the 10× objective.

Two problems arise if you use the millimeter ruler to determine the width of the field of view when using the high power objective. The first problem is the thickness of the ruler; it is too thick to fit into the space between the objective and the slide. This space is called the **working distance.** Information in the fifth column of Table A-1 shows that the working distance decreases as the magnification increases. What is the term for this kind of relationship?

Table A-1 Specifications of the compound light microscope

Ocular power	Objective power	Total magnification	Width of field of view (mm)	Working distance of objective (mm)
10	Scanning or 4×			28.3
10	Low or 10×			6.3
10	High or 40×			0.4
10	Oil immersion or 97× or 100×		0.18	0.1

22

(The short working distance of the higher power objectives is one reason why you should not use the coarse focusing knob to adjust focus with these objectives.) The second problem is the field of view is less than one millimeter, so you do not have two marks in the field of view indicating the millimeter distance to which you can compare the diameter of the field of view. Instead, use proportional reasoning and the inverse relationship between the change in magnification and the diameter of the field of view to calculate the diameter of the field of view at the higher magnifications. The equation for this calculation is the following:

$$\frac{\text{Magnification at lower power}}{\text{Magnification at higher power}} = \frac{\text{Diameter at higher power}}{\text{Diameter at lower power}}$$

Double check your measurement of the diameter of the field of view at 100× using this equation. Calculate the 100× diameter using your measurement of the field of view at 40×. What is the calculated field of view at magnification of 100×?

23

Compare this answer to the measurement you made using the ruler slide. If these two correspond, calculate the diameter at 400× and record it in Table A-1. If these two calculations do not correspond, seek guidance from your teacher.

Parfocal

The microscope only needs small focusing adjustments after switching to the next higher objective. This feature is called **parfocal.** If the specimen was in sharp focus when using the adjacent shorter objective, each objective will clear the slide without rolling the stage down *unless* the specimen or cover slip is very thick.

PART TWO Some aspects of microscope use

Useful magnification

Observe the series of photographs of a ruler in Figure A-3. Each photograph in the series is an enlargement of the preceding photograph. List two factors that change in the series.

24

What limits magnification? Actually there is no upward limit to the magnification in light microscopy; there is only a limit to the **useful magnification.** The basic limitation is not magnification but **resolving power,** the ability to observe the fine detail of the object. Does the last photograph in the series give you more details about the ruler than the preceding photograph?

25

If the object has been magnified to the point that its image is becoming fuzzy or indistinct because of the limited resolving power, further magnification does nothing but make the image larger and less distinct. Increasing magnification without increasing resolving power gives no additional detail. Such useless increase in magnification is called *empty magnification.*

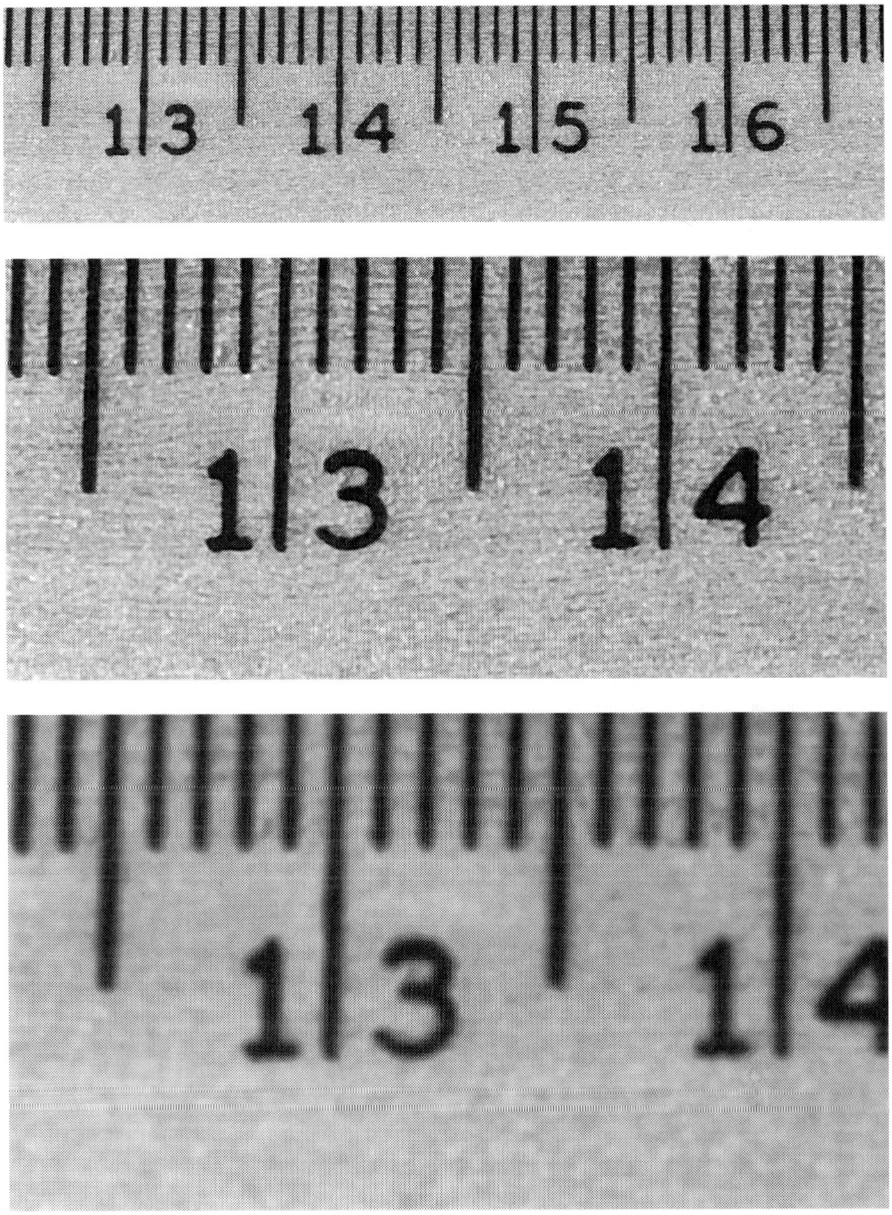

Figure A-3
Enlarging a photograph.

Resolving power

Make two chalk dots on the board very close together. Now move across the room from the board. What appears to happen to the two dots?

26 ______________________________

When the two dots blur together and can no longer be distinguished from each other, the limits of the resolving ability of your eyes have been exceeded. Resolving power is represented by the shortest distance between two points that still permits you to distinguish them as two points instead of viewing them as one. This is another inverse relationship: as the distance between the dots decreases, resolving power increases. Humans can usually distinguish two bright pinpoints of light that are 1 mm apart if the lights are not more than 10 meters away.

Increasing the resolving power of the microscope is based on several factors. Two of these are (1) using light of a shorter wavelength and (2) increasing the number of light rays from the specimen that reaches the objective lens. The oil used with the oil immersion objective increases the number of light rays that enters the oil objective. This increase in light rays gives this objective greater resolving power than the objectives that are designed to use air between the objective and specimen. The electron microscope is designed to use very short wavelengths to illuminate the specimen. These shorter wavelengths increase the resolving power of the electron microscope.

Illumination

Effect of specimen on light

If the lamp on your microscope has an intensity control, adjust it to low intensity. Keep the substage condenser at its highest position and place the scanning objective into viewing position without any slide on the stage. While looking through the ocular, move the lever of the iris diaphragm over its maximum range. Describe the results.

27 ______________________________

Put the "R" slide on the stage with the unmarked portion of the paper in position for viewing. Rotate the focusing controls to obtain the sharpest image possible on the paper fibers, first with the 4× and then the 10× objectives. When the fibers are in sharp focus at 100 magnification, move the iris diaphragm over its range. What happens to your ability to see details of the paper fibers under the lowest light conditions?

28 ______________________________

What happens under the very highest light conditions?

29 ______________________________

Does there seem to be an amount of light between these extremes that gives the greatest clarity?

30 ______________________________

Now move the dark ink of the letter into the field of view. Again move the iris diaphragm lever over its range. Which needs more light for maximal visibility, the dense, inked area or the less dense, noninked areas?

31 ______________________________

Effect of objectives on light

Focus on the paper fibers using 4× objective with the best lighting conditions. Turn the 10× objective into place without changing lighting conditions. Is the amount of light reaching your eye the same as when you used the 4× objective?

32 ______________________________

If not, does the amount of light need to be increased or decreased as magnification increases?

33 ______________________________

Maximizing your success as a microscopist

The adjustment of the illumination system is the most important way you can control the performance of the microscope. In theory, the greater the amount of light that enters your eye from the microscope, the more detail you can see. However, when the parts of the specimen have little difference in density (low **contrast**), additional light decreases this contrast and "washes out" the image. In general, you increase the light slightly when moving to a more powerful objective and when you have a more dense specimen.

There is no hard and fast rule to follow for light adjustment. In time, you will gain experience in adjusting illumination as you move from one level of magnification to another and between specimens of varying densities, thicknesses, and contrasts. With this experience, you will quickly adjust the lighting for maximal visibility as you observe each part of a specimen.

Several ways of controlling the illumination in light microscopes follow:

1. Iris diaphragm: The best adjustment for optimal light is to open the diaphragm until the image of the diaphragm opening is slightly larger than the field of view. If the specimen has low contrast and there is no intensity control on the microscope, you may need to decrease the diaphragm opening slightly or lay thin paper over the lamp to decrease the light coming through the diaphragm.
2. Substage condenser: In general, keep the upper surface of the substage condenser close to the surface of the stage. It should form a sharp image of the lamp iris at the level of the specimen.
3. Light intensity control: This control decreases the amount of light without decreasing the size of the cone of light reaching the specimen and objectives.

Most beginning students have a tendency to close down the microscope's illumination too much. Ask your teacher to check your lighting adjustments.

The light path

View the film, *Microscope Illumination: the Light Path,* and read the written information on the container. Using these sources of information and the drawing in Figure A-4, describe the effect that each of the following parts has on the beam of light:

34 Lamp filament ______________________________

____

Iris diaphragm ______________________________

____

Substage condenser ______________________________

____

Objective lens system ______________________________

____

Ocular lens system ______________________________

____

Figure A-4
The light path in a compound microscope.

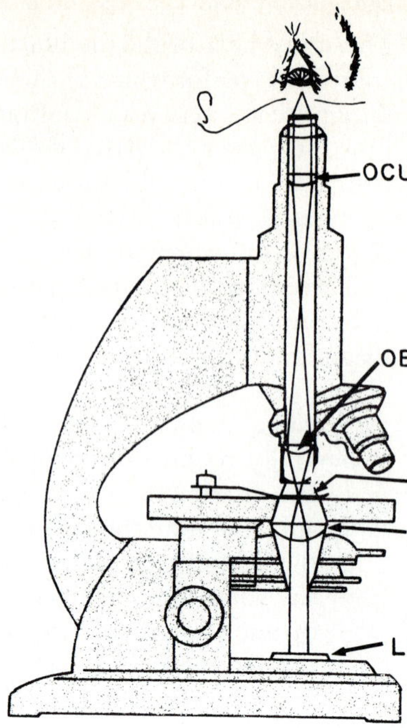

What is a possible explanation for the inverted and reversed image of the light microscope?

35

Rules for using the compound light microscope

1. Always carry the microscope in a vertical position with two hands.
2. Use only special lens paper when cleaning the lenses. Always clean lens surfaces when first starting to use a microscope; include the surfaces of the condenser and lamp lenses.
3. Always lay the specimen on a glass slide before placing it on the stage.
4. Observe liquid materials only after covering with a slip.
5. Observe an object under high power only after you have first focused on it with each of the lower powers and centered the object in the field of view.
6. Use the coarse objective knob only when the scanning or the low power objective is in place. Do not use the coarse adjustment knob with 40× or 100× objectives. You may use the fine adjustment knob with any objective.
7. Use the high power objective to observe details that cannot be seen with low power.
8. Always lower the stage after using 40× or 100× objectives before you remove a slide.
9. Promptly remove any fluid or other material that spills on a lens surface or the stage. Seek help from your teacher if you have any problems in thoroughly removing the foreign substance.
10. Return the microscope to the cabinet with (1) the 4× objective in viewing position, (2) the stage close to the body tube, (3) the condenser at its highest position, and (4) all surfaces clean.

APPENDIX B

Preparation of material for microscopic study

There are at least five types of preparations of materials for microscopic study. These are wet mounts, whole mounts, smears, squashes, and sections. When you first use this appendix, you will prepare a wet mount of baker's yeasts. Later in the course you will use commercially prepared slides of sections and whole mounts and make your own squash preparations.

Wet mounts

Materials that can be suspended in a drop of liquid and are thin enough to let light rays pass through them are easily prepared for temporary use in a **wet mount.**

Procedure for making a wet mount

Obtain a thin glass slide and clean glass coverslip. (Wash the slide with detergent and water even if it is new. Holding the slide by its edges, thoroughly rinse and dry the slide.) If the material you are observing is already in a liquid, place a drop of the liquid in the center of the slide. If the material is not suspended in a liquid, add a drop of liquid (such as water) to the slide and place the material in the drop.

Place one edge of the coverslip at the edge of the drop. Tilt the coverslip until the angle between the coverslip and the slide is about 45 degrees (Figure B-1). Allow the liquid to run along the edge of the coverslip. Lower the coverslip onto the drop of liquid. You can use a dissection needle to aid this process.

Frequent problems with wet mounts

Air bubbles. If you see an air bubble after lowering the cover slip, center the air bubble on the microscope and observe it. What characteristics distinguish magnified air bubbles from other material in the wet mount?

1 ______________________________

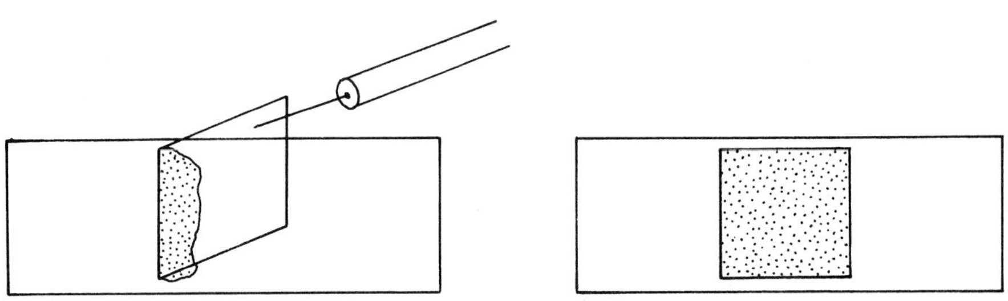

Figure B-1
Making a wet mount.

If you are uncertain whether you have an air bubble in the field of view, ask your teacher to view your slide. What happens to your observations of organisms when an air bubble is next to or over them?

2 ____

Two precautions help prevent trapping of air under cover slips:

1. Use clean slides and slips.
2. Lower the coverslip slowly from a 45-degree angle.

Amount of liquid. Too much liquid or too little liquid causes problems. If the liquid on your slide is more than the amount needed to just fill the space between the coverslip and slide, view your slide through the microscope. Are you having problems with liquid flowing through the field of view?

3 ____

If so, what effect does this flowing liquid have on the organisms you are trying to view?

4 ____

Liquid spilling outside of the coverslip area is easily moved by air currents. This movement causes movement of the liquid in the field of view. In addition, if some of the liquid gets on the upper surface of the slip, it will interfere with light to the objective and adversely affect resolving power. However, if the liquid does not fill the space between the slip and the slide, the slip will settle down on the material, possibly crushing it.

When making wet mounts, completely fill the space between the slide and coverslip without overfilling it.

Preparing and examining a wet mount of yeasts

Make a wet mount of a drop of baker's yeasts suspended in water, which is available at the supply table. When completed, view the small oval yeast with your compound light microscope. First, view the material using the scanning objective and adjust the light and focus mechanisms for clearest viewing. If you have trouble finding the yeast, focus on the edge of the cover slip and turn the coarse focusing knob slightly to find the yeasts. Second, increase magnification by rotating the 10× objective to viewing position. After the necessary adjustments for best viewing at 10× magnification, use the 40× objective. When you have the interior of the yeast under the best lighting conditions and the sharpest focus you can obtain, ask your teacher to check your wet mount and your use of the microscope.

After checking your technique of viewing the cells' interiors, the teacher may ask you to adjust the microscope to view the edge of the yeast. If needed, the teacher will give you tips for improving your technique.

Your teacher may also ask you to dye or **stain** the yeasts in your wet mount. A small amount of the stain can be added to the liquid of the wet mount before adding the coverslip. To stain material in a completely prepared wet mount, touch a small amount of stain at the end of a toothpick to one edge of a coverslip; then let the stain move under the slip. You can hasten the movement of the stain by touching a paper towel to the liquid at the edge of the coverslip opposite the addition of the stain. What effect does the stain have on your ability to see the yeasts and their internal structures?

5 ____

Sections

Many specimens are too big to put on a glass slide. Consequently, only a thin slice, or **section,** of the specimen can be placed on the slide for study. Preparing these sections is an involved procedure and is usually done in a way that permits the preparation to be used for a long time. These sections are frequently stained with dyes to increase contrast between parts of the material.

Squashes

In a **squash,** a small portion of the material suspended in a liquid is pressed between the slide and the coverslip, spreading the material to a single layer. Do you think that this technique is used if you want to preserve the structure of and the relationships between the specimen's parts?

6 ____

Explain your answer.

7 ____

____

Whole mounts

Some organisms are small enough to be placed on a slide without sectioning. These preparations are stained and prepared for long-term use. These are **whole mounts.**

Smears

In a **smear,** the suspended material is spread out on the surface of a slide. The material is then attached to the slide, frequently by warming the slide.

APPENDIX C

Use of the stereomicroscope

Some low-power microscopes have two lens systems, one for each eye. Only with this binocular (*bi* refers to "two") vision can you perceive the third dimension of a specimen. These **stereomicroscopes,** or **dissecting microscopes,** are extremely useful for viewing opaque objects at low magnifications or when manipulating a specimen. These microscopes are valuable instruments and should be given the same care as compound light microscopes.

To observe a specimen, place it in a dish, on a glass slide, or on a paper towel to avoid getting dirt and other debris on the microscope. Place the container in the center of the stage.

A lamp on a separate base provides light. Having a separate lamp enables you to change the position of the lamp on the table and alter the angle at which the light shines on the specimen. Some dissecting microscopes have a tunnel in the arm for inserting the bulb case of the lamp. Some have a mirror below the stage, which you can adjust to reflect light through the specimen from below. Plug in the lamp and rotate the dial on the lamp's base to turn on the light. By varying the position of the dial, you can change the intensity of the light. Turn off the light when you finish your observations.

The magnification control of the objective lenses is either a dial on the top or side of the microscope or a ring inserted in the body tube. Rotate this control to vary the magnification. Always begin at the lowest magnification. While looking through the oculars, rotate the focusing knob on the arm of the microscope until the specimen comes into view. Is the image of the specimen inverted and reversed as in the compound microscope?

1 ______________________________

How does the orientation of the specimen's image help you with the dissection of small organisms?

2 ______________________________

Proper adjustment of the oculars is necessary for stereoscopic viewing. Adjust the distance between the oculars to match the distance between your eyes by pushing the oculars together or pulling them apart. The distance should be such that you see the same field of view with each eye and the fields completely overlap. Also adjust the focal distance of the left ocular to compensate for any focusing difference between your right and left eyes. Change the focal distance by following these steps: (1) Close your left eye and view the specimen with your right eye. (2) Bring the image of the specimen into sharp focus with the focusing knob. (3) Now close the right eye and look through the left ocular with the left eye. (4) Do *not* turn the focusing knob; instead, rotate the ridged focusing ring at the base of the left ocular until the left eye has a sharply focused image. Once the oculars are properly adjusted for your eyes (both receiving the same sharp image), record the distance between the oculars as given on the distance scale (located between them) and record the focal setting of the left ocular. What is the distance setting?

3 ______________________________

What is the left ocular setting?

4 ______________________________

With these numbers, you can quickly reset the microscope to the proper positions for your eyes each time you use the stereomicroscope.

After making proper adjustments for stereoscopic viewing, you are ready to examine the specimen at greater magnification. To do this, simply rotate the magnification knob and adjust the focusing knob (do not change the focal distance for the left eye). Although the stereomicroscope has a zoom system of lenses (similar to the zoom lens of a camera), the depth of field decreases as magnification increases. Critical focusing therefore requires careful adjustment with the focusing knob. Is this microscope parfocal?

5 ______________________________

How does the working distance compare to the working distance in the compound light microscope?

6 ______________________________

To determine the magnification of an object, apply the same principles as for the compound microscope. What is the greatest possible magnification of the optical systems of your stereomicroscope?

7 ______________________________

What is the lowest magnification?

8 ______________________________

Become familiar with the use of the stereomicroscope by examining various objects in the laboratory.

APPENDIX D

Metric system

Scientists in the United States and people in most countries use metric units when measuring. Table D-1 gives the metric equivalent used by biologists of some commonly used measuring units in the United States.

You can easily manipulate measurements in the metric system because it is a system based on powers of ten. Each property of matter has a basic measuring unit such as a gram or meter. For each property, measuring units that are smaller than the basic unit are obtained by *dividing* the basic unit by 10 or a power of 10 (for example, 100 or 1000). Measuring units that are larger than the basic unit are obtained by *multiplying* the basic unit by 10 or a power of 10. Each unit is designated by a prefix that is used with all types of basic measuring units (see Table D-2). For example, the prefix *milli* identifies the measuring unit as 1/1000th of the basic unit: milliliter is 1/1000th of a liter, milligram is 1/1000th of a gram, and millimeter is 1/1000th of a meter.

The exponential equivalent is frequently used in scientific notation. The exponent of the number $10^{exponent}$ refers to the number of times that 10 must be multiplied by itself to equal a specific number. For example, the number 10^4 is the same as $10 \times 10 \times 10 \times 10$ or 10,000. The number 50,000 can be written as 5×10^4 or $5 \times 10 \times 10 \times 10 \times 10$. The number 3.2×10^{-3} is equivalent to $3.2 \times \frac{1}{10} \times \frac{1}{10} \times \frac{1}{10}$ or 0.0032.

Practice using the metric system and expressing the measurements in the smaller and larger units. Determine the following:

1 8 ounces = ________ pounds (16 ounces = 1 pound)

2 pounds = ________ grams

Table D-1 Metric equivalents

Property of matter	Common unit in United States	Metric equivalent	Basic metric unit of measure (symbol)
Mass	1 pound	454 grams*	Gram (g)
Distance	1 foot	0.30 meters	Meter (m)
	1 inch	2.54 centimeters	
Volume	1 quart	0.95 liter	Liter (l or L)

*A pound is a unit of weight and gram is a unit of mass; weight and mass are not equivalent terms even though they are frequently interchanged when referring to objects on Earth. Weight measures the force of gravity acting on an object. Mass is the quantity of matter as measured by its inertia.

Table D-2 Standard metric prefixes and equivalents

Prefix	Symbol	Numerical equivalent	Word equivalent	Scientific notation
mega	M	1,000,000	One million	10^6
kilo	k	1,000	One thousand	10^3
hecto	h	100	One hundred	10^2
deka	da	10	Ten	10 or 10^1
basic unit		1	One	1 or 10^0
deci	d	0.1	One tenth	10^{-1}
centi	c	0.01	One hundreth	10^{-2}
milli	m	0.001	One thousandth	10^{-3}
micro	μ	0.000001	One millionth	10^{-6}
nano	n	0.000000001	One billionth	10^{-9}
angstrom	Å	0.0000000001	One ten-billionth	10^{-10}
pico	ρ	0.000000000001	One thousand-billionth	10^{-12}

50 grams = ________ kilogram

2.3 kilograms = ________ grams

1567 yards = ________ mile (1760 yards = 1 mile)

1 yard = ________ meter

347 meters = ________ kilometer

2500 micrometers = ________ millimeters

1.5 quarts = ________ liter

120 liters = ________ milliliters

1 μl = ________ ml

30 nm = ________ μm

A piece of typing paper measures 8½ inches by 11 inches. What is its size in centimeters? ________ In meters? ________ In millimeters? ________ In nanometers? ________

APPENDIX

E

Seeing two dimensions, thinking three dimensions

Using sections

Using the compound light microscope to observe living materials requires that the materials be small and thin. Most living organisms are not small enough to place over the opening in the microscope's stage or thin enough for light to pass though. To examine the microscopic structure of most organisms, biologists prepare thin sections or slices of these organisms for examination.

Preparing sections

Slicing soft tissues of living organisms to light-penetrating thinness is difficult. To prepare these sections, biologists first submerge (embed) small pieces of the specimen in a supporting material such as wax. After embedding, the wax block is sliced with a cutting instrument called a *microtome*. Controls on the microtome regulate the thickness of the sections. The sections are transferred to glass slides, attached (fixed) to each slide, stained to increase contrasts between parts of each section, and protected with a coverslip. Sometimes the entire series of sequential sections is placed on the slide; at other times, only one or a few of the sections from different parts of the organism are placed on the slide.

Working from sections

The problem that sectioned material presents students is the task of reconstructing the three-dimensional organism or its parts from the two-dimensional sections. Essentially the mind has to build the three-dimensional form by sequentially "stacking" sections from the specimen and filling in its form between sections. (In one way this activity is similar to children connecting numbered dots to form a picture. However, in this case the lines between the dots, or points on the sections, remain in the mind of the observer and the observer is thinking in three dimensions.)

Try reconstructing a container that is commonly found on the table of many neighborhood restaurants. Assume that the container is sectioned in planes parallel to the surface of the table. Five of these sections are shown on the right half of Figure E-1. The first diagrammed section in Figure E-1 is near the top of the container and the last one is through its base, which rests on the table. The other three sections are in order (top to bottom). A reconstruction of the container, a water pitcher, is shown on the left side of Figure E-1.

Sections from perpendicular cutting planes

Now that you have a three-dimensional mental picture of the water pitcher, proceed one step further. In your mind, view the pitcher from the point indicated by the larger arrow in the upper left corner of Figure E-1. Draw the pitcher from this viewpoint on the left side of Figure E-2.

Figure E-1
Reconstructing a container from sections.

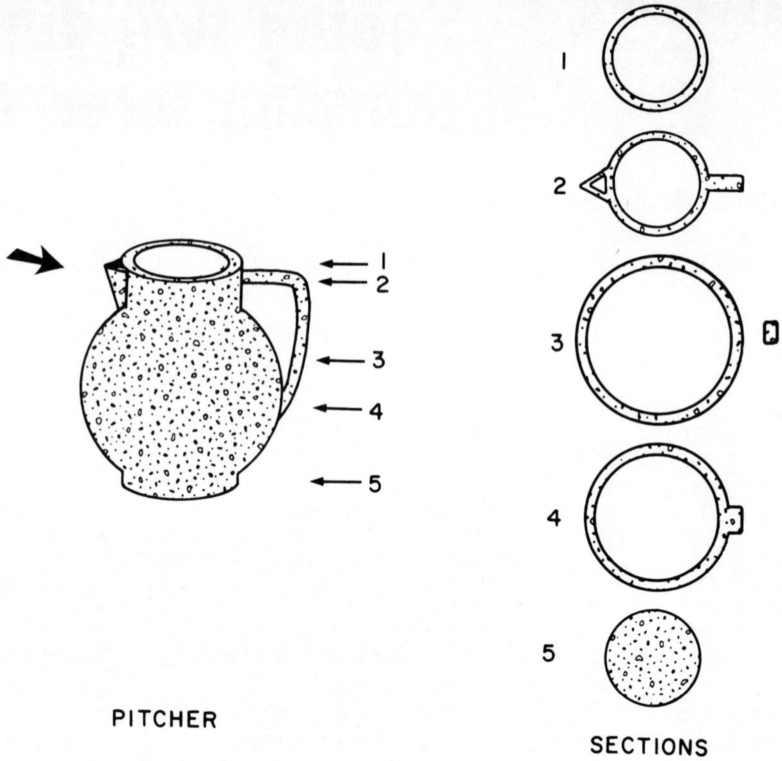

Figure E-2
Mental formation of pitcher and sections from a different perspective.

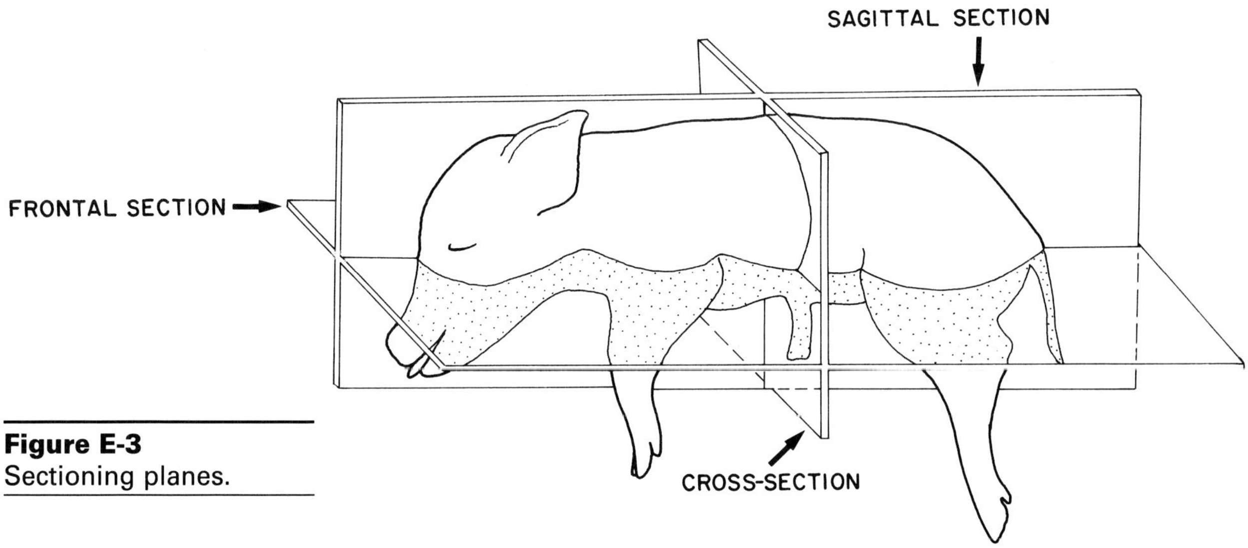

Figure E-3
Sectioning planes.

What are the expected shapes of the pitcher's sections if (1) they are cut perpendicular to the plane of the table instead of parallel to the plane of the table, (2) the first section is cut from one side of the pitcher, and (3) the remaining sections are parallel to the first and continue through to the other side? In Figure E-2, draw five of these sections that are spaced along the container from one side to the opposite side.

Sectioning and terminology

Unless the sectioned structure is exactly round, the different planes of sectioning result in sections with different shapes. For example, a middle section along the long axis of a cylinder is rectangular, with the rectangle as long as the length of the cylinder and as wide as the diameter of the cylinder. In contrast, a section resulting from a cut across a cylinder is a circle with its diameter the same as the diameter of the cylinder.

Biologists use terms to identify the orientation of the sectioning planes to the organism. The terms for the planes of sectioning are shown in Figure E-3. The plane of sectioning is always given on the label of the slide.

The different planes of sectioning separate different parts of the organism. If a pig is used as a convenient example of an organism, then the following apply:

1. Cross-section or transverse section (x.s. or c.s.) is a cut across the pig separating the head end (cranial or cephalic) from the tail end (caudal).
2. Saggital section (sag.) is a longitudinal cut that separates the right side from the left side.
3. Frontal section (frontal or f.s.) is a longitudinal cut that separates the back (dorsal) from the belly (ventral).

Mental exercise in visualizing sections

Determine the effects of different sectioning planes on the shape of the resulting sections. In Figure E-4, draw a middle section from each of the three sectioning planes of the "spool dog." Place arrows on the diagram of the spool dog to show where the cuts should be made. Label each section with its plane of sectioning.

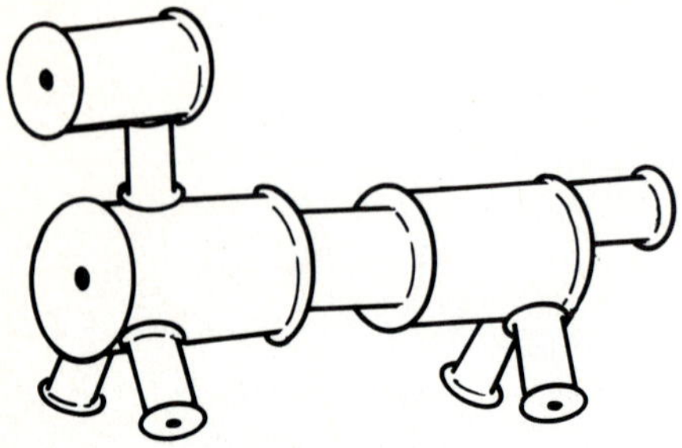

Figure E-4
Sections of "spool dog."

APPENDIX F
Using a pH meter

A pH meter is an instrument that precisely measures the acid content (the hydrogen ion concentration) of solutions. There are many models of pH meters. In general, using any model requires two procedures. The first procedure calibrates the instrument to the temperature of the solution and its general pH range. The second procedure measures the pH of the solution.

Figure F-1 shows the external parts of a typical pH meter. The electrodes are fragile. To care for the electrodes, you should store them in a container of buffered solution, protect them from bumping against other objects, and rinse them with distilled water as you move them from one solution to another.

Calibrating the pH meter

1. Take a beaker of buffer solution of known pH to a pH meter.
2. With a thermometer, measure the temperature of the buffer.
3. Turn the *STDBY/pH* knob to *pH*.
4. Lift the electrodes from their storage solution. Over a water-collecting beaker, wash them with distilled water from a squirt bottle.
5. Lower the electrodes into the buffer.

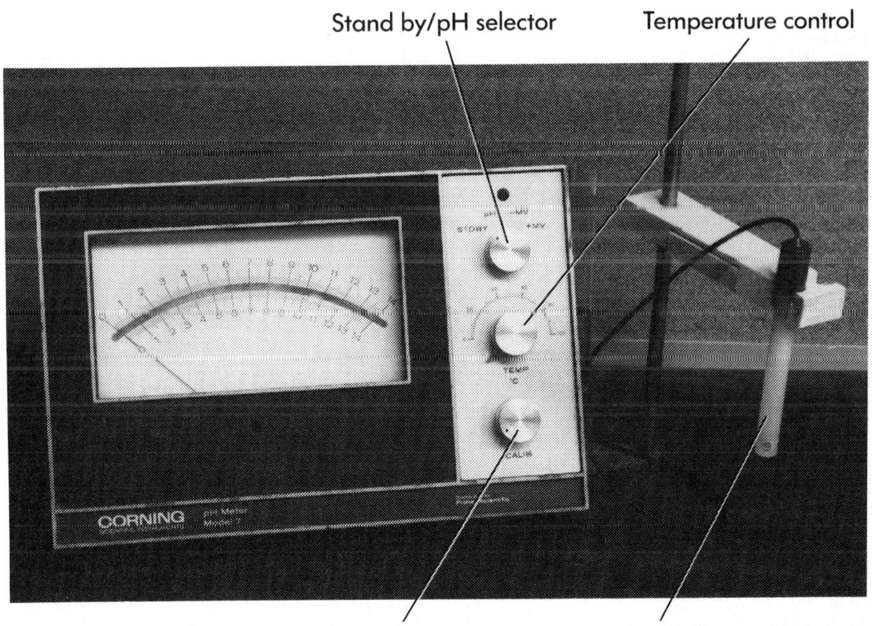

Figure F-1
External view of a pH meter.

6. Turn the *TEMP* knob to the temperature of the buffer.
7. Turn the *CALIB* knob until the meter reads exactly the same as the pH of the buffer.
8. Raise the electrodes and rinse with distilled water. Store the electrodes in the storage solution and turn the operations knob to *STDY* or proceed with the measuring of the pH of solutions.

Measuring the pH of a solution

1. Take a beaker of the solution to be measured to a calibrated pH meter.
2. Raise the electrodes from the buffered solution and wash them with distilled water from a squirt bottle over a water-collecting beaker.
3. Turn the operations knob to *pH*.
4. Lower the electrodes into the solution of unknown pH.
5. Read the pH from the pH display. Record the reading.
6. Raise the electrodes and rinse with distilled water. Repeat steps 4 through 6 to measure the pH of other solutions.
7. Return the electrodes to the holding buffer and turn the operations knob to the *STDBY* position.

APPENDIX
G

Spectrophotometry

Basic concepts and use of the Spectronic 20

The concepts of spectrophotometry

Some scientists measure the amount of light energy absorbed by a substance in solution to determine its light-absorbing properties or its concentration and to identify it. They use a spectrophotometer to make these absorption measurements. The spectrophotometer is an instrument that measures the intensity of light passing through a substance and compares it to the amount of light illuminating the substance. It indirectly measures the amount of light absorbed by the substance.

As light shines on substances, three things can happen: light can be reflected, transmitted, or absorbed (see Figure 9-3). **Reflected light** is the light that is turned back by the substance. **Absorbed light** is the light energy that is incorporated into the molecules of the substance. **Transmitted light** is the light that passes through the substance.

The difference between the light that shines on the substance and the transmitted light is the result of light absorbed by the substance *and* light reflected by the substance. In spectrophotometry, the operator uses a **reference blank** to distinguish the amount of light absorbed by the substance from the light absorbed by other molecules (such as the tube and solvent molecules) and the reflected light. The reference blank is a tube with all of the molecules in the solution *except* the substance that the investigator is interested in measuring. The blank is placed in the beam of light, and the operator adjusts the spectrophotometer to read 100% transmittance (100% T). This adjustment sets the amount of light coming through the reference blank as the standard for comparison. In effect, setting the reference blank to 100% T negates the loss of light to reflection and the absorbance of light by the tube molecules, the solvent molecules, and any other substances that are in the reference blank. Once the reflected light and the light absorbed by the reference blank are negated, the operator places a tube with the substance to be measured in the beam of light. As the substance molecules absorb the light, the percent of transmitted light decreases. This change in the transmitted light compared to the transmittance of the reference blank (100% T) is the measure of light absorbed by the substance.

Relationship between transmittance and absorbance

Based on your knowledge of the definitions of transmitted and absorbed light, what happens to the percent of light transmitted through a tube as its contents absorb more light?

1 ____

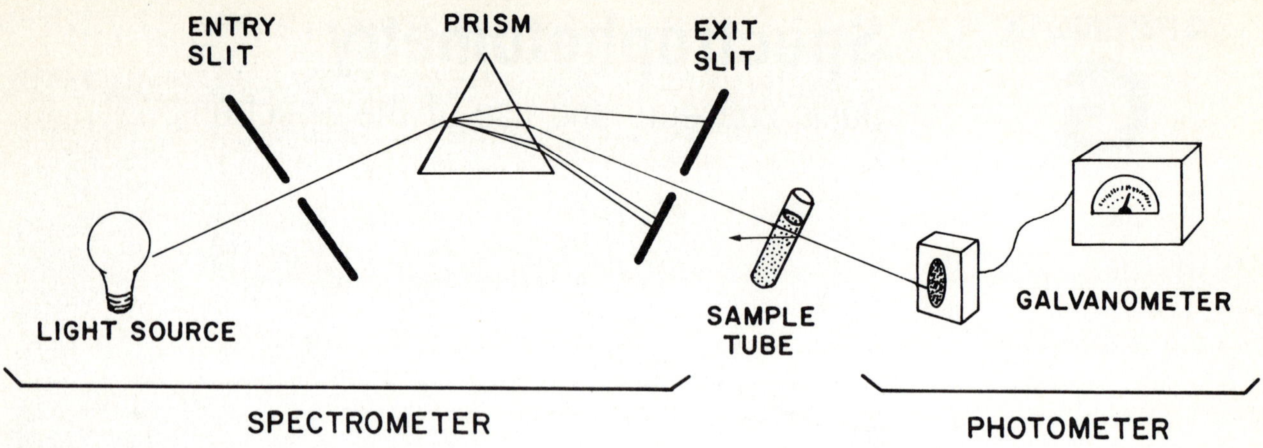

Figure G-1
Basic components of a spectrophotometer.

What kind of a relationship exists between transmitted and absorbed light?

2 ______________________________

In addition to this relationship, absorbance (optical density, or OD) is expressed as a logarithmic function.

The instrument

The spectrophotometer frequently used in student laboratories is the Spectronic 20. It consists of two major parts: a spectrometer and a photometer (Figure G-1). A **spectrometer** consists of a light source, a prism, and a narrow slit. The prism and the slit limit light rays from the lamp to those of specific wavelengths. The light that passes through the slit shines on the tube. A **photometer** measures the light transmitted through the tube. It has two components: a light detector, which is a photoelectric cell sensitive to the emitted wavelengths of light, and a galvanometer, which records the output of the detector on a double scale. The double scale consists of both a percent transmittance scale (from 0% to 100%) with evenly spaced unit markings and a logarithmic absorbance scale (from 2+ to 0) with unequally spaced marks.

Figure G-2 shows the external parts of the Spectronic 20. Match the following regulating or associated external structures of the Spectronic 20 to the basic components of the spectrophotometer shown in Figure G-1:

1. Power switch/0 control: Rotate this knob to turn on the machine and to adjust the intensity of light shining on the tube.
2. Wavelength control: Turn this knob to select the specific wavelength of light that shines on the tube.
3. Light control: Rotate this knob to regulate the amount of light passing to the phototube.
4. Sample holder: Place the reference blank or the sample tube in this well, which is in the path of the light.

Spectrophotometer tubes and their care

You will be working with a set of special tubes. Because the spectrophotometer measures the amount of light passing through the tube, all tubes must be *clean and without scratches.* Handle the tubes carefully. Prevent scratches by setting tubes only in wooden or plastic racks. To thoroughly clean the spectrophotometer tubes, shake detergent water in them, rinse with tap water, and then rinse with distilled water. Just before placing the tube in the spectrophotometer, remove all fingerprints with a laboratory tissue such as Kim-wipes.

Measurements are made in a set of **matched tubes.** Each tube has its own optical properties; therefore only tubes with *similar optical properties* should be

Figure G-2
Spectronic 20, external view.

used together. Usually, tubes that transmit light within one percentage point of each other are used in a set. Keep matched tubes together. Each matched set has its own identification.

Directions for use of the Spectronic 20

1. Plug in the instrument. Rotate the power switch clockwise until you hear a click. The pilot light will glow. Allow at least 15 minutes for the instrument to warm up.
2. With the power turned down but not off, adjust the wavelength control to the proper setting. (Adjusting wavelength with power turned up can damage the photo unit.)
3. With the cover of the sample holder closed and with no tubes in the machine, adjust the power switch/0 control until the meter needle is set exactly at 0% light transmission. (Stand directly in front of the spectrophotometer as you make all adjustments and take readings.)
4. Fill one clean matched tube to its label with the solvent and all substances *except the substance to be measured.* This is the reference tube, or blank. Wipe the outside of the blank to remove fingerprints.
5. Place the filled blank in the sample holder with the label of the tube aligned with the mark on the sample holder. Close the cover.
6. Adjust the light control knob until the meter reads 0 absorbance (100% T). This negates the effects of reflected light and the absorption of light by the tube and solvent system.
7. Without resetting any controls, remove the reference blank and replace it with a matched tube containing the substance and solvent. Read the absorbance and the transmittance from the double scale. This is the absorbance and transmittance of the substance.
8. If you are making a series of measurements, frequently check the meter for drift. If no tube is in the sample holder, the cover is closed, and the needle has drifted from 0% transmittance, readjust the 0 control knob. If the reference blank does not give a 100% transmittance reading, readjust the light control.
9. If you make readings using different wavelengths of light, return to step 2 and repeat all adjustments for each new wavelength.

APPENDIX H

Equipment that measures physiological responses

Instruments that measure and record physiological responses have several components. Some manufacturers put most of these components into one cabinet or unit. Other manufacturers package each component separately. A description of these components follows.

Transducer

A transducer is any instrument that converts one form of energy into another form. There are many kinds of transducers. One example is a radio's speaker system, which converts electrical energy into acoustical (sound) energy. In scientific measuring, most transducers convert various forms of energy into a proportional amount of electrical energy. Two examples of transducers used in physiological measurement are the following:

1. **Force transducer.** This transducer converts mechanical force (pulling) energy into electrical current. As a muscle pulls against the transducer, the force of the contraction is converted into a proportional electrical signal.
2. **Displacement transducer.** This transducer converts the mechanical energy of movement into an electrical signal. The signal is proportional to the amount of movement (displacement) of the transducer's lever. This type of transducer is used to measure how much a muscle shortens when it contracts.

Electrodes

These flat metal plates or circular wells pick up weak electrical signals from the biological source. These sources may be contracting muscles or nerves carrying messages.

Amplifier

An amplifier is an electronic system that increases the strength of a weak electrical signal. Two amplifiers are frequently used for physiological measurements in the student laboratory:

1. **Bio-amplifier/supply and isolated amplifier:** This amplifying system increases the weak electrical signals received directly from biological materials. The electrocardiogram (ECG) of the heart, electromyogram (EMG) of skeletal muscle, records of nerve action potentials, and electroencephalagram (EEG) of the brain use this amplifying system.
2. **Bio-amplifier/supply and strain gauge preamplifier:** This amplifying system boosts the electrical signal resulting from the conversion of mechanical energy. This combination is used with a force or displacement transducer.

Controls on the amplifiers can do the following:
1. Change the amount of amplification
2. Reject extraneous DC current
3. Filter out unwanted electrical frequencies above 25 Hertz (Hz) (such as 60 Hz household current)

Recording or display devices

These devices display an electrical signal as a line on paper or as a visible image on a screen. The height of the line is proportional to the electrical signal received from the amplifying system. Some of these devices can display two incoming signals at a time (dual-channel devices). The three devices most frequently used are the following:

1. **Strip recorder:** The strip recorder converts the electrical signal to the movement of a pen against a moving paper strip. In some strip recorders, a warmed pen marks thermal-sensitive paper, and in others a pen releases ink on the paper. The result is a line on the paper with its height corresponding to the strength of the electrical signal received at the time the pen was touching that section of the paper. The operator can adjust both the amount of pen movement for a specified electrical signal and the speed of the moving paper. This instrument provides a permanent copy of the biological activity.
2. **Oscilloscope:** The oscilloscope converts the electrical signal to an image of a point of light moving horizontally across the display screen. As the electrical signal varies in strength, the image moves vertically. A second device provides a permanent copy of the image's path.
3. **Computer, video monitor, and printer:** More recently, recording and display equipment utilize computer-driven monitors and printers. With computer-interfaced equipment, the electrical signals are converted to images displayed as a line on the monitor. The line varies in height as the signal varies in strength. Various software packages manipulate measurement data and direct the printing of a hard copy.

Figure H-1
Components for measuring mechanical responses.

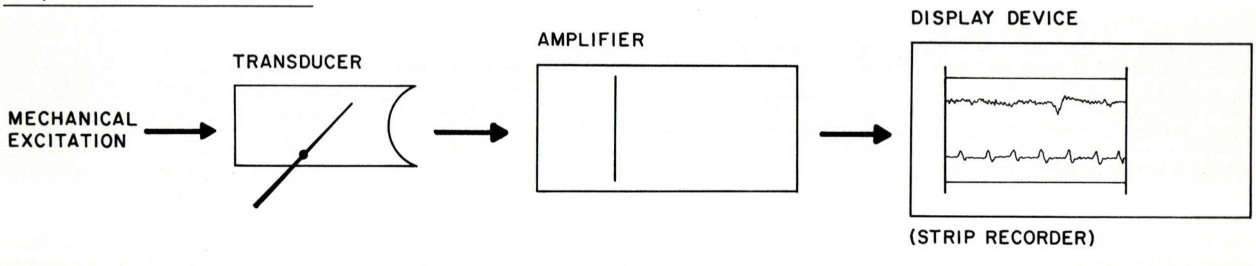

Figure H-2
Components for measuring electrical responses.

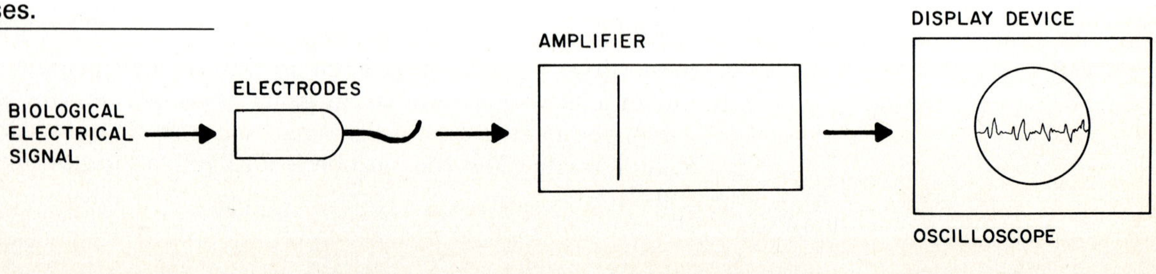

Stimulator

The stimulator produces a controllable electrical current of varying strength (amplitude-volts), duration (width-milliseconds), and frequency (pulses/second). It is used to initiate biological activity in nerves and some muscles.

Figures H-1 and H-2 present the general arrangement among the instruments when measuring biological phenomena. Figure H-1 illustrates the arrangement when measuring mechanical responses such as muscle contraction, and Figure H-2 shows the arrangement for measuring electrical activity in muscles and nerve cells.

Each manufacturer's equipment varies in the specific ways the components are set up and adjusted. Your teacher will provide directions for the specific use of the equipment in your laboratory.

Credits

1-1 South Dakota Tourism.

2-1 William Ober.

4-3 Left to right J.J. Candamse & B.C. Pugasheti/Tom Stack & Associates; David M. Phillips/Visuals Unlimited; Ed Reschke.

5-5 Barbara Cousins (art); Carolina Biological supply (photos).

6-2 Carlyn Iverson.

6-8 Nadine Sokol after William Ober.

17-2 Barbara Cousins.

18-1 Barbara Cousins.

19-1 Raychel Ciemma.

19-3 Michael Schenk, R.T. Hutching.

19-4 Barbara Cousins.

19-5 Raychel Ciemma.

20-4 Lisa Chuck/Michael Schenk.

21-2 Kevin A. Sommerville.

21-4 Copyright © Joel Gordon.

22-4 William Ober.

27-1 Nadine Sokol after William Ober.

27-2 Nadine Sokol after William Ober.

Plate 13 Ed Reschke.

Plate 14 Left to right: Triarch/Visuals Unlimited; John D. Cunningham/Visuals Unlimited.